Sustainable Development Goals Series

The **Sustainable Development Goals Series** is Springer Nature's inaugural cross-imprint book series that addresses and supports the United Nations' seventeen Sustainable Development Goals. The series fosters comprehensive research focused on these global targets and endeavours to address some of society's greatest grand challenges. The SDGs are inherently multidisciplinary, and they bring people working across different fields together and working towards a common goal. In this spirit, the Sustainable Development Goals series is the first at Springer Nature to publish books under both the Springer and Palgrave Macmillan imprints, bringing the strengths of our imprints together.

The Sustainable Development Goals Series is organized into eighteen subseries: one subseries based around each of the seventeen respective Sustainable Development Goals, and an eighteenth subseries, "Connecting the Goals", which serves as a home for volumes addressing multiple goals or studying the SDGs as a whole. Each subseries is guided by an expert Subseries Advisor with years or decades of experience studying and addressing core components of their respective Goal.

The SDG Series has a remit as broad as the SDGs themselves, and contributions are welcome from scientists, academics, policymakers, and researchers working in fields related to any of the seventeen goals. If you are interested in contributing a monograph or curated volume to the series, please contact the Publishers: Zachary Romano [Springer; zachary.romano@springer.com] and Rachael Ballard [Palgrave Macmillan; rachael.ballard@palgrave.com].

Ya Ping Wang · Keith Kintrea ·
David Everatt · Debolina Kundu
Editors

New Drivers of Division

Urbanisation and Spatial Inequality in Africa and Asia

Springer

Editors
Ya Ping Wang
School of Geography and Tourism
Shaanxi Normal University
Xian, China

Keith Kintrea
School of Social and Political Sciences
University of Glasgow
Glasgow, UK

David Everatt
School of Governance
University of the Witwatersrand
Jonhannesburg, South Africa

Debolina Kundu
India Habitat Centre
National Institute of Urban Affairs
New Delhi, Delhi, India

ISSN 2523-3084 ISSN 2523-3092 (electronic)
Sustainable Development Goals Series
ISBN 978-981-96-5628-8 ISBN 978-981-96-5629-5 (eBook)
https://doi.org/10.1007/978-981-96-5629-5

This Springer imprint is published by the registered company Springer Nature Singapore Pte Ltd.
The registered company address is: 152 Beach Road, #21-01/04 Gateway East, Singapore 189721, Singapore

Foreword

The Sustainable, Healthy and Learning Cities and Neighbourhoods (SHLC) project, led by a team at the University of Glasgow, is the platform for this book. The project ran from 2018 until 2023, having been seriously interrupted by the global coronavirus pandemic. I was privileged to chair the International Advisory Committee for the project. There were eight other universities, from seven African or Asian countries, who were partners in the work.

The SHLC project should be seen as a landmark in urban studies. The UN's Sustainable Development Goals (SDGs) and the New Urban Agenda (NUA) set an aspirational framework globally for human settlements for the period from 2016 to 2030. In particular, SDG 11 was to 'Make cities and human settlements inclusive, safe, resilient, and sustainable'. While there was recognition that local action was necessary to deliver on this and the other SDGs, there was surprisingly little knowledge of the barriers and opportunities at neighbourhood level. Similarly, the ambitions in the SDGs and in the NUA were necessarily generalised and high level, but would be interpreted in culturally specific ways and implemented through a range of different governance systems. Hence the need for systematic comparisons across spatial scales, from the global to the very local, and between larger and smaller cities in and between countries. The SHLC research framework was designed to explore this multi-comparative challenge.

The SHLC team probed in empirical detail the nature of inequalities at neighbourhood scale across 14 cities, large and small, stretching from Cape Town to Manila. Some 15,000 questionnaire interviews were conducted, as well as 74 focus groups. This book tells that story, giving the reader unique insights into the diverse spectrum of place-making that is happening at pace, as both slums and gated communities are created, managed and lived in. It shows the benefits that urbanisation can bring, but also the challenges of providing adequate shelter and services for all. In creating wealth, these dynamic cities also embed new inequalities, and entrench old ones.

The research breaks new ground by drilling down to neighbourhood scale in a systematic and comparative way. Despite the widespread advocacy of evidence-informed and data-driven policymaking, and the belated international recognition of the significance of urbanisation, comparisons are frustrated by the lack of common definitions and data sources, and we still know relatively little about what is happening at neighbourhood level. Indeed, 'neighbourhood' is itself a problematic concept, which does not translate universally. Inter-governmental agencies such as the United Nations are focused on national-level data and policy prescriptions; but national averages obscure important spatial differences within and between cities, and can elide crucial cultural and governance factors that shape the delivery of policy on the ground.

Another key feature of the SHLC project, and also of this book, is that it makes the connections between health, well-being, education, economic opportunity, and sustainable cities and communities. These inter-relations are often recognised rhetorically and intuitively, but rarely researched in the scale and depth needed to illuminate the web of urban inequalities. In academia, administration and policymaking, silos remain entrenched, wasting scarce resources and imposing perceptions of needs and solutions that do not accord with the more integrated reality experienced in daily lives on the street. Similarly, much of the research on urbanism in what is called the Global South is done by academics based in the Global North. Partnership and capacity building was a central feature of the SHLC project, and this publication benefits greatly from the inputs of those members of the SHLC team based in their universities in Africa and Asia.

The widening inequalities revealed here, and their relation to land and housing markets should be a concern to all those working in public service in an urban context. We now have a well-researched snapshot of rapid urbanisation from the bottom up in the early 2020s, and across some very different places. Inequality shapes neighbourhoods, and neighbourhoods reproduce inequalities. Each city is unique, shaped by its history, culture, religions, economy, and environment, yet a broad picture emerges. In particular, public services, such as health care, education, or urban planning can ameliorate, or exacerbate exclusion and disadvantage. What is happening in neighbourhoods in these 14 cities and across the rapidly urbanising countries is already spilling out to affect places elsewhere, and seems likely to do so more and more. *New Drivers of Division: Urbanisation and Spatial Inequality in Africa and Asia* gives readers an essential understanding of how rapid urbanisation is remaking our diverse and complex 21st-century urban world.

Edinburgh, UK Cliff Hague

Cliff Hague *was awarded an Officer of the Most Excellent Order of the British Empire, a British Honour (OBE), by the United Kingdom for his services to urban and regional planning. He is Professor Emeritus in Planning and Spatial Development at Heriot-Watt University, and a past President of the Royal Town Planning Institute, and of the Commonwealth Association of Planners.*

Acknowledgements

This book is based on work of the Centre for Sustainable, Healthy and Learning Cities and Neighbourhoods (SHLC), which was supported by the UK Economic and Social Research Council (ESRC) grant number ES/P011020/1. SHLC was one of the 37 large Growing Research Capacity projects funded by the UK Research and Innovation (UKRI) Global Challenges Research Fund.

This book and its sister book, *Urban Policy for Sustainable Development in the Global South: New Strategies for New Challenges*, report the main findings of the SHLC project. The five-year project involved research teams in eight countries and more than 100 researchers and project managers through its lifetime. Not all team members were involved in the production of these books. However, we would like to thank all SHLC team members for their hard work over the project period and beyond, and we are delighted to see so many of them are continuing the research in this important area and becoming research leaders in the field. We would like to thank the SHLC project management team led by Ms Gail Wilson and her successors, as they were critical to the project's functioning.

Apart from the team members, the project involved many local interviewers gathering data in 14 cities, often working in different languages and under the very difficult conditions of the COVID-19 pandemic. More than 15,000 individuals participated in the neighbourhood survey, and many residents and local community workers participated in the 74 neighbourhood focus group interviews. The success of the project owes a great deal to the hard work of interviewers and the generous support of residents.

SHLC's Advisory Committee, chaired by Professor Cliff Hague, provided continuous and professional support and advice over the project period. We would like to thank all the committee members for their expert insights and practical advice. The project benefited enormously from their experience and expertise.

As the project addresses three of the UN 2030 Sustainable Development Goals (SDGs), we worked very closely with UN-Habitat. Dr Remy Sietchiping, Chief of the Policy, Legislation and Governance Section, Urban Practices Branch Global Solutions Division, and his staff provided valuable support and policy advice during the project. We organised several joint online and offline policy discussion and dissemination events. We would like to thank UN-Habitat for its strong support.

We would like to thank our host institutions. The University of Glasgow as the overall project lead and grant holder provided strong and professional support in project management, finance, legal matters, and ethical approval. The Research Office of the College of Social Sciences, in particular, provided essential management support and staff cover during a very hectic period in response to the impact of Covid-19. Similarly, we would like to thank the host institutions of SHLC's international partners— Ifakara Health Institute in Tanzania, Khulna University in Bangladesh, Nankai University in China, National Institute of Urban Affairs in India, the Human Sciences Research Council and the University of Witwatersrand in South Africa, the University of Rwanda, and the University of the Philippines—for their professional support and management, which ensured the project ran successfully and smoothly in its respective locations.

Finally, we would like to thank the National Institute of Urban Affairs in India, Ms Pragya Sharma and Dr Arpita Banerjee, in particular, for ensuring the publication quality. Their contributions have been instrumental in bringing this work to fruition.

Ya Ping Wang
Keith Kintrea
David Everatt
Debolina Kundu

Contents

Introduction

1

Ya Ping Wang, Keith Kintrea, David Everatt, and Debolina Kundu

Abstract

Urbanisation in the twenty-first century has resulted in more spatially fragmented and socially divided cities in developing countries. Based on a UK government supported international collaborative project (Sustainable Healthy Learning Cities and Neighbourhoods—SHLC), this book brings a fresh understanding of recent trends in urbanisation and urban spatial inequality in Africa and Asia, using the neighbourhood as the unit of analysis. The authors reveal new patterns of urban and neighbourhood inequalities and driving forces for division under different economic, historical, cultural conditions and political systems. In this introduction chapter, the editors review recent theories and debates about urban and spatial inequalities. They summarise recent trends in urbanisation and their implications for inequalities and international policy responses, before providing some background information about the research project and an outline of the book's chapters and main findings.

Keywords

Urbanisation · Urban spatial inequality · Developing countries · SHLC project · Sustainable Development Goals (SDGs)

1.1 Background and Aim

Fast urbanisation in the twenty-first century has resulted in a dual process of urban expansion in most developing countries. New, well planned, and serviced areas for the emerging middle class are constructed by professional builders, while unplanned, informal, and off-grid settlements and slums make up much of the simultaneous urban expansion. Inside existing urban areas, traditional neighbourhoods and residence divisions are gradually replaced by new socio-economic and political statuses created through political change,

Y. P. Wang (✉)
School of Geography and Tourism, Shaanxi Normal University, Xian, China
e-mail: yaping.wang@snnu.edu.cn; yaping.wang@glasgow.ac.uk

Y. P. Wang · K. Kintrea
School of Social and Political Sciences, University of Glasgow, Glasgow, Scotland, UK
e-mail: keith.kintrea@glasgow.ac.uk

D. Everatt
Wits School of Governance, University of the Witwatersrand, Johannesburg, South Africa
e-mail: David.everatt@wits.ac.za

D. Kundu
National Institute of Urban Affairs, New Delhi, India
e-mail: dkundu@niua.org

Y. P. Wang et al. (eds.), *New Drivers of Division*, Sustainable Development Goals Series,
https://doi.org/10.1007/978-981-96-5629-5_1

large-scale migration, and economic development. In many parts of the world urbanisation has become more spatially fragmented and socially divisive, with the rich and poor living under very different conditions.

Responding to the dualities of urbanisation and inequality requires an understanding of the complex relations between economic, social, political, and environmental developments in cities and specifically at the neighbourhood level. Yet, recent urban policies for developing countries emanating from supranational agencies tend to operate at a very general level; research and understanding of urbanisation are fragmented and mainly focused on living conditions and poverty in undifferentiated 'slum areas'. Debates about sustainable and just development tend to concentrate on higher levels such as national government, 'the city', or 'the mega city', and emphasise the physical and environmental aspects of urbanisation at macro- rather than neighbourhood levels and micro-scales.

This book brings a fresh understanding of recent trends in urbanisation and urban spatial inequality in Africa and Asia, using the neighbourhood as the main unit of analysis. Through a set of well-coordinated theme- and city-based studies that follow an overall framework, we address the following questions about cities in developing countries:

- What do socio-economic and spatial inequalities look like in cities in Africa and Asia under the processes of 21st-century urbanisation?
- To what extent do urban neighbourhood differences reflect social inequalities and how do they manifest spatially in different countries?
- What are the main driving forces of increasing spatial inequalities and how are these inequalities produced and sustained in relation to history, culture, political system, class, race, ethnicity, and religion, as well as systems of health and education provision?
- What are the implications of these new forms of inequality for urban and neighbourhood sustainability and policy development?

By addressing these questions, the book makes a contribution to progressing the United Nations 2030 Sustainable Development Goals (especially SDGs 3, 4, 10, and 11) and the New Urban Agenda by enhancing policy and theoretical debates on how we understand Asian and African cities.

1.2 Urban and Spatial Inequality

Inequality has accompanied the development of human society throughout time, and spatial inequality describes the unequal distribution of income, wealth, and resources across geographical regions and areas. Inequality in the pre-Industrial Revolution period was represented by the different lifestyles between the rulers and the ordinary people. While the rulers enjoyed a luxurious life in their castles and palaces, most ordinary people lived in poverty with an income similar to today's poorest countries (Roser 2017). The first major widening of income inequality in the world started in Britain between 1750 and 1850, as the gains from the Industrial Revolution went to the urban and rural middle class, but not to the urban and rural poor (DeLong 2016). Since then, industrialisation has been a major driver both of development and of urban inequality in the world. As the world and many of its countries become richer, inequalities between regions and among people grow. The speed of growing inequality only slowed slightly between 1930 and 1980 in some western countries where governments promoted progressive taxation and welfare provision systems. Under neo-liberal policies, income and wealth inequality have grown again substantially in many countries over the last four decades (Piketty 2014; Musterd 2023).

Urban inequality has long been an important theme of social science research. It is widely acknowledged that urban inequality is a multi-faceted phenomenon with economic, social, environmental, and political dimensions; variations across class, race, ethnicity, age, gender, or citizenship status are also considered as important factors at national level (Galster and Sharkey

2017; Hamnett 2019; Storey et al. 2020; Pardo and Prato 2021; Pryce et al. 2021). Inequalities exist in all aspects of cities, such as housing, education, health, transportation, infrastructure provision, and social and commercial services. Different dimensions of urban inequality are also intertwined and intersectional and related to geographical factors, including urban space, scale, zoning, environment, and accessibility to various urban facilities (Wei et al.2017; Musterd et al. 2016). Inequality trends have also been exacerbated recently by economic globalisation, international migration, and neo-liberal government policies. The rise of information-based economy has resulted in new socio-economic and spatial inequalities in cities and among urban populations. Income gaps have widened, inter-city disparities have grown, and suburbs have been re-sorted into a wide array on the basis of class and race or ethnicity. Crime rates are higher in places with more inequality, and people in unequal cities are more likely to be unhappy (Nijman and Wei 2020; Piketty 2020; Mela and Toldo 2019). High inequality, as revealed by the Covid-19 pandemic undermined social cohesion, eroded public trust, and deepened political polarisation, all of which negatively affect governments' ability and readiness to respond to crises (Doucet et al. 2021).

Debate about the drivers of urban inequality in developed western countries has shifted in the last twenty years from poverty to the presence of the wealthy (Galster and Sharkey 2017; Wilkinson and Pickett 2009; 2020; Piketty 2014). As early as the 1980s, researchers found that income inequality was increasingly exacerbated by inequality in wealth, especially the appreciation of property values (Goodman 1988; Piketty 2014). Skyrocketing housing prices and the problem of housing affordability have remained a challenging issue globally. Housing costs and markets are intertwined with urban inequality in multiple ways. The segregation of housing markets and the problem of housing affordability have been regarded as new drivers of urban inequality (Baker et al. 2016; Piketty 2014; 2020; Wei and Ewing 2018; Pryce et al. 2021). London, for example, has been captured by a new wealthy elite that is pushing out not only the working class from the city, but also old money from its enclaves, and changing the fabric of the city (Atkinson 2020). In developing countries, housing conditions in different types of neighbourhoods are a direct reflection of the wealth or poverty levels of residents compounded by government policies—both neo-liberal economic policies, and those meant to manage urbanisation.

Many aspects of inequality derive from where people live, work, grow up, go to school, and retire (Roser 2017). Urban spatial inequality can be observed between different areas and neighbourhoods. People living in different neighbourhoods do have very different life outcomes even though those neighbourhoods are in the same area of the city. Spatial inequality also exists because of unequal distribution of resources, job opportunities, public services, green spaces, etc. Apart from local geographical and environmental features, the distribution of components of public infrastructure, land use, and planning policies and government investments in services such as health care, education, and transport are determining factors for urban spatial inequality as they directly affect the life quality and well-being of people. People living in neighbourhoods with poor infrastructure and public services, such as informal settlements and slums, are at a greater risk of poor health and well-being. Such impacts compound over time and over generations, leaving individuals to become more susceptible to future health problems and illnesses (McVie et al. 2019; Storey et al. 2020). Spatial inequality when aligned with political, religious, or ethnic tensions can be dangerous to social and political stability. As such, city planning and the provision of public infrastructure and services remain essential to public policy considerations for rapidly urbanising communities. The case for policy interventions to ensure a more spatially equitable and efficient allocation of infrastructure and public services is important for urban sustainability (Shirazi and Keivani 2019).

The incremental spatial transformation associated with neo-liberal urban development is often characterised by unfettered competition, spatial segregation, and injustice (Hubbard 2004; Brenner and Theodore 2005; Mela and Toldo

2019). The combined ramifications have been socio-spatial divisions and cities that provide highly unequal spaces of abode (Musterd et al. 2016; Van Ham et al. 2021; Soliman 2021). This is why spatial inequality is commonly understood through a spatial and social injustice lens (Soja 2009). Thus, rising income and wealth associated with changes in the labour and financial markets coupled with austerity and neo-liberal policies drive inequality, social exclusion, and poverty (Donald et al. 2014; Piketty 2014; Zwiers et al. 2016) while undermining urban governance. In the developing world, informality simply fills the vacuum created when local (and national) governments continue to invest in existing formal suburbs and under-provide for the urban poor.

As centres of economic production and consumption, cities have increasingly become places of innovation and accumulation where reorganisation and re-engineering of the urban ecosystem occur at different scales (Martínez et al. 2022). The processes that play out at the global and city levels often manifest at the neighbourhood level. This positions the neighbourhood as a key spatial unit for understanding the internal dynamics of the city (Wang and Kintrea 2019). Growing socio-economic inequalities continue to shape neighbourhood formation. Neighbourhoods are different and are often shaped by multiple factors, including housing stock, income status, education, race, religion, and occupation, among others. Income intersects with other differentiating factors such as race and education levels to produce and sustain socio-spatially segregated areas (Marcuse and Van Kempen 2000; Kazepov 2005; Unceta et al. 2020).

In developing countries, urban inequalities are at their most acute in part due to fast urbanisation, rural-to-urban migration, policy failures, racial discrimination, and exclusion, neo-liberal market economic reforms and, more recently, the impacts of the Covid-19 pandemic. Many also have a distinct post-colonial shape, resulting from decades of segregation under colonialism (Soliman 2021). However, urban inequalities and spatial segregation remain under-studied. Beyond some general observations on the striking division between the rich and the poor, between slums and gated communities, we know little about the incidence and the impacts of socio-economic and spatial inequalities in the fast-urbanising cities, nor about the impact of urban spatial inequality on the people living in their midst.

We explore these observations further in the following chapters and present some fresh understanding of the current speed of urbanisation, and to what extent neighbourhood inequalities in African and Asian cities resemble or differ from those found in western cities. While taking the view that the land market has been the main driver of spatial inequality, we argue that understanding the forms, causes, implications of, and solutions to inequality in cities in the Global South requires us to rethink the way we understand urban inequality. We need to place more emphasis on national and local historical background, stages of development, different cultural and political values, and forms of governance, racial, ethnic, and religious composition, and informality. We also need to pay particular attention to the roles that public services such as education and health care have played in the creation and amelioration of inequalities (Table 1.1).

1.3 Urbanisation, Inequality, and Development Policies

Cities are the sites where diverse social and economic resources are concentrated, and that concentration itself can generate substantial social and economic benefits in the form of innovation and income growth (Jacobs 1969). There is also widespread recognition of the importance of cities for global sustainability and how we build them will determine humanity's future (PEAK Urban 2023). Over recent history, it has generally been assumed that urbanisation goes hand-in-hand with industrialisation and 'development' (Kamete et al. 2002). It has been argued that industrialisation and urbanisation are two of the most significant processes affecting societies in the late twentieth century and beyond (Devas and Rakodi 1993; Potter et al. 2004). Mera (1978, 1981) stresses many positive links between urbanisation, particularly between the growth of large

Table 1.1 Main drivers of urban and spatial inequality

Drivers	Western and developed countries	Developing countries
Urbanisation	Suburbanisation, polycentric development and diversification	Rural to urban migration, sprawl and informal development
Industrialisation	Post-industrial cities, gentrification, rich and poor segregations	New divisions of labour and creation of social–economic groups and neighbourhoods
Poverty	Persistent but being controlled at a relatively low level and among a smaller group	Large-scale poverty and urban poor
Wealth and property market	Housing price inflation and property wealth distribution	Access to land and shelter; informal housing; growing inequality
Urban governance and services	Stronger, but diminishing social welfare provision, policy intervention, professional urban planning	Lack of effective planning and intervention policies; limited access or lack of public services, especially health and education
Colonial legacy	Past colonisers	Influence being embedded socially, culturally, physically, and spatially in major cities
Disasters and pandemics	More resources and relatively quick reaction, but not necessarily offering equal protection	Lack of resources; very unequal reaction and impacts with the urban poor suffering serious consequences

Source Authors, 2025

cities and social indicators such as school enrolments, literacy, declining infant mortality, nutritional intake, life expectancy, and falling birth and death rates. However, the relationship between urbanisation, industrialisation, and development has become more questionable in relation to developing countries. Some of them, many in Africa, have become more urbanised without much industrialisation at all (Gollin et al. 2016). There is also generally growing apprehension about population increase, high levels of unemployment, poverty, slums, soaring infrastructure demands and costs, and ecological dangers in cities, and about further decline of rural areas and 'lagging' regions (Potter et al. 2012). Responding to these trends, politicians and planners often focus on what they perceive as 'urban problems', including 'excessive' in-migration and the limits to urban absorptive capacity (Gilbert and Gugler 1992).

For a long time, researchers and policymakers around the world have been searching for answers to the question of how the urbanisation process can be managed to avoid 'urban problems' and especially how the large number of urban poor in developing countries can be housed. During the immediate post-war period, many newly independent states followed the policy and practice of industrialised countries by continuing to adopt urban policies and regulations which had been authored by the now-departed colonial authority. As part of a general strategy of modernisation and industrialisation, subsidised public housing was built in many countries on cheap suburban land (Jenkins et al. 2007; Renaud 1981; Wakely 1988). This policy proved ineffective. In the late 1960s a different approach to shelter and urban development in the form of 'self-help' proposed by John Turner and others such as De Bono was promoted by the United Nations and the World Bank through a series of demonstration projects in the 1970s. At the same time there was a shift in overall development strategies from modernisation to the idea of 'basic needs' and 'redistribution with growth' (Harris and Giles 2003). Further changes occurred in the 1980s. With the emergence of neo-liberal development strategies globally, urban development and human settlement policies became closely related to macro-economic policies and structural adjustment (Burgess 1992; Pugh 1995). This ‘support approach’, complemented by the growing interest in urban management and neo-liberal tendencies towards privatisation, was taken up by the United Nations and the World Bank through ‘enabling policies’, which facilitated and encouraged the private and voluntary sectors to respond to urban development and shelter demands, and limit state intervention to the provision of legislative, institutional, and financial frameworks (Pugh 1997; Rakodi and Lloyd-Jones 2002). Despite these policy initiatives, urban poverty in developing countries deepened and research in development studies came to the so-called ‘impasse’ period (Arsel and Dasgupta 2015).

From the 1990s onwards, the United Nations, especially its Human Settlements Programme (UN-Habitat) played an important and leading role in global urban policy development and implementation. The Second UN Conference on Human Settlements (Habitat II, also known as the City Summit) in 1996 advocated for socially just and environmentally sustainable cities. Instead of framing urbanisation as inherently problematic, it considered urban issues as challenges that can be overcome by seeking best practices and models to replicate. The adopted Habitat Agenda emphasised the need to create sustainable and inclusive cities with adequate shelter for all residents (United Nations 1996). Inspired by the radical assertion of the right to the city, the Habitat Agenda advocated for inclusivity, which encouraged participation of all sectors, including the urban poor, in the production of urban life. The goal of building ‘shelter for all’ considers the need to address issues of inequality alongside the importance of recognising differences and diversity in cities (Jenkins et al. 2007).

The goals of the Habitat II Agenda were later reaffirmed in 2001 through the UN Declaration on Cities and Other Human Settlements in the New Millennium, which led to the publication of the Millennium Development Goals (MDGs) to reduce extreme poverty by 2015 (United Nations 2000). The MDGs called for a significant improvement in the lives of at least 100 million slum dwellers by the year 2020.[1] While focusing on ways to improve life within slums, the report recognised that such areas also provide important forms of affordable shelter, especially for the growing proportion of urban informal sector workers. They also were seen as being the basis for positive social and cultural movements. The report stressed that efforts to resolve living conditions in such areas had declined since the 1980s when they were a major development focus, and that slum upgrading or eradication programmes had failed to address underlying causes of the existence of these areas, mainly poverty (Jenkins et al. 2007; UN-Habitat 2003). It was claimed by the then Secretary-General of United Nations Ban Ki-moon that ‘The global mobilisation behind the Millennium Development Goals has produced the most successful anti-poverty movement in history’. He also recognised that for all the remarkable gains, ‘inequalities persist and that progress has been uneven. The world’s poor remain overwhelmingly concentrated in some parts of the world’ (United Nations 2015a: 5).

Based on experiences and evidence from the efforts to achieve the MDGs, in 2015 United Nations Member States adopted The 2030 Agenda for Sustainable Development which was intended to provide a shared blueprint for peace and prosperity for people and the planet (United Nations 2015b). At the heart of the new agenda are the 17 Sustainable Development Goals (SDGs) which are an urgent call for action by all countries—developed and developing—in a global partnership. The goals recognise that ending poverty and other deprivations must go hand-in-hand with strategies that improve health and education, reduce inequality, and spur economic growth—all while tackling climate change and working to preserve oceans and forests. Compared to the MDGs, the SDGs single out urbanisation as one

[1] Slums were defined as places without access to adequate drinking water, sanitation, quality of housing and security of tenure, and were seen as being the product of two main processes: rapid urbanisation and the urbanisation of poverty (UN-Habitat 2003).

of the key issues in achieving sustainable development through SDG 11. This emphasis on urbanisation speaks to the fact that more than half of the world's population lives in urban areas, and that cities are the important terrains of multiple social, political, economic, cultural, and ecological issues.

To consolidate the SDG urban policies, a New Urban Agenda (NUA) was adopted and ratified in 2016 at the UN-Habitat III conference. The NUA reflects key sustainability challenges that need to be addressed by cities all over the world and provides a new framework that lays out how cities should be planned and managed to best promote sustainable urbanisation (UN-Habitat 2017). The NUA outlines principles of focusing on urban design and planning, sub-national, and local governments, the role of participatory and 'bottom-up' practices and innovation, data and technology. It adopts a city-wide approach to development with strategies and actions, introducing funding mechanisms and effective means of monitoring. It seeks to create a mutually reinforcing relationship between urbanisation and development, with the aim that they become parallel vehicles for sustainable development.

In the SDGs and NUA declarations, there is a clear progress over previous policies. Firstly, there is a recognition that urbanisation has an important role to play as it is 'intrinsically linked to and can function as a transformation driver of sustainable development' (UN-Habitat 2017). Secondly, it is recognised that the way in which cities are governed and the way in which the urban challenges have hitherto been addressed should be overhauled in order to build 'cities that can serve as engines of prosperity and centres of cultural and social well-being while protecting the environment' (United Nations 2016). In a critical review of the approach to urban policy Kintrea and Baffoe (2025) found that policies developed by the main supranational agencies lean heavily towards their potential role in urban economic growth, and most of the vehicles promoted by UN-Habitat and the Organization for Economic Cooperation and Development (OECD) to carry forward urban policy globally have been national policies. In the UN's formulation, urban policy is 'national' because it argues for the need for countries to have an integrated national approach and not just a series of local policies driven bottom-up by individual city governments. Grover et al. (2022) argue for what they call 'place based policies' but it is clear that they consider that cities are the most important object of such policies because they represent the highly concentrated economic activity that is seen as the key to greater prosperity. The SDGs and the NUA still rely heavily on national governments for implementation. They also emphasise the important roles of local and municipal governments and the private sector in achieving sustainable urban transformation and development.

The SDGs and NUA set very ambitious goals and the vision of 'cities for all', 'equal use and enjoyment of cities and human settlements', and 'inclusive urban development' (UN-Habitat 2016). In some sense they read like a long-term and utopian manifesto that underestimates the huge differences and inequalities between the world's regions, countries, cities, and more importantly, urban communities. Although they emphasise implementation and follow-up reviews and progress monitoring, the 'one size fits all' approach may not provide a clear route map for sustainable urban development, especially in the developing countries. Taking a closer look at the realities of inequalities in cities located in the Global South, it is evident that the technical prescription of 'readdressing the way cities and human settlements are planned, designed, financed, developed, governed and managed' (UN-Habitat 2016: iv) is far from enough to inform the actions and policies required to end poverty and hunger in all its forms and dimensions and to reduce inequalities. Many parts of cities throughout of the world have been built simultaneously without much professional planning or design. Large-scale informal settlements are still emerging somewhere every day. Urban planning practice in some African cities is underdeveloped in comparison with that in other parts of the world, which in some suburban new areas only involves some simple street markings on the ground and the provision of a few electricity poles and water taps. The technical planning, design, and financing approach based mainly on developed countries promoted by the New Urban Agenda does not seem capable of addressing the political and socio-economic dimensions of urban planning in the developing country context.

It is also important to note that, in common with the other recent urban policies developed by the main supranational agencies which emphasise urban economic growth, the SDGs lack focus on socio-spatial inequalities, except for a very minor mention of uneven development at regional level within countries. Spatial inequality may be a sub-set of general inequalities but it requires a very specific focus and set of solutions. This book (and the associated research project) is positioned on these understandings and aims to demonstrate that understanding of urban spatial inequalities, the divisions and differences of urban neighbourhoods is critically important in achieving the UN SDGs and the NUA.

1.4 The SHLC Project

The book draws on recent work by the Centre for Sustainable, Healthy and Learning Cities and Neighbourhoods (SHLC), a UK Research and Innovation-The Global Challenges Research Fund (UKRI GCRF)—supported multi-national research consortium. SHLC teams have been studying the internal structures and social-spatial divisions of cities in Africa and Asia since 2017. They conducted systematic studies of urbanisation and the formation and differentiation of neighbourhoods in urban areas in order to address the challenges associated with urbanisation and large-scale rural-to-urban migration and understand the new patterns of social and spatial inequalities. The project design and selection of countries reflect the diversities in Africa and Asia in relation to economic development, regional characteristics, cultural and historical background, level of industrialisation and urbanisation, and enable cross-country comparison (Robinson 2016, 2022). Case study countries include two broad categories: the large and emerging economies (South Africa, India, and China) and the others (Rwanda, Tanzania, Bangladesh, and the Philippines). To give a better understanding of the regional complexity in urbanisation within each country, two cities were selected as case studies: one large city, or the country's capital, and one regional city, as shown in Table 1.2.

In each city, major types of neighbourhoods were identified, investigated, and categorised into five groups, according to economic and wealth conditions ranging from slums to rich residential areas/gated communities. The design allows comparative analysis across several dimensions: within country and region, between countries and continents, and between different economic development levels and types of cities and, more importantly, between neighbourhoods.

The core SHLC project includes three major primary data collection activities: neighbourhood study (fieldwork and auditing), a household survey, and neighbourhood focus group interviews. Firstly, for each of the five main groups of neighbourhoods identified in the city, one (or two) typical ones were chosen for detailed study. In each city we examined between 6 and 12 neighbourhoods depending on the size of the city and neighbourhood complexity. For each of the selected neighbourhoods, a detailed study through on-site auditing and fieldwork was conducted. Areas of research and data collection covered a range of topics and indicators, including physical size, history, population and demographic features of residents, housing, built environment, health, education, infrastructure and service provision, geographical conditions and environmental hazards, economic opportunities and job availability, local politics, governance and management, etc. This neighbourhood study brought fresh understanding of the internal structures of these cities and generated a major inter-medium report for each city, and the summary of research findings which was released at the time at the centre's website (https://www.centreforsustainablecities.ac.uk/). These neighbourhood studies and the related internal reports are the main source for the chapters included in this book.

The quantitative household survey data was collected through mainly face-to-face interviews using a common questionnaire designed,

Table 1.2 Countries and cities covered in the SHLC project

Region	Country	Case study cities
Africa	South Africa	Cape Town (national parliament seat) and Johannesburg (economic centre/capital of South Africa)
	Tanzania	Dar es Salaam (old national capital and financial centre) and Dodoma (new national capital)
	Rwanda	Kigali (national capital) and Huye (regional city)
South Asia	India	Delhi (national capital) and Madurai (regional city)
	Bangladesh	Dhaka (national capital) and Khulna (regional city)
East Asian	China	Chongqing (one of the four cities under central government control) and Datong (a provincial capital and regional city)
	Philippines	Manila (national capital) and Batangas (regional city)

Source Authors, 2025

discussed and agreed together by all teams and implemented in local languages in the case study cities. Interviewees were a random adult member of the sampled household. The questionnaire covered socio-economic and demographic profiles of population living in different types of urban neighbourhoods and their experiences of the neighbourhood in relation to health, education, neighbourhood facilities and governance, economic opportunities and income, as well as the impact of Covid-19. The database includes over 15,000 responses (around 2000 from each country). Primary qualitative data was collected through focus group interviews, which covered views and opinions about neighbourhoods as well as health and educational needs, issues and problems. In total, 74 focus groups were organised in 12 of the case study cities.[2] The China team was unable to do any focus groups and the Chongqing survey was also delayed until summer 2023 due to Covid-19 lockdowns, which lasted longer than in many other countries.

Due to disruptions from Covid-19 lockdowns, data collection from the survey and focus groups was completed only in the last few months of the project's funding period; neither of these datasets have been fully analysed by the SHLC research teams. The city-focused chapters in this book were written by the partner teams in each country and draw heavily on the neighbourhood study. The analysis also focuses mainly on the larger city in each country. The contextual chapters, including those on health and education, were prepared by team members specialised in the relevant topic areas. The overall chapter structure and arguments were guided by the editorial team and based on the common research framework of SHLC.

1.5 Chapter Outline and Main Findings

This book includes 13 further chapters which are organised into several sections. Chapters 2 and 3 set the context for urban and spatial inequalities. Kintrea's chapter (2) considers what we know about urban socio-spatial segregation from the Global North literature and explores the ways in which urban inequalities might be different in developing countries. It argues that social and spatial inequalities largely originate from economic divisions between people, and divisions that continue to widen in many contexts by the competition for urban land and housing, sometimes mediated by the role of the state. It argues that such divisions are pervasive, mainly harmful to poor residents but also to society as a whole, but at the same time difficult to change even when public policy is sympathetic. With respect to developing countries, extensive existing knowledge of the extent of poverty

[2] SHLC data from the household survey and focus groups has been published by UK Data Service at ReShare. The data files and accompanied research tools can be accessed and downloaded from the ReShare website (https://reshare.ukdataservice.ac.uk/855998/).

and the weakness of public policy in mitigating housing and living conditions suggests that the extent and negative impact of socio-spatial segregation is more exaggerated even than in the Global North.

Wang, Baffoe, and Kamete continue the discussion about urban and neighbourhood inequality in the context of the Global South in Chapter 3. Under the general trend of urbanisation, they find very different modes of urban development in the study countries. Cities in China benefit from strong government control and public ownership of land, and urban development is dominated by professionally planned high-rise and high-density areas. In Bangladesh and India, urban development is driven by industrial development and migration, and the development process is led by small housing cooperatives and private housing renewal partnerships, and development mainly occurs on privately owned city or suburban land. Small plots of land and traditional private ownership dramatically increase the urban land use intensity and population density. Cities in the Philippines come in between the Chinese and South Asia development models. In places where there is a concentration of higher income groups, including expatriates or people of Chinese origin, larger apartment buildings and sizeable expensive suburban middle class new estates are built. However, other locations in Manila show a similar development pattern to that observed elsewhere in South Asia. Chapter 3 also shows that urbanisation in Africa follows very different routes. In South Africa, the emergence of large-scale informal settlements in its major cities further magnifies spatial and racial inequalities. In the central African countries of Tanzania and Rwanda, urban expansion is dominated by family-led private housing development in suburban areas. These very different processes of urbanisation create many different types of neighbourhoods, which are characterised by huge differences in income levels of residents, family wealth, access to public and social services, as well as very unequal conditions in respect of housing quality and the general living environment.

Chapters 4 and 5 examines urban spatial and neighbourhood inequalities from the perspectives of education and health, drawing on both qualitative and quantitative data collected by the Centre for Sustainable, Healthy and Learning Cities and Neighbourhoods. Education and urban education in particular, is often viewed as a leveller of inequalities and a key to social, economic, and personal development. Yet, education is riven with inequalities that perpetuate broader socio-economic inequalities and injustices. In Chapter 4, Nesterova, Osborne, and Schweisfurth explore the ways that spatial inequalities, particularly neighbourhood inequalities, shape educational opportunity in cities in the Global South, and how participation in schooling and adult education differs between neighbourhoods of different socio-economic status. They find that the 'urban advantage' in education masks vastly different levels of access, participation, retention, and quality between city neighbourhoods, particularly in relation to formal schooling. Yet while urban neighbourhoods shape educational advantage, and therefore life chances, they do not fully determine them. Their analysis shows there are some predictable patterns of educational advantage in wealthier neighbourhoods across countries, but also some outliers where poorer neighbourhoods perform better than expected, which are worthy of further exploration.

Bhandari and Mitchell, in Chapter 5 examine the role of urban neighbourhoods in shaping health outcomes within low- and middle-income countries (LMICs) amidst rapid urbanisation and pronounced social inequalities. They explore how neighbourhood attributes contribute to health disparities but also consider the potential for fostering healthier urban lifestyles. They argue that while the importance of the social and environmental determinants of health is relatively uncontested, the rapid unplanned urbanisation in developing countries is a challenge because, even if planners and politicians know what kinds of neighbourhoods are required, they are not generally in control of what kinds of neighbourhoods are actually created. Poor or absent urban planning can create neighbourhoods that promote and foster poor health behaviours, as well as those that threaten health directly via toxic or unsafe environments and hamper access to services. Bhandari and Mitchell underscore the influence of physical, social, and economic environments within neighbourhoods on the well-being of urban residents. By leveraging insights from the 14-city household survey, their chapter highlights the intricate ways in which neighbourhood factors can exacerbate or mitigate health disparities, emphasising the potential of localised, neighbourhood-level interventions in advancing the health-related Sustainable Development Goals (SDGs).

The book also focuses on urban and spatial inequalities in cities in Africa. South Africa is among the most unequal countries in the world. Chapters 6, 7 and 8 therefore focus on Cape Town and Johannesburg. In Chapter 6, Visagie, Turok, and Scheba address the question of whether Cape Town experienced racial desegregation since the transition to democracy in 1994. They examine the changing pattern of racial segregation in the city, which experienced a large decline in share of the coloured and white population in contrast to strong growth in the population share of Africans. They find that Cape Town has become less segregated, but mainly because of city-wide population growth among Africans, with limited evidence of deeper social integration or structural reduction in spatial inequalities. The sorting of households based on income and affordability continues to limit the options available to poorer groups who remain mostly black. It is middle-income (or lower middle class) neighbourhoods that are most conducive to racial integration but these neighbourhoods are still unaffordable for the average black South African. They also show there is a limit to growth of the black middle class, with a poor economic outlook for the country in the foreseeable future. These patterns reflect how the legacy of apartheid continues to be reproduced through residential sorting in the housing market. They conclude that in post-apartheid Cape Town, urbanisation has led to densification in poorer areas away from the core city rather than socio-spatial integration.

Lynge, Everatt, and Katumba measure quantitatively the hidden local exposure to economic inequality in the city of Johannesburg in Chapter 7. Knowing about inequality is one thing but experiencing it on a daily basis is something else. While the effects of the former have been examined through surveys and controlled experiments, the effects of the latter remain largely unexplored primarily due to the difficulties in methodology. Lynge and colleagues make a first attempt to measure local exposure to economic inequality. They calculate and map Gini-like coefficients for 5800 neighbourhoods across Johannesburg, drawing on household income data extracted from census data. They measure the degree of inequality not just within each neighbourhood but also in a specified radius around them. Their main objective is to test this methodology and create a measure of local exposure to economic inequality and lay the groundwork for future research. Their analysis finds that local context matters: two similar neighbourhoods (neighbourhoods with similar socio-economic characteristics) may be exposed to economic inequality in very different ways, due to their location in the city. They conclude that the engines of change seem to lie in neighbourhoods with high Gini-like coefficients, suggesting that exposure to economic inequality is an important and under explored variable in neighbourhood research.

In Chapter 8, Abrahams examines socio-spatial inequality in neighbourhoods in Johannesburg from a qualitative perspective. Drawing on focus group data from five different types of neighbourhoods, she looks specifically at three concepts which come to define place-making, namely, sense of place, belonging, and agency. Her analysis finds several clear juxtapositions where neighbourhood inequality is evident—lamentation and resignation at the poor quality of life in deprived neighbourhoods, and relative safety and conviviality in those well-off; a sense of belonging in a secure living environment and a sense of being forgotten or uncared for in those places which are less secure; and direct, mediated or marketised action or exercised agency for those in deprived, middle class, and wealthy areas respectively. These translate into the unequal ways that residents live in the city, can directly influence their own well-being or have a say in the way they live. Urban life in Johannesburg is then characterised by the experience of life and interaction with others, and public infrastructure, services, and amenities are only the stage upon which these interactions and associations play out. Life in the city is about how life is lived, how agency is exercised, and whether people have the freedoms to be able to exercise neighbourliness or live a good life. More importantly, Abrahams finds that it is about whether residents in neighbourhoods have the ability to stave off the harsh effects of market-like displacement and the slide into precarity and anomie.

Nduwayezu, Manirakiza, Malonza, Mugabe, Nsabimana, Rutayisire, Nzayirambaho, and Kato analyse in Chapter 9 the extent of city expansion in the last three decades in Kigali, Rwanda, focusing on the main categories of residential neighbourhoods—planned, unplanned, and mixed or semi-planned. Drawing on the SHLC household survey data, they investigate inequality by comparing indicators related to housing, income, employment, household amenities, health, and education across five income bands of neighbourhoods. Their analysis shows that Kigali has experienced substantial demographic and spatial growth in the last 30 years, driven primarily by internal migration and natural population increase. Rapid urbanisation is causing urban sprawl and extensive informal settlements that are overloading the capacity of city authorities to ensure adequate housing and equal distribution of basic amenities to all neighbourhoods, despite notable developments such as income diversification activities, new business infrastructure, and improved housing facilities. They also observe that the high rate of home ownership and the predominantly detached houses of Kigali can be understood as an indicator of relative liveability and high quality of life. However, the form of development is a threat to sustainable urban planning, as it is the cause of rapid spatial expansion.

The next section of the book focuses on Asian cities. In Chapter 10, Kundu and Debnath compare neighbourhood inequalities and access to housing, amenities, and services in Delhi and Madurai. Unlike most other empirical research on India cities that is based on ward level data, this study examines neighbourhood level inequality that shapes residents' everyday lived experiences. Based on the findings of the SHLC household survey of 21 neighbourhoods in Delhi and Madurai, the authors examine the nature and extent of inequalities between different types of neighbourhoods. They find stark inequality across neighbourhoods of different socio-economic status in both cities, especially on the basis of access to housing, amenities, and services, and particularly in respect of networked infrastructure such as public sewerage networks. Inter-city comparison reveals a higher level of inequality in Delhi in case of access to amenities and services compared to Madurai. Distribution of amenities and services, except for piped drinking water, is more inclusive in the regional city of Madurai. However, there was high inequality on the access to piped drinking water in Madurai across and within neighbourhoods of similar socio-economic status. Moreover, Madurai exhibits higher levels of inequality in terms of access to adequate living space.

Roy, in Chapter 11, examines neighbourhood inequalities in Dhaka, Bangladesh, and explores the challenges for urban planning. A significant share of neighbourhoods in Dhaka has evolved organically; they are flexible to change, and welcome migrants. Yet, the city is rapidly becoming unequal because of organic and uncontrolled changes. The distribution of health and education facilities and service provision varies considerably and results in spatial inequalities across neighbourhoods. Institutions, partisan politics, power, and the market are all responsible for the widening gap. Inadequate resources for meeting developmental needs, weak institutional capacity for guided urban growth, disproportionate distribution of the available resources due to market forces, and elitist planning are all identified as prime forces in making Dhaka an extremely unequal city. The prevailing nature of governance with colonial roots, ineffective urban policies and the neo-liberalisation of planning and governance have all contributed to sustained neighbourhood differentiation in Dhaka. Roy argues that it is urgent to promote the distribution of resources in ways that contribute to tackling the spatial disparities in the city. Innovative approaches to people-centred neighbourhood planning, acknowledging the internal structure and differentiation of neighbourhoods would make Dhaka a relatively equal city.

In Chapter 12, Liu, Li, Yin, Zhan, Sun, Zhai and Sun report on their study of urbanisation and neighbourhood diversification in the fast-changing Chinese mega city of Chongqing. They analyse how the state and the market have interacted to drive the city's urbanisation and how the previous socialist organisation of residence and work-unit system has been transformed since the 1980s through housing privatisation and commercial property development into a diversified and complex urban neighhourhood arrangement. They find that the city now is made up of very different types of neighbourhoods, including work-unit, commercial housing estate neighbourhoods, government-subsidised housing neighbourhoods, and urban villages. Each type exhibits some differences in terms of residents' socio-economic characteristics, housing conditions, internal spatial arrangements, accessibility to public services, as well as commercial facilities. Predicated upon both topographical and historical determinants, the city's residential districts are characterised by a polycentric and clustered spatial configuration. Distinct residential clusters preserve a degree of spatial segregation, concurrently facilitating relatively autonomous employment opportunities and public amenities within their respective precincts. Consequently, disparities have emerged among these clusters, contingent upon variances in educational institutions, health care facilities, income brackets, and property valuations.

Rivera, Gamboa, Delos Reyes, Racoma, and Manlapas examine the spatial inequality in Manila, the Philippines in Chapter 13. Their study takes a different approach by taking the notion of the ethics of (un)care as a spatially produced and reinforced phenomenon. They show that the contrasting yet profoundly similar living conditions in different neighbourhoods tell a tale of a city that cultivates, on one hand, a compassionate, humane, and caring environment, and on the other, signifies a global city that can be forgetful, restraining, and uncaring. Learning from these two tendencies, the chapter reflects on the potential for a bottom-up, solidarity-based, and empathy-enabled manner of urban citizenship to navigate long-standing spatial inequalities among neighbourhoods. The discussion serves as a backdrop for planning cities with a place-based caring ethic.

Earlier chapters focus either on a specific subject or on an individual country. In the final section of the book, Chapter 14, Ahmad, Sowgat, and Wang examine income, employment, and housing inequalities across all case study countries, using the household survey data collected from various neighbourhoods in the 14 Asian and African cities. Their analysis shows that cities in the Global South face significant levels of inequality and the unique social and political contexts of different countries have significant impacts on inequality. In general, Asian countries, particularly China, exhibit better income and employment stability compared to African countries; satisfaction with housing is higher in Asian countries, though a large number of households in these countries engage in informal employment and live in overcrowded dwellings compared to African countries. Their analysis also shows that major cities perform better on many aspects than smaller regional cities within a country. Disparities in housing and income are prevalent across cities but become more pronounced at the sub-city level, where households in low-income neighbourhoods face significant disadvantages in both housing and income compared to those in affluent neighbourhoods. Living in affluent neighbourhoods provides better housing and a superior built environment, contributing to overall well-being; conversely, residents of less wealthy neighbourhoods face heightened exposure to economic inequality. Analysis also shows that there is a substantial presence of high-income households in poor neighbourhoods, while affluent neighbourhoods have a significant share of low-income households. This indicates that Asian and African cities are currently experiencing early stages of socio-spatial exclusion when compared to developed Western countries. Their findings highlight the importance of neighbourhood data for understanding fast growing cities in the Global South.

References

Arsel M, Dasgupta A (2015) Critique, rediscovery and revival in development studies. Dev Chang 46(4):644–665. https://doi.org/10.1111/dech.12183

Atkinson R (2020) Alpha city: how London was captured by the super-rich. Verso Books

Baker E, Bentley R, Lester L, Beer A (2016) Housing affordability and residential mobility as drivers of locational inequality. Appl Geogr 72:65–75

Brenner N, Theodore N (2005) Neoliberalism and the urban condition. City 9(1):101–107

Burgess R (1992) Helping some to help themselves: third world housing policies and development strategies. In: Mathey K (ed) Beyond self-help housing. Mansell, London

DeLong JB (2016) A brief history of (in)equality. Public lecture, Online resources at: https://www.socialeurope.eu/brief-history-inequality

Devas N, Rakodi C (eds) (1993) Managing fast growing cities: new approaches to urban planning and management in the developing world. Longman Scientific and Technical

Donald B, Glasmeier A, Gray M, Lobao L (2014) Austerity in the city: economic crisis and urban service decline. Camb J Reg Econ Soc 7(1):3–15. https://doi.org/10.1093/cjres/rst040

Doucet B, Melik RV, Filion P (2021) Global reflections on COVID-19 and urban inequalities—Volume 1: community and society. Bristol University Press, Bristol

Galster G, Sharkey P (2017) Spatial foundations of inequality. Russell Sage Found J Soc Sci 3(2):1–33

Gilbert A, Gugler J (1992) Cities, poverty and development, urbanization in the third world. Oxford University Press, Oxford

Gollin D, Jedwab R, Vollrath D (2016) Urbanization with and without industrialization. J Econ Growth 21(1):35–70

Goodman AC (1988) An econometric model of housing price, permanent income, tenure choice, and housing demand. J Urban Econ 23(3):327–353

Grover A, Lall S, Maloney W (2022) Place, productivity, and prosperity: revisiting spatially targeted policies for regional development. World Bank, Washington, DC

Hamnett C (2019) Urban inequality. In: Schwanen T, van Kempen R (eds) Handbook of urban geography. Edward Elgar, Northampton, MA, pp 242–254

Harris R, Giles C (2003) A mixed message: the agents and forms of international housing policy, 1945–1973. Habitat Int 27(2):167–191

Hubbard P (2004) Revenge and injustice in the neoliberal city: uncovering masculinist agendas. Antipode 36(4):665–686

Jacobs J (1969) The economy of cities. Random House, New York

Jenkins P, Smith H, Wang YP (2007) Planning and housing in the rapidly urbanising world. Routledge, London and New York

Kamete AY, Tostensen A, Tvedten I (2002) From global village to urban globe: urbanisation and poverty in Africa: implications for Norwegian aid policy. Christian Michelsen Institute, Bergen

Kazepov Y (ed) (2005) Cities of Europe, changing contexts, local arrangement and the challenge to urban cohesion. Wiley-Blackwell

Kintrea K, Baffoe G (2025 forthcoming) What is the case for urban policy? In: Kundu D, Kintrea K, Everatt D and Wang YP (eds) Urban policy for sustainable development in the global south – new strategies for new challenges (Chapter 2), Springer.

Marcuse P, Van Kempen E (2000) Globalizing cities: a new spatial order? Blackwell

Martínez E, Rodríguez A, Altuzarra A, Álvarez I (2022) Remaking urban divides: shifting patterns of neighborhood differentiation in Bilbao, Spain. J Urban Aff. https://doi.org/10.1080/07352166.2022.2057319

McVie S, Understanding Inequality (UI) Project Team (2019) Summary report: the impact of inequalities in the early years on outcomes over the life course. Report published at the Understanding Inequality project website: https://www.understanding-inequalities.ac.uk/

Mela A, Toldo A (2019) Socio-spatial inequalities in contemporary cities. Springer, Cham

Mera K (1978) On urban agglomeration and economic efficiency. In Bourne LS, Simmons JW (eds) Systems of cities. Oxford University Press, New York

Mera K (1981) City size distribution and income distribution in space. Reg Dev Dialogue 2(1):105–120

Musterd S (2023) Advanced introduction to urban segregation. Edward Elgar, Cheltenham, UK

Musterd S, Marcińczak S, van Ham M, Tammaru T (2016) Socioeconomic segregation in European capital cities. Increasing separation between poor and rich. Urban Geogr 38(7):1062–1083. https://doi.org/10.1080/02723638.2016.1228371

Nijman J, Wei YD (2020) Urban inequalities in the 21st century economy. Appl Geogr 117(2020):1–8

Pardo I, Prato GB (2021) Urban inequalities: ethnographically informed reflections. Springer Nature

PEAK Urban (2023) Global report—PEAK Urban: unleashing sustainable cities of the future. Report published at PEAK Urban website: https://www.peak-urban.org/publications/global-report

Piketty T (2014) El capital en el siglo XXI [Capital in the 21st century]. Fondo de Cultura Económica

Piketty T (2020) Capital and ideology. Harvard University Press, Cambridge, MA

Potter RB, Binns T, Elliott JA, Smith D (2004) Geographies of development, 2nd edn. Pearson Education Limited, Harlow

Potter R, Conway D, Evans R, Lloyd-Evans S (2012) Key concepts in development geography. Sage, London

Pryce G, Wang YP, Chen Y, Shan J, Wei H (eds) (2021) Urban inequality, segregation & integration in Europe & China. Springer (Open Access)

Pugh C (1995) The role of the World Bank in housing. In: Aldrich B, Sandhu R (eds) Housing the urban poor. Policy and practice in developing countries. Zed Press, London

Pugh C (1997) The changing roles of self-help in housing and urban policies 1950–1996: experience in developing countries. Third World Plan Rev 19(1):91–109

Rakodi C, Lloyd-Jones T (2002) Urban livelihoods: a people-centred approach to reducing poverty. Earthscan, London

Renaud B (1981) National urbanization policy in developing countries. Oxford University Press, New York

Robinson J (2016) Comparative urbanism: new geographies and cultures of theorizing the urban. Int J Urban Reg Res 40(1):187–199

Robinson J (2022) Comparative urbanism: tactics for global urban studies. Wiley

Roser M (2017) The history of global economic inequality. Published online https://ourworldindata.org/the-history-of-global-economic-inequality

Shirazi MR, Keivani R (eds) (2019) Urban social sustainability, theory, policy and practice. Routledge, London and New York

Soja E (2009) The city and spatial justice. Justice Spatiale/Spat Justice 1(1):1–5

Soliman AM (2021) Urban informality—experiences and urban sustainability transitions in Middle East cities. Springer, Cham

Storey AD, Sheehan M, Bodoh-Creed J (eds) (2020) The everyday fife of urban inequality: ethnographic case studies of global cities. Lexington Books, London

Unceta PM, Hausleitner B, Dąbrowski M (2020) Socio-spatial segregation and the spatial structure of 'ordinary' activities in the Global South. Urban Plan 5(3):303–318

UN-Habitat (2003) The challenge of slums: global report on human settlements 2003. Earthscan, London

UN-Habitat (2016) The new urban agenda. UN-Habitat, Nairobi

UN-Habitat (2017) The new urban agenda. Nairobi: UN-Habitat. Available at https://unhabitat.org/about-us/newurban-agenda

United Nations (1996) The Habitat Agenda: The Istanbul Declaration on Human Settlements. Available at https://unhabitat.org/sites/default/files/2014/07/The-Habitat-Agenda-Istanbul-Declaration-on-Human-Settlements-20061.pdf (Accessed 23 February 2023)

United Nations (2000) United Nations Millennium Declaration. Available at https://documents-dds-ny.un.org/doc/UNDOC/GEN/N00/559/51/PDF/N0055951.pdf?OpenElement (Accessed 26 February 2023)

United Nations (2015a) Foreword to The Millennium Development Goals Report 2015. United Nations, New York

United Nations (2015b) Transforming our world: the 2030 Agenda for Sustainable Development. Available at https://documents-dds-ny.un.org/doc/UNDOC/GEN/N15/291/89/PDF/-N1529189.pdf?OpenElement.).

United Nations (2016) The New Urban Agenda: key commitments. Available at https://www.un.org/sustainabledevelopment/blog/2016/10/newurbanagenda/

Van Ham M, Tammaru T, Ubareviciene R, Janssen H (eds) (2021) Urban socio-economic segregation and income inequality. A global perspective. Springer

Wakely P (1988) The development of housing through the withdrawal from construction: changes in third world housing policies and programmes. Habitat Int 12(3):121–131

Wang Y, Kintrea K (2019) Sustainable, healthy and learning cities and neighbourhoods. Environ Urban ASIA 10(2):146–150. https://doi.org/10.1177/097542531985912

Wei YD, Ewing R (2018) Editorial: urban expansion, sprawl and inequality. Landsc Urban Plan 177:259–265

Wei YHD, Li H, Yue W (2017) Urban land expansion and regional inequality in transitional China. Landsc Urban Plan 163(2017):17–31

Wilkinson R, Pickett K (2009) The spirit level: why equality is better for everyone. Penguin Books

Wilkinson R, Pickett K (2020) The inner level: how more equal societies reduce stress, restore sanity and improve everyone's well-being. Penguin Press

Zwiers M, Bolt G, van Ham M, van Kempen R (2016) The global financial crisis and neighbourhood decline. Urban Geogr 37(5):664–684. https://doi.org/10.1080/02723638.2015.1101251

2 Urban Inequality and Socio-Spatial Segregation

Keith Kintrea

Abstract

This chapter considers what we know about urban socio-spatial segregation, mainly from the extant, mainly Global North literature. It argues that socio-spatial segregation is longstanding and near universal. While acknowledging authors who favour segregation, it is largely considered to be harmful, both to residents of the most disadvantaged areas and to urban society as a whole. The chapter than explores how urban inequalities might be different in developing countries. It suggests that the extent and negative impact of socio-spatial segregation is more exaggerated even than the Global North, driven by diverse, complex, and multifaceted emergent processes of urbanisation that are shaped by situated competitions for living space and development profits. The chapter is intended to provide essential context for interpreting the empirical, city-focussed chapters that follow.

Keywords

Segregation · Socio-spatial inequality · Cities · Global South · Urban policy

K. Kintrea (✉)
School of Social and Political Sciences, University of Glasgow, Glasgow, Scotland, UK
e-mail: keith.kintrea@glasgow.ac.uk

2.1 Introduction

Socio-spatial segregation is the division of the urban population both socially (between income groups and/or between groups with different socioeconomic, ethnic or religious characteristics) and spatially (between different residential neighbourhoods). It is typically identified empirically by the percentage of people with a particular characteristic that live in sub-areas of a city compared to the average distribution across the city as a whole.[1] Each socio-economic characteristic has its own distribution, so, for example, segregation by ethnicity may be stronger than segregation by income. Exact measures of segregation depend on the unit of analysis. The smaller the unit of area, the more likely it is that segregation will be found. The lack of common spatial units between cities internationally therefore makes it difficult to compare across countries.

Socio-spatial segregation is found wherever there are divisions in society that derive from socioeconomics and/or culture and is apparent to some degree in every city and large settlement in the world. Hamnett (2019) provides a useful overview of the forms and dimensions of urban inequalities. It is also as longstanding as cities themselves. The urban historian Nightingale (2012) reports that segregated cities existed

[1] There are many ways of measuring segregation; see Veneri et al. (2021) for a recent discussion.

Y. P. Wang et al. (eds.), *New Drivers of Division*, Sustainable Development Goals Series,
https://doi.org/10.1007/978-981-96-5629-5_2

in ancient Mesopotamia around 4000 BCE where, in the world's first cities, citadel zones around temples were set aside for the religious and political classes, for purposes of both defence and propaganda. He identifies similar 'sacropolitical' districts in ancient Africa, Peru and Cambodia, and in China, where the roots of the Imperial (or 'forbidden') City appeared in what became Beijing during the Shang dynasty not long after 2000 BCE. Staying with the ancient world, in 360 BCE Plato observed '*any city, however small, is in fact divided into two. One the city of the poor and the other of the rich; these are at war with each other*' (Marinoff 2019: 186).

In later centuries, planned segregation of cities became more diverse in intent and manifestation (Nightingale 2012). Political-religious citadels were extended during the medieval era but city neighbourhoods were also set aside for foreign traders and merchants. Sometimes foreigners were corralled into their own separate zones according to origin by the city authorities such as in Constantinople. But, increasingly, the foreigners led segregation themselves as their power grew. After the English conquest of Ireland, so as to keep the Irish themselves 'beyond the pale', the enclave of Dublin was created in the 1400s. It was the forerunner of a mode of planning, which developed across the globe that separated settlers from natives and encouraged further spatial sorting by colour, race, and religion, and the legacy of colonial spatial planning continues to shape cities today. As the economy of cities became more differentiated, distinctions between city dwellers by social class and status thrived. The separation of income groups and occupations also has ancient roots, often reflecting racial and religious differences, although in medieval and early modern cities it was frequently as much vertical within buildings as spatial, or highly variegated within tight areas of cities.

With the rise of industrial cities and the accompanying migration of labour from rural areas that gave a huge boost to urban growth, new forms of market-influenced segregation appeared. The global context for this was a new system of exploitation of people and resources based on colonial power and backed by violence as required, which shaped the world order. Merchants, manufacturers, and members of the new professional classes were enriched and they escaped air pollution and filth by commandeering cities' most salubrious and advantaged locations, often at a height and upwind of industrial sites, where they built fabulous mansions in imitation of the established noble classes (Simpson and Lloyd 1977). At the same time, a land market shaped by profit-oriented property owners and developers provided highly differentiated housing products and services for the bulk of the population. The subtle gradations of class and status in the industrial city were deliberately easy to read off in the built environment from its densities, plot sizes, dwelling sizes, and building design and ornamentation. However, the poorest usually were priced out of the new industrial housing and typically were left behind in the slums of the old medieval city cores, which had been otherwise abandoned.

When social policy and then urban studies emerged as academic fields in the late nineteenth and early twentieth century, the principal work of key figures such as Charles Booth in London and the leaders of the Chicago School was to describe and explain the separation of classes of people in urban space in works such as *The Gold Coast and the Slum* (Zorbaugh 1983). Meanwhile, new state-led segregation emerged as governments in the developed world responded to the political demands of organised labour by funding or directly providing state housing (see, e.g. Harloe 1985). Distinctive in its built form and characterised by large-scale construction as 'estates', state housing not only marked out those who benefitted from its bricks-and-mortar subsidies but it also contained strong social divisions, between 'rough' and 'respectable' tenants, whose effects are still apparent today (Damer 2018).

From the days of the pioneers of urban studies, but particularly in the period since the 1960s when conflict with significant racial dimensions emerged in segregated cities in the USA and later in Europe, research on urban socio-spatial segregation has exploded and has never died down. It continues to be a pressing matter in the Global North, both in terms of social equity and national security, as the disturbances in French cities following the killing of a young person of Algerian background in the Paris suburbs in the summer 2023 show (Observer 2023). At the time of writing, a search for 'urban segregation' on *Google Scholar* leads to over 1.5 million items, while related topics such as urban deprivation and gentrification each generate hundreds of thousands of results.

There is often acknowledgement in this literature that urban segregation is a global phenomenon but it is apparent that such academic studies are dominated by North American and other English-speaking researchers and that most of the empirical literature on segregation is likewise derived from Global North case studies and data. In spite of the recent massive growth of cities in the Global South, neighbourhood research as a whole is relatively undeveloped there, with the exception of China (Baffoe and Kintrea 2022, 2023). This Global North 'overburden' has caused Garrido (2021) to argue that the research field of urban socio-spatial segregation as it applies to the South has been stunted or even closed down, thereby distinctive dimensions of segregation in the South have been missed, or not discovered. The rationale of this book supports Garrido's position but it is hard to ignore the vast body of extant work which has generated important ideas about the origins of social inequality, how it translates into socio-spatial inequality, why such inequality matters and what -if anything- might be done about it by public policy, even if this must be carefully re-interpreted in the context of the Global South.

2.2 Unequal Societies and Unequal Places

A necessary starting point for any consideration of socio-spatial inequalities must be the existence and persistence of inequalities between groups in the population within particular societies, particularly in terms of income and wealth. The *World Inequality Report* (Chancel et al. 2022), using ratios between the top 10% and the bottom 50% of the global population, shows that inequalities in locations across the world grew steeply in the nineteenth century and then flattened out but with the income of the top 10% of the population globally being between 38 and 53% of the bottom half of the population during the period 1900–2020. Across the same period, the share of total income going to the top 10% has been between a half and 60%, and of the bottom 10% somewhere between 5 and 15%. So income inequalities are large and have been highly persistent across an extended period when there has been rapid societal, political, economic, and technological change, although the report also shows that global income inequality has declined somewhat in the last 20 years (Chancel et al. 2022: Figure 2.1, 2.2 and 2.3).

Considering wealth (including both financial and non-financial assets), globally the top 10% has held at least half of the world's wealth for the last 200 years, with their share growing during the period of industrialisation and colonialisation up to 1920 to 60% and fluctuating since. Since 1995 there has been a notable tendency for those wealthy people in the developed world at the very top of the distribution (the top 1%) to capture an even greater share of wealth, along with some groups in the middle of the hierarchy, representing the rise of the middle class in emerging economies. However, the bottom half's share of global wealth fell from about 15% to 10% over the 160 years to 1980 but has risen slightly since (Chancel et al. 2022: Figure 4.2).

If the global picture of inequality between individuals appears fairly stable, even if the evolution of patterns of income and wealth inequality continues, inequalities between countries appear to be lessening somewhat, as more of the world develops. But within countries, inequalities have reached a record high level (Chancel et al. 2022: 28), which is directly relevant to our concerns with socio-economic segregation. In 2021, the people in the bottom 50% of the income distribution in the countries included in the SHLC study on which this book draws all have less than 18% of their country's income. This ranges from 17.1% in Bangladesh down to 5.8% in South Africa. Meanwhile, the people in the top 10% command over 40% of the income in each SHLC country, the range running from Bangladesh on 42.4% to

South Africa on 65.4%. The extent of inequality in developing countries is far stronger than in Europe (the world's least unequal region). For example, the comparable figures for the UK are 20.3% and 35.8% and for France 23% and 31.2%. It is also apparent that, although the countries in the SHLC study have all benefitted from economic growth in the last 40 years, the share of income going to the bottom 50% has fallen in several countries, most notably in China, India, and South Africa. It has risen in only one country- the Philippines. Simultaneously, over the same period, the top 10% in every SHLC country has benefitted from a rising share of income, except again for the Philippines. (All data from the World Inequality Database 2022.)

Such extreme and deepening inequalities are driven substantially by unequal growth. They provide global and national moral philosophical challenges on their own terms and arguments can be made for more equal distributions drawn from perspectives on social justice, human rights and 'the right to the city'. But inequalities also have clear impacts on social welfare. Wilkinson and Pickett's (2009) study *The Spirit Level* demonstrated that the more unequal a country is the more likely people will experience, among other things, worse physical and mental health, lower life expectancy, greater obesity and higher levels of violence, and citizens are also less likely to enjoy trust between each other. Therefore, they conclude that, in unequal societies after a certain point, economic growth falters in its task of delivering greater population welfare, while a policy strategy to reduce inequality is likely to have widespread benefits.

These deep inequalities also provide the raw material of socio-spatial segregation. However, it is not possible to read off the extent or nature of socio-spatial segregation from the data on income or wealth divisions. Broadly, segregation comes about as a result of the translation of economic transformations in cities, such as changes to the economic base and the labour market, into socio-spatial transformations. In the contemporary era, Sassen (1991) suggested that the more connected cities were to the global economy, the more likely they were to be highly segregated, with well-paid knowledge workers separated both economically and spatially from a much less well-rewarded and less well-housed service class. Two recent comparative international studies (Musterd et al. 2017, building on Tammaru et al. 2016, and Veneri et al. 2021), provide some insights into explanations for different patterns of segregation. The work by Musterd and Tammaru and their colleagues showed that European capital cities were becoming more economically unequal in the first decade of the twenty-first century and at the same time more segregated, especially between the highest and lowest income groups.

Segregation is mediated principally through the housing system but it is also influenced by exclusion and discrimination and the exercise of political power, usually involving the direct collusion of the state in the form policies and practices towards land use planning and housing. Hence there is extensive interest in gentrification as a process of socio-spatial transformation in post-industrial Global North cities,[2] including its 'state-led' forms (e.g. Watt 2009; López-Morales 2019). These cities have seen the rise of the so-called 'hourglass' labour market, with increasing distance between the knowledge workers at the top of the labour market and the routine service workers at the bottom, as well as revived preferences for city centre living among the middle class, which has the strong support of city planners who are pursuing policies such as 'compact cities' and '20 minute neighbourhoods' in pursuit of climate change and liveability goals.

A critical lesson for fast developing cities is that once set, the socio-spatial structure of cities is not something that can be quickly changed. The historical class and status associations of particular neighbourhoods tend to persist for decades, even centuries. Perhaps this is most evident where the state has defined areas of residence for particular classes or cultural or ethnic groups, such as with colonial-era planning, or where areas have been designated for low-status social housing. In a spatial demonstration of the reproduction of social status, such designations thrive in the social value of particular sites (or the lack of it) long after

[2] Gentrification research is now global.

the policy that created them has been superseded. In one documented case, the low social status of a neighbourhood in the twenty-first century was shown to have originated hundreds of years prior from encampments of the poor beyond a city's walls (Robertson et al. 2010).

What is more, the operation of the land market typically creates 'spatial lock-in', which means that large adjustments in financial value (and hence significant social change) in neighbourhoods) are rare and can only be achieved through significant intervention such as complete physical destruction and rebuilding of a neighbourhood (for example Nygaard and Meen 2013). For example, the basic socio-spatial structure of central London survived a century of industrialisation and deindustrialisation, aerial bombing and reconstruction, extensive state housebuilding, financialisation of the housing market and successive waves of immigration (Orford et al. 2002).

This is important knowledge in the context of fast-urbanising cities. Experience suggests that solutions to the patterns of human settlement that are allowed or encouraged in the early twenty-first century are likely to be around for many decades, even centuries, ahead.

2.3 In Defence of Segregation?

While segregation in cities is pervasive, arguments about its benefits and disbenefits are not fully settled, which in part accounts for its popularity as a research topic. Arguments in defence of segregation are heard much less often, and centre around the benefits that people derive from the proximity of other people who are like themselves. During the apogee of international interest in mixed-income neighbourhoods as a part of planning and housing policy (Arthurson 2012; Kintrea 2013), the economist Paul Cheshire was notoriously sceptical about the international acceptance of socially mixed neighbourhoods as a policy aim (Cheshire 2007; Cheshire et al. 2008). In making his pitch against social mix policies, he identified the benefit of what he called 'specialised neighbourhoods', that is those that house people from particular income groups. He argued that such neighbourhoods had benefits in they provided income-appropriate services, which, in poor neighbourhoods, were of direct benefit to lower-income people, such as affordable food shops and pawn shops. They also hosted linguistically and culturally appropriate services for disadvantaged migrant groups. Cheshire also suggested that social interaction with neighbours who share characteristics also brought benefits in the form of useful social networks and, drawing on Luttmer (2005), claimed that low-income residents would benefit by feeling less bad about their own poverty if they were surrounded by others in the same situation.

The philosopher Michael Merry (2013) makes some similar but perhaps more sophisticated arguments in his 'defence of separation'. Merry seeks to challenge accepted views on segregation (and separation more widely) and points to the limitations of what he calls 'integrationism', that is policies that attempt to bring together people from diverse groups, although he does not tackle 'social mix' policies directly. He argues that in a situation where there are excluded and stigmatised groups (and the urban poor would come into that category), they may usefully seek 'voluntary segregation' on their own terms. While not denying there are some poor neighbourhoods that compound their residents' disadvantage, he argues that spatial concentrations of disadvantaged people provide the possibility of the community acting in its own interests, as he illustrates elsewhere through the story of one village's housing campaign (Merry et al. 2016).

2.3.1 The Case Against Segregation

In spite of the cases put by Cheshire and Merry, most of the literature focuses on the undesirability of segregation. There are two lines of argument. The first is that segregation is particularly bad for those people that have the fewest choices, that is those at the bottom end of the income distribution whom segregation confines to neighbourhoods that are dominated by poor people. Such neighbourhoods are typically labelled by planners using language such as 'deprived', 'disadvantaged' or 'excluded' or even officially called 'slums', a term that has been condemned as discriminatory and even dangerous (e.g. Gilbert 2007). The second broader line of argument is that segregation is bad not just for residents in disadvantaged areas but also for the wider society.

Taking the first line of argument, it has three elements. The first concentrates on spatial processes and it holds that disadvantaged areas further harm their residents because poor areas have inferior or absent services. Considering services provided for profit, the accentuation of disadvantage come from simple market forces reacting to weak demand. So, for example, shops may be scarce in a disadvantaged area and the few shops that exist are may offer low quality goods. Considering public services, there is often unconscious or sometimes deliberate steering in the allocation of resources away from deprived areas (e.g. Hastings et al. 2017). Disadvantaged areas may be poorly provided, for example, with health services, schools, and street cleaning. In developing countries, poor areas they may also lack access to clean water, sewerage and energy services and to the publicly maintained roads that are found in rich areas. People in disadvantaged neighbourhoods may also find it hard to access jobs and economic opportunities they are mainly located in city centres and more affluent suburbs.

The second line of argument is that people who live in impoverished areas suffer placed-based or territorial stigma in the world beyond their neighbourhood, based purely on their home location. In consequence they suffer discrimination in their access to education, work, health care, and social relationships, with subsequent impacts on their wellbeing. This phenomenon has been evidenced by a large number of studies, both recent and historical, across the Global North and Global South, with territorial stigma often co-existing with ethnic, racial, or religious discrimination. Global North studies include those by Wacquant (1993), Jacobs and Flanagan (2013) and Butler-Warke (2020) while recent examples of Global South studies on the same themes include Devika (2016), Fattah and Walters (2020) and Dev et al. (2023).

The third part of the argument that poor areas further disadvantage their residents' points to the negative 'neighbourhood effects' that operate though social process at a neighbourhood level (for reviews see, e.g. Galster 2012; Sampson 2019). Such ideas are generally traced to the work of Wilson (1987) who studied Black neighbourhoods in the USA that had become geographically isolated from their urban economies through deindustrialisation and spatial relocation of economic activity and in consequence suffered extensive worklessness and the outward migration of the middle class. Neighbourhood effects subsequently become a major multi-disciplinary research enterprise in the UK and Europe, however it has yet to make much impact in developing countries.

Briefly summarised, the theory of neighbourhood effects rests on a series of four propositions (Atkinson and Kintrea 2002; see also Musterd et al. 2019). First, disadvantaged neighbourhoods function as a defining social world, and second, that in such neighbourhoods, people mainly mix with others who are like themselves. So, in a disadvantaged context, social interactions within the neighbourhood are among the most important that people have and most of the interactions are between people who share a common experience of disadvantage. The third proposition is that people develop shared beliefs and expectations (for example about work, education, and culture) based on their common experiences of poverty and exclusion. Fourth, limited social networks and the kinds of beliefs and expectations that develop around poverty are ultimately harmful to life chances, as they serve to limit the possibilities of residents' lives.

Relating these propositions to theories of social capital, it could be said that in poor neighbourhoods, there is a surfeit of 'bonding' social capital (the kind that helps people 'get by') but a shortage of bridging social capital (the kind that helps people 'get on'). Hence a central part of the case for more socially mixed neighbourhoods rests on the wider and deeper social networks that might become available. Taking the three parts of the argument together (territorial stigma, urban services and neighbourhood social networks), a useful insight is that neighbourhoods represent a 'spatial opportunity structure' (e.g. Galster and Sharkey 2017) that serves to negatively affect outcomes for residents, such as educational attainment, employment status and health.

The case against segregation centres also centres on its ill effects on the wider society and the urban economy, particularly due to its potential to contribute further to social divisions and to undermine social cohesion. Social cohesion is a slippery concept but at its heart is the capacity of diverse groups in the city to recognise each other and to achieve mutual accommodation through social and political institutions (Cassiers and Kesteloot 2012). Cohesion is put at risk by many features of modern cities, including a lack of economic security for many individuals and very wide inequalities in income and wealth, as well as the distrust and conflict that are fomented by the rise of populist politics. However, socio-spatial segregation plays a key role in undermining cohesion. Place attachment and identity, that is where there is a sense of belonging to particular place and an intertwining of personal identity and place, is one of five domains of social cohesion identified in the classic analysis of Forrest and Kearns (2001). Where cities are sharply divided into zones that are significantly exclusive, the cohesion of the whole city is at risk, as it breaks the everyday links between groups. Spatial divisions have the power to undermine empathy for the poor and make redistributive policies even less popular as they intensify mutual distrust between people in different economic situations. Such distrust also creates a ratchet effect so that the rich seek to barricade themselves within gated enclaves and engage in 'cocooning' in their travel, consumption and leisure behaviours, all designed to create distance from others in society, far away from the so-called 'no go' areas of the poor (e.g. Atkinson 2006; Boterman and Musterd 2016).

While well-off people may be more attached to global networks but cut themselves off from their city, disadvantaged residents often exhibit deep place attachment and strong community solidarity, which as Merry (2013) argues, may be of great benefit. But considering social cohesion, inward-looking isolated communities also represent a risk. Many of the violent disturbances (or uprisings) that have occurred in European cities have originated in socially isolated, disadvantaged areas, usually with high proportions of migrants and minority groups, where younger residents feel a powerful sense of separation and injustice. For example, the waves of disturbances across France in the summer 2005 and again in 2023 originated mainly in poor, segregated enclaves of social housing (Fieschi 2023). The same phenomenon was apparent in several episodes of disturbances in Swedish suburbs in 2022, where segregation was identified by the justice minister as a very serious cause of the problems, as well as in other locations internationally. Ultimately, inequality, socio-spatial segregation and a lack of social stability can make cities a less attractive places to live, repel investment and affect their competitiveness (Tammaru et al. 2016).

2.3.2 Public Policy Versus Socio-Spatial Segregation

On the basis that socio-spatial segregation is harmful, or perhaps more often because it is so emblematic of the failure to produce outcomes that are evidently fair for all citizens, especially when disturbances break out, national and city governments across a wide range of countries have from time to time prioritised it. Andersson (2006) neatly sums up the three main policy options in the typical situation where there is a gradient of segregated neighbourhoods ranging from rich to poor. The first option is to generally redistribute income and wealth so that the extent of inequality in society between the ends of the distribution is lessened. In principle, this might be achieved by fiscal policy (tax and public spending), redistributing resources in favour of the poorer citizens who are most likely to live in disadvantaged neighbourhoods. As we have seen, in most Global North countries, this is the opposite of what has been happening, with the distance between the rich and poor widening somewhat in the period since 1980 as the state has withdrawn from a strong welfare role in favour of distribution through the market.

The second and third options are in the realm of urban policy. The second one is longstanding, since at least the 1960s in the UK and USA, for example, and became very widespread in the Global North in the 1990s and 2000s (Smith 1999). It aims to flatten out the gradient of neighbourhood's right at the bottom end of the distribution through diverse types of area-based initiative (ABI). ABIs typically target the 'worst' (i.e. the most disadvantaged) neighbourhoods and through public-sector-led partnership agencies they aim to improve housing and neighbourhood conditions and services, often alongside initiatives to reduce unemployment and other forms of worklessness. While they often improve material conditions through targeted spending, studies show they have they are not able to do much fundamentally change the disadvantaged status of such areas and their position at the bottom of the socio-economic ranking (see, e.g. Lawless and Beatty 2013).

The third approach to the problem of socio-spatial segregation are policies on social mix, which became extremely prominent internationally in the first decade of the twenty-first century (see, Arthurson 2012; Kintrea 2013). Social mix as a policy was intricately connected to theories about the lack of opportunity structures in poor areas and rested on the idea that wider social networks and role models had the potential to be transformative of people's life chances, and that higher income residents would improve local services through spending power and political advocacy. Like ABIs, social mix policy positioned poor areas as the problem to be solved and, while planners promoted some newly built areas as socially mixed communities, there was never any sense that social mix polices would be aimed at neighbourhoods dominated by the well-off.

There were very many interesting social mix experiments and initiatives in several different countries, some quite successful at a basic level, showing that people from a wide range of income groups could share the same neighbourhood. However, evidence of beneficial effects on the life chances of disadvantaged people was slight at best (e.g. Briggs et al. 2010), and social mix policy was never developed on such a scale that it could overcome the power of the market- based division of residential areas based on preferences to live apart, if possible.

2.4 Implications for Cities in Developing Countries in Africa and Asia

The analysis above of socio-spatial segregation is of its longstanding development, its rootedness in the economic structure that shapes inequalities of income and wealth, the damage that it causes, its persistence over time and of the difficulty of doing anything fundamental to change it. This account is drawn mainly from the Global North experience. But while it is easy to be fatalistic, there still are positive decisions that can be made about income and wealth redistribution, at least given sympathetic politics. The public sector can also take a role in planning, regulation of the built environment and service provision, which has the possibility of at least of ameliorating inequalities between neighbourhoods. Local communities, even disadvantaged ones, can also sometimes organise to defend their own neighbourhood interests. So, given the subject of this book, what knowledge about segregation can be carried forward into the context of fast developing cities in Africa and Asia?

Unfortunately, it is difficult to muster a great deal of optimism. Almost all the signs point to socio-spatial segregation in the Global South being more extreme, more damaging in terms of 'opportunity structures', and even less likely to be countered by public policy. Considering, first, the urban economic conditions under which segregation develops, the countries in the SHLC study are far more unequal in terms of income and wealth than the Global North countries where the study of segregation has a history of more than 120 years. Importantly, they are also developing in a quite different economic era. While industrialism and colonialism in the north widened the income and wealth span, in the twentieth century they also offered a range of economic opportunities in the middle range (skilled manual jobs) that helped build a society of 'affluent workers'. In turn, these workers were able to exert strong political power that helped to shape welfare states. Welfare institutions have in recent decades been under political attack, and the labour market in the north has become more polarised but a considerable legacy of welfare remains in key fields such as education, health, and social security. But cities in developing countries are bypassing the industrial stage, or seeing it pass quickly as in China, and moving straight into service economies that require residents to have formal credentials to get the few high paid jobs. While this helps to create a new, small middle class, this kind of economy leaves behind employees in manufacturing (outsourced from the developed world), which now requires less and mainly unskilled labour, as well as those in low-paid personal services jobs and a very large the informal economy (Randolph and Storper 2023). In many developing country cities there is also the phenomenon of jobless growth (e.g. Tejani 2016). Prompted by low and falling living standards in rural areas, often as a consequence of conflict and climate change, migrants move to cities but without even insecure low paid employment to go to. All this means that the raw material of segregation, i.e. the divisions of income and wealth, is particularly stark, and many citizens lack economic value and therefore political clout.

Then there is the question of the urban fabric within which neighbourhoods develop. Many cities in Africa and Asia suffer spatial legacies where colonial planning has scripted a pattern of differential neighbourhood social status, even if urban growth has gone far beyond colonial-era boundaries. But it is sprawl and peripherality that characterise a great deal of current development, especially in Africa, and give rise to huge, predominately poor neighbourhoods which, lacking any reasonable transport infrastructure, are disconnected from the jobs that exist in the more affluent areas and in city centres (e.g. Bloch 2018). Many of these impoverished areas also have few regularised streets where more formal economic activity can coalesce. So poor neighbourhoods form that are spatially distanced from economic activity.

The physical conditions of neighbourhoods in developing countries also show huge disparities between social groups, as evidence in this book shows. The better off live in neighbourhoods that have been planned, at least to some degree, and are well ordered and serviced with utilities. In contrast, poor neighbourhoods often emerge incrementally on unused land and are wholly informal and self-constructed, and most are still waiting for a full complement of utilities. Therefore, the wide social distance between neighbourhoods is matched by enormous disparities in housing quality, sanitation and energy availability that are much greater than those in the Global North, even the north of 100 years ago.

The role of government in the Global North has been important in shaping neighbourhood segregation and the quality of neighbourhoods at different points on the socio-economic spectrum. Although there are many critics of the state, including by those who point to the state doing the work of capital in shaping cities for the benefit of more affluent people, governments at least retain some capacity to plan residential areas and to regulate construction activity and environmental standards. As discussed, some maintain an interest in segregation and have policies to counter it, albeit weak ones. It helps that nearly all residents have some enforceable rights of occupancy. However, urban policy in many developing countries, including some included in this study, is severely lacking. UN-Habitat submits that that many governments in developing countries lack the data and the professional skills to develop urban policies, and the knowledge and resources to effectively implement them (UN-Habitat 2022).

It does not help that the United Nations' Sustainable Development Goals, and specifically Goal 11 which is about 'making cities and human settlements safe, resilient and sustainable' (United Nations 2015), have absolutely no concern about socio-spatial inequality, even though it poses a clear challenge to sustainability in all of its forms. Even more surprisingly is the same silence within the *New Urban Agenda (NUA)*, which is UN-Habitat's declaration on 'sustainable cities and human settlements for all' (UN-Habitat 2017: 2).

The *NUA* calls on UN member countries to readdress the ways in which cities are planned and managed and this has prompted a wide interest in urban policy (Kintrea and Baffoe, forthcoming) but there is no encouragement at all to use urban policy to challenge socio-spatial inequalities.

In order to defend their interests, residents of disadvantaged areas require a certain level of rights. However, in developing countries universal land registration systems rarely prevail and security of tenure, especially in self-built neighbourhoods, is often absent. A significant strand of the evidence focuses on urban policy that is punitive towards the poor. The precarity of property rights and sometimes of citizenship rights, and the lack of capacities to be involved in the formal economy, accentuates the vulnerability of residents in disadvantaged areas. So, in very many documented cases, rather than being a force for integration, urban policy involves dispossession in the poorest areas and relocation to ever more marginal sites (e.g. Doshi 2015; Ortega 2020).

The commentary above suggests that problem of socio-spatial segregation in developing counties will share characteristics with the Global North but is likely to be even more problematic in terms of the wide social distance between the rich and the poor and its impact on residents. But even if some fundamentals are similar, an important question concerns whether the nature of socio-spatial segregation in some respects or in some places is emerging in fundamentally different forms compared to the Global North.

Increasingly, writers are suggesting that the idea that the socio-spatial structure of cities can be read off from the Global North model or from universal theories of global cities should be rejected (Robinson 2016). Relatedly, Schmid et al. claim evidence exists for a 'multitude of urban forms in a variety of contexts' (2018: 22) and argue that cities no longer constitute 'units that can be delimited in a clear and easy way, but are highly dynamic, multifaceted and complex' (2018: 31), a situation that arises from multi-dimensional and somewhat unpredictable urbanisation processes. They also stress that key concepts in urban studies, such as gentrification, suburbanisation and even informality as applied to settlements in developing countries, are essentially rooted in western experiences.

Researchers are beginning to respond to these kinds of ideas in tandem with a desire to decolonialise urban research, to take three recent examples. In Metro Manila, Garrido (2021) finds that informal settlements are interspersed throughout the conurbation, with multiple locations in each administrative area leading to the inevitable collision of areas of strongly contrasting social status. The equivalent of the 'Gold Coast and the slum' are not in separate zones but in close proximity. In Manila, this all significantly driven by the competition for space. Rather than the social and economic isolation of the poor, he suggests there is considerable interaction, even if it is unequal between classes. If this tendency is widespread beyond the Philippines, it tends to suggest that the impacts of segregation, including the social isolation and the limited opportunity structures that have fuelled the concept of and the evidence on neighbourhood effects in the Global North (especially the USA) may be somewhat less in developing country cities, even if the housing opportunities, physical conditions and sanitation in many poor neighbourhoods may be objectively much worse than their equivalents in the north.

Similar observations have been made in the context of Latin American cities where urban fragmentation has been developed as a key idea, sitting alongside residential segregation and territorial inequality to explain divisions within cities (Santiago et al. 2022). Fragmentation, they argue, has been enabled by the development of an urbanisation model that is driven, on one hand, by a lack of regulation and therefore the flexible use of land, and on the other by the development of what they call 'satellite-auto contained units' such as gated communities and shopping centres, which are products of a neo-liberal development process. It has also been aided by the breakdown in the clear distinction between the ex-colonial planned centres and the unplanned peripheries as the housing market have become more vital and complex.

A third example of research that stresses new forms of socio-spatial structure is Meth et al.'s (2021) work on mainly residential African urban peripheries. While peripheries have a set of common characteristics, including distance to central cores and labour markets, and relatively cheap land, on the basis of the emerging literature and their own research, they find very great diversity in terms of wealth and poverty, connectivity, and economic dynamism. They propose five 'logics' of urbanisation that produce different archetypes of peripherality, which essentially concern the main drivers of urbanisation, including speculators, self-builders and the state.

These examples all have in common an emphasis on explanations for urban form and socio-spatial structures that lie with diverse emergent processes of urbanisation. These are substantially contingent on locally-shaped competitions for living space and for development profits, leavened or encouraged by the role of the state, albeit competitions that are substantially more multifaceted and complex than the divisions that emerged in Global North cities amid industrialisation. But so far it is not so clear to what extent they cut across the fundamentals of class, race, and power that have generated segregation since the inception of cities.

2.5 Conclusion

This chapter has situated socio-spatial divisions in cities historically, and in the well-researched context of Global North, mainly western, urban studies. It has set out that they largely originate from economic divisions between people, divisions that continue to widen in many contexts, and which are translated into spatial segregation by the competition for urban land and housing, sometimes mediated by the role of the state. It has argued that such divisions are pervasive, mainly harmful to poor residents but also to society as a whole but at the same time difficult to change even when public policy is sympathetic. With respect to developing countries, extensive existing knowledge of the extent of poverty and the weakness of public policy in mitigating housing and living conditions, suggests that the extent and negative impact of socio-spatial segregation is more exaggerated even than the Global North. New epistemologies and methodologies in urban research are emerging that challenge western understandings and some appear to show that urban forms in developing countries are more variegated than those provided by existing accounts, however socio-spatial segregation is not often at the centre of these new enquiries. The empirical chapters in this book, each focussed on a different city, provide new data on emerging segregation to add to this debate about twenty-first century urban forms.

References

Andersson R (2006) Breaking segregation: rhetorical construct or effective policy? The case of the metropolitan development initiative in Sweden. Urban Stud 43(4):787–799

Arthurson K (2012) Social mix and the city: challenging the mixed communities consensus in housing and urban planning policies. Csiro, Clayton

Atkinson R (2006) Padding the bunker: strategies of middle-class disaffiliation and colonisation in the city. Urban Stud 43(4):819–832

Atkinson R, Kintrea K (2002) Area effects: what do they mean for British housing and regeneration policy? European Journal of Housing Policy 2(2):147–166

Baffoe G, Kintrea K (2022) Towards understanding the landscapes of neighbourhood research: an insight from bibliometric analysis. Sustainability 14(5):3115

Baffoe G, Kintrea K (2023) Neighbourhood research in the global south: what do we know so far? Cities 132:104077

Bloch R (2018) Africa's new suburbs. In: Hamel P, Keil R (eds) Suburban governance: a global view, Chapter 12. Toronto: University of Toronto Press, pp 253–277

Boterman WR, Musterd S (2016) Cocooning urban life: exposure to diversity in neighbourhoods, workplaces and transport. Cities 59(1 November):139–147

Briggs X, Popkin S, Goering J (2010) Moving to opportunity: the story of an American experiment to fight ghetto poverty. Oxford University Press, London

Butler-Warke A (2020) Foundational sigma: place-based stigma in the age before advanced marginality. Br J Sociol 71(1):140–152

Cassiers T, Kesteloot C (2012) Socio-spatial inequalities and social cohesion in European cities. Urban Stud 49(9):1909–1924

Chancel L, Pitketty T, Saez E, Zucman G (2022) World inequality report 2022. World Inequality Lab, Paris

Cheshire P (2007) Segregated neighbourhoods and mixed communities. Joseph Rowntree Foundation, York
Cheshire P, Gibbons S, Gordon I (2008) Policies for "mixed communities": a critical evaluation. SERC Policy Paper 2. London School of Economics, London
Damer S (2018) Scheming: a social history of Glasgow Council Housing, 1919–1956. Edinburgh University Press, Edinburgh
Dev S, Duval J, Galivanche A, Shitole T, Sawant K, Shitole S, Patil-Deshmukh A, Lincoln A, Subbaraman R, Weinstein L (2023) Spatializing stigma-power: mental health impacts of spatial stigma in a legally-excluded settlement in Mumbai, India. PLOS Global Public Health 3(7):e0001026
Devika J (2016) Aspects of socio-economic exclusion in Kerala, India: reflections from an urban slum. Critical Asian Studies 48(2):193–214
Doshi S (2015) Rethinking gentrification in India: displacement, dispossession and the spectre of development. In: Lees L, Shin H, Lopez-Morales E (eds) Global gentrifications: uneven development and displacement. Policy Press, Bristol
Fattah K, Walters P (2020) 'A good place for the poor!' Counter narratives to territorial stigmatisation from two informal settlements in Dhaka. Soc Incl 8(1):55–65
Fieschi C (2023) These French riots are different and far more disturbing. Politico. https://www.politico.com/news/magazine/2023/07/07/french-riots-banlieues-00105143 Accessed 14th July 2023
Forrest R, Kearns A (2001) Social cohesion, social capital and the neighbourhood. Urban Stud 38(12):2125–2143
Galster G (2012) The mechanism(s) of neighbourhood effects: theory, evidence, and policy implications. In: Van Ham M, Manley D, Bailey N, Simpson L, Maclennan D (eds) Neighbourhood effects research: new perspectives. Springer, Dordrecht
Galster G, Sharkey P (2017) Spatial foundations of inequality: a conceptual model and empirical overview. Russell Sage Found J Soc Sci 3:1–33
Garrido M (2021) Reconceptualizing segregation from the global south. City Community 20(1):24–37
Gilbert A (2007) The return of the slum: does language matter? Int J Urban Reg Res 31(4):697–713
Hamnett C (2019) Urban inequality. In: Schwanen T, van Kempen R (eds) Handbook of urban geography. Edward Elgar, London
Harloe M (1985) The people's home?: Social rented housing in Europe and America. Wiley, London
Hastings A, Bailey N, Bramley G, Gannon M (2017) Austerity urbanism in England: the "regressive redistribution" of local government services and the impact on the poor and marginalised. Environ Plan A: Econ Space 49(9):2007–2024
Jacobs K, Flanagan K (2013) Public housing and the politics of stigma. Aust J Soc Issues 48(3):319–337
Kintrea K (2013) Social mix: International policy approaches. In: Van Ham M, Manley D., Bailey N, Simpson L, Maclennan D (eds) Neighbourhood effects or neighbourhood based problems? A policy context. Springer, Dordrecht
Kintrea K, Baffoe G (forthcoming). What is the case for urban policy? In: Kundu D, Everatt D, Kintrea K, Wang Y-P (eds) Urban policy for sustainable development in the global south: new strategies for new challenges. Springer, Dordrecht
Lawless P, Beatty C (2013) Exploring change in local regeneration areas: evidence from the new deal for communities programme in England. Urban Stud 50(5):942–958
López-Morales E (2019) State-led gentrification. In: Orum A (ed) The Wiley Blackwell encyclopedia of urban and regional studies. Wiley, London
Luttmer E (2005) Neighbors as negatives: relative earnings and well-being. Q J Econ 120(3):963–1002
Marinoff L (2019) On human conflict: the philosophical foundations of war and peace. Hamilton Books, Lanham, MD
Merry M (2013) Equality, citizenship, and segregation: a defense of separation. Springer, Dordrecht
Merry M, Manley D, Harris R (2016) Community, virtue and the white British poor. Dialogues Hum Geogr 6(1):50–68
Meth P, Goodfellow T, Todes A, Charlton S (2021) Conceptualizing African urban peripheries. Int J Urban Reg Res 45(6):985–1007
Musterd S, Marcińczak S, van Ham M, Tammaru T (2017) Socioeconomic segregation in European capital cities. Increasing separation between poor and rich. Urban Geogr 387:1062–1083
Musterd S, Andersson R, Galster G (2019) Neighbourhood effects on social outcomes. In: Schwanen T, van Kempen R (eds) Handbook of urban geography, Chapter 18. London: Edward Elgar, pp 281–296
Nightingale C (2012) A global history of divided cities. The University of Chicago Press, Chicago
Nygaard C, Meen G (2013) The distribution of London residential property prices and the role of spatial lock-in. Urban Stud 50(12):2535–2552
The Observer (2023) The Observer view on the riots in France: a grim tale of the growing gulf between haves and have-nots. The Observer, 2 July
Orford S, Dorling D, Mitchell R, Shaw M, Smith G (2002) Life and death of the people of London: a historical GIS of Charles Booth's inquiry. Health & Place 8(1):25–35
Ortega A (2020) Exposing necroburbia: suburban relocation, necropolitics, and violent geographies in Manila. Antipode 52(4):1175–1195
Randolph G, Storper M (2023) Is Urbanisation in the Global South fundamentally different? Comparative global urban analysis for the 21st century. Urban Stud 60(1):3–25
Robertson D, McIntosh D, Smyth J (2010) Neighbourhood identity: the path dependency of class and place. Hous Theory Soc 27(3):258–273
Robinson J (2016) Thinking cities through elsewhere: comparative tactics for a more global urban studies. Prog Hum Geogr 40(1):3–29
Sampson R (2019) Neighbourhood effects and beyond: explaining the paradoxes of inequality in the changing American metropolis. Urban Stud 56(1):3–32

Santiago C, Acuña R, Monje-Hernández Y (2022) Long-term features of cities in Latin America and the Caribbean: socio-residential segregation, territorial inequality, and spatial fragmentation. In: González-Pérez JM, Irazábal C, Lois-González R (eds) The Routledge handbook of urban studies in Latin America and the Caribbean: cities, urban processes, and policies. Routledge, London

Sassen S (1991) The global city: New York, London, Tokyo. Princeton University Press, Princeton

Schmid C, Karaman O, Hanakata NC, Kallenberger P, Kockelkorn A, Sawyer L, Streule M, Wong KP (2018) Towards a new vocabulary of urbanisation processes: a comparative approach. Urban Stud 55(1):19–52

Simpson M, Lloyd T (eds) (1977) Middle class housing in Britain. David and Charles, Newton Abbot

Smith G (1999) Area-based initiatives: the rationale and options for area targeting. LSE STICERD Research Paper No. CASE025. London School of Economics, London

Tammaru T, van Ham M, Marcińczak S, Musterd S (eds) (2016) Socio-economic segregation in European capital cities: east meets west. Routledge, Abingdon

Tejani S (2016) Jobless growth in India: an investigation. Camb J Econ 40(3):843–870

UN-Habitat (2017) New Urban Agenda. UN-Habitat, Nairobi

UN-Habitat (2022) Global State of National Urban Policy 2021. Accessed 10 November 2022. https://unhabitat.org/global-state-of-national-urban-policy-2021

United Nations (2015) Sustainable Development Goals: Goal 11: Make cities inclusive, safe, resilient and sustainable. Accessed 6 December 2023. https://www.un.org/sustainabledevelopment/cities/

Veneri P, Comandon A, Garcia-López M-A, Daams M (2021) What do divided cities have in common? An international comparison of income segregation. J Reg Sci 61(1):162–188

Wacquant L (1993) Urban outcasts: stigma and division in the Black American ghetto and the French urban periphery. Int J Urban Reg Res 17(3):366–383

Watt P (2009) Housing stock transfers, regeneration and state-led gentrification in London. Urban Policy Res 27(3):229–242

Wilkinson R, Pickett K (2009) The spirit level: why more equal societies almost always do better. Allen Lane, London

Wilson W (1987) The truly disadvantaged: the inner city, the underclass and public policy. University of Chicago Press, Chicago

World Inequality Database. Accessed 26 July 2023. https://wid.world/data/

Zorbaugh H (1983) The gold coast and the slum: a sociological study of Chicago's near north side. University of Chicago Press, Chicago

3 Urbanisation and Neighbourhood Inequality in the Global South

Ya Ping Wang, Gideon Baffoe, and Amin Kamete

Abstract

This chapter continues the discussion about urban inequality and provides further context for the case study chapters. It looks at neighbourhood development and socio-spatial inequalities and the trend and style of urbanisation in the case study countries. It then discusses neighbourhood typology and its implication for socio-spatial inequalities. Fast urban expansion and sprawl in Asia and Africa is resulting in new patterns of inequality and informality. Neighbourhood differentiation is an important indicator of urban inequality, especially spatial inequality. There are huge differences between rich and poor neighbourhoods in relation to income levels of residents, family wealth, housing condition, as well as general living environment and access to public services. In all cities, the most striking spatial differences exist between slum settlements and new middle-class areas.

Keywords

Urbanisation · Urban development · Neighbourhood inequality · Spatial inequality · Sustainable Development Goals (SDGs)

3.1 Introduction

Chapter 2 introduced what we know about urban socio-spatial segregation from the Global North literature, and explored briefly how urban inequalities might be different in developing countries. It pointed out that a necessary starting point for any consideration of socio-spatial inequalities must be the existence and persistence of inequalities between groups in the population within particular societies, particularly in terms of income and wealth. Citing Goodman (1988), Nijman and Wei point out 'that income inequality had been increasingly exacerbated by inequality in wealth, especially the appreciation of property values' (Nijman and Wei 2020). They highlight the problems of rising housing prices and housing unaffordability as issues which have

Y. P. Wang (✉)
School of Geography and Tourism, Shaanxi Normal University, Xian, China
e-mail: yaping.wang@snnu.edu.cn; yaping.wang@glasgow.ac.uk

Y. P. Wang · A. Kamete
School of Social and Political Sciences, University of Glasgow, Glasgow, Scotland, UK
e-mail: Amini.Kamete@glasgow.ac.uk

G. Baffoe
Department of Environment and Geography, University of York, York, UK
e-mail: gideon.baffoe@york.ac.uk

Y. P. Wang et al. (eds.), *New Drivers of Division*, Sustainable Development Goals Series,
https://doi.org/10.1007/978-981-96-5629-5_3

become global challenges. Housing costs and markets are intertwined with urban inequality in multiple ways. The segregation of housing markets and the problem of housing affordability are widely regarded as drivers of urban inequality (see Baker et al. 2016; Wei and Ewing 2018). In developing countries, housing conditions in different types of neighbourhoods are a direct reflection of income and wealth inequality among residents. This chapter continues the discussion of urban inequality, especially neighbourhood and spatial inequality, with a close focus on the case study countries and cities covered by the Sustainable Healthy Learning Cities (SHLC) project. The chapter aims to explore neighbourhoods and socio-spatial inequalities, trends in urbanisation globally, and diverse urbanisation processes that shape urban and spatial inequalities.

This chapter begins by examining the pivotal role of neighbourhood development in influencing socio-spatial inequalities. This will be followed by an outline of the general trends of urbanisation and different styles of urban development in the SHLC case countries. The chapter then discusses the importance of urban neighbourhoods in understanding cities in the Global South, and neighbourhood differentiation and spatial division in Africa and Asia. The important differences between traditional and new urban neighbourhoods, planned and unplanned (spontaneously developed) neighbourhoods, formal and informal neighbourhoods, as well as regulated and unregulated settlements are highlighted in the context of inequality. The final section concludes the chapter by summarising key insights and proposing directions for future research and policy.

3.2 Neighbourhood and Socio-Spatial Inequality

To define the concept of neighbourhood in one city is a challenge; to define a commonly accepted urban neighbourhood concept for several different countries is even more difficult. Neighbourhoods in the central historical area of a city, for example, are always very different from suburban housing estate areas; neighbourhoods created under different geographical, political, and economic conditions vary in size, architectural and construction style, and the social and economic profile of residents. Popular neighbourhood typologies tend to follow subjective, administrative, and physical attributes of neighbourhoods (Hopkins 2016). For instance, drawing on census data, Jackson-Smith et al. (2016) created typologies that share unique characteristics with the immediate natural environment, built environment, and social structures. Similarly, Lynge et al. (2022) used census data to develop typologies that reflect the socio-economic and environmental conditions in South Africa.

Factors such as inner-city proximity, race and ethnicity, employment and income level, service provision, housing types, and availability of natural resources, therefore, go a long way to define a particular neighbourhood. These factors also define the life chances of residents (Atkinson and Kintrea 2004; Department of Health and Human Services [HHS], Washington, USA 2021), and the existence of 'neighbourhood effects' has been studied extensively by scholars in the Global North (see van Ham et al. 2012). More recently, the neighbourhoods where people live have been observed to play a major role in how they experienced and coped with the Covid-19 pandemic, with affluent neighbourhoods doing better as compared to their poor counterparts (see Lynge et al. 2022). But how different neighbourhood types play out in shaping socio-economic inequalities, particularly in the Global South, continues to remain a subject of investigation in the field of urban studies (Lynge et al. 2022; Baffoe 2019; Baffoe et al. 2020).

Neighbourhoods can be broadly categorised into several types based on their socio-economic characteristics, planning history, and access to urban amenities. Understanding these typologies is essential for grasping the underlying dynamics of inequality within cities. The factors that shape socio-spatial inequalities in neighbourhoods are as follows:

- *Socio-Economic Dynamics*: Socio-economic characteristics, such as income levels and employment types, significantly impact neighbourhood quality and residents' life chances (Mouratidis 2020; Nutakor et al. 2023). In many cities in the Global South, informal economies play a critical role in shaping neighbourhood dynamics. Informal settlements, often characterised by poor infrastructure and lack of services, are home to large segments of the urban poor (Lombard 2014; Campos et al. 2022). These areas contrast sharply with gated luxury communities, which offer high-quality amenities and security but contribute to social segregation (Manzi and Smith-Bowers 2005).
- *Environmental and Spatial Factors*: Access to natural resources and environmental conditions also contribute to socio-spatial inequalities. For instance, neighbourhoods with better access to green spaces and clean water often attract wealthier residents, while poorer communities are left in areas prone to flooding or pollution (Kruize et al. 2019). Spatial segregation and connectivity within cities further exacerbate these inequalities. Poorly connected neighbourhoods limit residents' access to employment opportunities, education, and health care services, reinforcing cycles of poverty (Cerf 2023).
- *Health and Education*: Health and educational services are unevenly distributed across different neighbourhood types, leading to significant disparities in residents' well-being and life opportunities (Zajacova and Lawrence 2018). Affluent neighbourhoods typically have better access to high-quality health care and education, while poorer areas struggle with under-resourced schools and clinics. The Covid-19 pandemic, for instance, highlighted these disparities, as affluent neighbourhoods could better cope with lockdowns and health care demands compared to their poorer counterparts (Shadmi et al. 2020; Organisation for Economic Co-operation and Development (OECD 2020).

3.3 Trends in Urbanisation and Urban Growth

The rapid pace of urbanisation and the shift of people from rural to urban areas has been a driving force for development and inequality in recent years. According to UN-Habitat, 47% of the world's population (2.86 out of 6.07 billion) lived in urban areas in 2000. By 2020, this had increased to 56.2%. Of the 4.4 billion urban population in the world, 77.1% lived in the less developed regions (13.4% in Africa and 53.9% in Asia). UN-Habitat also projected that the world's urban population would reach 5.56 billion (62%) by 2035 with some regional variations (UN-Habitat 2022). In the Global North, urbanisation in the past used to go hand in hand with industrialisation, and major world industrial powers had high levels of urbanisation, but this relationship between urbanisation and industrialisation has become weak recently (Voumik and Sultana 2022; Fox and Goodfellow 2022). The rapid pace of urbanisation is happening in regions without major industrial development (Randolph and Storper 2023). Consequently, in much of Africa and Asia, there is a disconnect between urbanisation and economic development (Kundu 2020).

Table 3.1 shows urban population and rate of change in the world and SHLC case countries, while Table 3.2 takes a closer look at the percentage of population living in urban areas and the rate of change in the world and in the SHLC case countries. As shown in Table 3.1 there are marked differences in the size and proportion of the urban population among different groups of countries and the case study countries included in this book. Between 2015 and 2020, the urban population in the more developed regions increased by around 0.5% every year; in the less developed regions, it grew by over 2.3% every year (UN-Habitat 2022).

All of the SHLC case study countries show some degree of rapid urban population growth in comparison with the more developed countries and regions. Between 2015 and 2020, Tanzania

Table 3.1 Urban population and rate of change in the world and SHLC case countries

Region, country	Urban population at mid-year, 2015–2035 (thousands)					Average annual rate of change of the urban population, 2015–2035 (in %)			
	2015	2020	2025	2030	2035	2015–2020	2020–2025	2025–2030	2030–2035
World	**3,981,498**	**4,378,994**	**4,774,646**	**5,167,258**	**5,555,833**	**1.90**	**1.73**	**1.58**	**1.45**
More developed regions	979,089	1,003,640	1,027,097	1,049,699	1,070,804	0.50	0.46	0.44	0.40
Less developed regions	3,002,409	3,375,354	3,747,549	4,117,558	4,485,029	2.34	2.09	1.88	1.71
Africa	**491,531**	**587,738**	**698,149**	**824,014**	**966,330**	**3.58**	**3.44**	**3.32**	**3.19**
South Africa	35,844	39,551	43,113	46,457	49,631	1.97	1.72	1.49	1.32
Tanzania	17,035	22,113	28,245	35,529	44,001	5.22	4.89	4.59	4.28
Rwanda	1977	2281	2660	3144	3769	2.86	3.07	3.34	3.63
Asia	**2,119,873**	**2,361,464**	**2,589,655**	**2,802,262**	**2,998,511**	**2.16**	**1.84**	**1.58**	**1.35**
China	775,353	875,076	956,554	1,017,847	1,059,619	2.42	1.78	1.24	0.80
Philippines	47,078	52,009	57,606	63,844	70,692	1.99	2.04	2.06	2.04
India	429,069	483,099	542,743	607,342	675,456	2.37	2.33	2.25	2.13
Bangladesh	55,305	64,815	74,838	84,689	93,958	3.17	2.88	2.47	2.08

Source UN-Habitat (2022)

Table 3.2 Urban population size and rate of change in the world and SHLC case countries (2022)

Region, country	Percentage of population at mid-year residing in urban areas, 2015–2035					Average annual rate of change of the percentage urban, 2015–2035 (in %)			
	2015	2020	2025	2030	2035	2015–2020	2020–2025	2025–2030	2030–2035
World	**53.9**	**56.2**	**58.3**	**60.4**	**62.5**	**0.82**	**0.75**	**0.71**	**0.67**
More developed region	78.1	79.1	80.2	81.4	82.7	0.24	0.27	0.30	0.32
Less developed regions	49.0	51.7	54.3	56.7	59.0	1.09	0.97	0.88	0.80
Africa	**41.2**	**43.5**	**45.9**	**48.4**	**50.9**	**1.09**	**1.08**	**1.06**	**1.04**
South Africa	64.8	67.4	69.8	72.1	74.2	0.76	0.71	0.65	0.59
Tanzania	31.6	35.2	38.9	42.4	45.9	2.16	1.96	1.76	1.56
Rwanda	17.0	17.4	18.3	19.6	21.5	0.50	0.96	1.40	1.82
Asia	**48.0**	**51.1**	**54.0**	**56.7**	**59.2**	**1.26**	**1.10**	**0.98**	**0.88**
China	55.5	61.4	66.5	70.6	73.9	2.03	1.58	1.21	0.91
Philippines	46.3	47.4	49.0	50.9	53.3	0.48	0.64	0.79	0.91
India	32.8	34.9	37.4	40.1	43.2	1.27	1.36	1.42	1.46
Bangladesh	34.3	38.2	42.0	45.6	49.0	2.14	1.90	1.67	1.44

Source UN-Habitat (2022) (World Cities Report)

and Bangladesh had the highest annual growth rates of 5.2% and 3.2%, respectively (UN-Habitat 2022). As shown in Table 3.2, in terms of the size and rate of change of the urban population, South Africa is the most urbanised country with 67.4% of people living in urban areas in 2020, followed by China at 61.4%. Apart from the Philippines (47.4% urban), the other four SHLC countries all have a relatively low proportion of urban population. Less than 18% of people in Rwanda live in urban areas. India, now the largest country in the world in terms of population, has a relatively lower percentage of urban population (34.9%) (Table 3.2; UN-Habitat 2022). These low levels of urbanisation make Africa and Asia the focal regions for future world urban development and expansion (World Bank 2023). Kundu et al. (2022) analysed the trend of global urbanisation. They found that Asia and Africa are indeed poised to be the epicentres of urbanisation, without a corresponding increase in their income levels. Much of the future urban growth will be concentrated in countries with very low urbanisation and income levels and poor infrastructural bases. In Asia, urbanisation is fuelled by a combination of factors such as rural–urban migration, in situ urbanisation, and enlargement of urban areas, while in Africa high fertility rates are driving urbanisation. Moreover, the two least urbanised regions in these two continents, that is, Eastern Africa and Southern Asia, are expected to experience the highest urban growth and rural–urban transition in the next three decades (Kundu et al. 2022).

The large Asian and African cities included in the SHLC study have experienced continuous and large-scale population growth. They received more migrants and experienced faster economic transformation. Delhi, for example, saw its population increase by 3.4 million to 16.3 million between 2001 and 2011, a 2.4% increase per annum; the city had experienced an even faster annual population growth of 4.2% in the previous decade (Kundu et al. 2020). The population in Cape Town increased from 2.6 million in 1996 to an estimated 4 million in 2016, although the annual growth rate fell from 3.3% between 2000 and 2010 to 1.5% since 2010 (Turok et al. 2020a). In Johannesburg, the population growth has remained high since 1996, ranging between 2.6 and 3.2% per annum in recent years, with the total population in 2020 at 5 million (Everatt et al. 2020). Dar es Salaam, a city with approximately 10% of the Tanzanian population, saw an increase from 4.4 million residents in 2012 to an estimated 6 million (Msuya et al. 2020).

Kundu et al. (2020) project that Asian urbanisation, which has been top-heavy with population concentration in big cities, will continue to be so even after the slowdown in urban growth in the region. By 2035, six of the ten most populated megacities of the world will be in Asia, and Delhi is projected to be the most populated urban agglomeration with 43.3 million people. Urbanisation in Africa, currently with few megacities, will follow the Asian experience with a rise in the number and population of megacities. However, contrary to the Asian counterparts, most of the new megacities of the continent will be 'places of consumption' in place of 'growth engines' (Kundu et al. 2020).

In the cities covered by the SHLC project, suburban sprawl and fragmented land development and transformation have created new subcentres and industrial districts, as well as ribbon development of residential areas along major roads. The city of Chongqing, for example, has transformed from a regional industrial centre in inland China to a huge metropolitan area of national significance in a short period of 20 years. Its built-up area expanded more than five times so that almost all the developable flat ground around the city has been built on (SHLC China Team 2020). Delhi's population growth has also been accompanied by urban sprawl; its built-up area expanded by 65% in the two decades preceding 2018 (Kundu et al. 2020). Kigali is another of the study cities that has seen a dramatic increase in its built-up area, from around 25 to 115 square kilometres in the last 30 years (Manirakiza et al. 2020).

Migration and population growth have also induced large-scale infill development, intensifying land use in existing built-up areas, and loss of green spaces and waterbodies within cities. In Dhaka, for example, the population density increased from 14,000 per square kilometre in

1991, to 28,900 in 2011 and 31,800 in 2015. Currently, 40% of the residential areas in the city have a density of over 99,000 people per square kilometre (Roy and Sowgat 2020). The city of Manila saw similar population density increases (Delos Reyes et al. 2020). In Africa, urban population density is low in comparison with these Asian cities but densities are rising. In Johannesburg, density increased from 2055 to 3281 people per square kilometre between 2002 and 2017 (Turok et al. 2020a, b). Kigali's new master plan aims to bring its overall population density up to 5198 persons per square kilometre (Kigali City Council 2020).

3.4 Characteristics of Urban Development

SHLC team members carried out comparative fieldwork in the case study cities and examined the ongoing process of urbanisation. The study found very different ways of urban development, despite the similar trends of urbanisation outlined above. Johannesburg and Cape Town in South Africa, for example, show very serious social and spatial inequality as a result of early apartheid policies and race segregation. The rich and middle-class white population enjoy a European and US style of suburban neighbourhoods (Mace 2018). Large houses hide away in gardens covered by mature trees; modern and expensive apartments enjoy high standards of amenities, infrastructure, and transport and road networks. New and well-designed expensive housing estates sit side by side with professional office blocks and shopping malls. New mixed-race middle-class neighbourhoods fill the gaps between the black townships and the rich districts. Small-scale house upgrading and redevelopment have happened in existing black and coloured townships such as Delft in Cape Town and Soweto in Johannesburg.

There are also areas experiencing serious decline. In Johannesburg's city centre, many white residents have moved out, but the poor cannot afford the rent and prices. Some high-rise housing and office blocks have become empty, abandoned, and taken over by criminal cartels. Illegal activities and crimes have made the city centre a major problematic and unsafe area to live in or visit. In some older white working-class districts, unemployment has resulted in poverty and neighbourhood deterioration. Overall, the post-apartheid economic and social progress and resorting of the urban population has blurred the white and coloured division only a bit; they have not fundamentally changed the racial segregation and spatial division and inequality created by apartheid. One of the dominant and striking features of recent urbanisation in South Africa is the rapid expansion of informal settlements in major cities (Turok and Borel-Saladin 2014). The post-apartheid government sought to ensure every citizen had a safe place to live in, but the system was unable to cope with the large volumes of migrants (from rural areas and neighbouring countries), which resulted in many organised and simultaneous informal squatter settlements on marginal areas, natural reserves, and other publicly owned land. Covid-19 lockdowns made the situation even worse. Sheds made of metal sheets and wood board can be found in their thousands along riverbanks, motorways, and public parks (Fig. 3.1). In Cape Town, it was estimated that about 20% of the population live in these types of shelter. These informal settlements (they were not referred to as slums in the country), lacking major public facilities including water and electricity, will be a major challenge to planners and policymakers in the future (Nassar and Elsayed 2018; Satterthwaite et al. 2020).

Tanzanian cities show a very different picture of urbanisation from South Africa. Rapid economic growth and international trade has attracted many people to Dar es Salaam and the new national capital, Dodoma. There is large-scale house rebuilding in the central business districts; traditional houses built under the influence of Arabs and Middle Eastern migrants were either rebuilt or expanded for more floor space. There are also large-scale slum neighbourhoods along the riverbanks or other marginal areas. Together with the urban villages, they provide low-income housing for the urban poor. These, however, are not the main features of urbanisation in the country. Population movement, cheap

Fig. 3.1 Informal settlement ('A Covid-19 Settlement') in Cape Town. *Note* These types of informal settlements emerged during and post the COVID-19 pandemic. Photo by Ya Ping Wang

land prices, and a very flexible and minimal planning and building regulation system have enabled large-scale urban expansion and sprawl in the two cities (Fig. 3.2). Huge areas of rural land were designated as new urban districts for residential development; current landowners divided their land into small plots and sold them to individuals coming from the city centre or elsewhere to build their own homes. Apart from major roads between districts which were paved, there is no road surfacing inside these development areas. Water pipes and electricity can reach each individual plot; but there are no sewage or drainage systems. As these houses are organised and individually built, there is no standard design or neighbourhood similarity as is often found in commercially developed estates; some houses are big and others are very small; some homes are very fancy, while others are very basic; some houses have been completed and occupied by their owners for several years, other plots are still empty. In Dar es Salaam, two new urban districts were created to accommodate such expansion. This type of urban development is not found only in Tanzania; it is very popular in other eastern and central African countries such as Rwanda. The huge size of these emerging middle-class areas has made them a major characteristic of central and eastern African urbanisation and they form a sharp contrast to the large-scale expansion of informal settlements in South Africa.

Urban development in South Asia shares some similarities with Africa, including private ownership of land, small plot and family organised construction, a few large and commercially built housing estates, and very basic planning and regulations resulting in various degrees of informality. There are also very important differences between them. African urban development is characterised by low-rise and horizontal expansion and sprawl with individual family houses at very low density (including the informal areas as shown in Cape Town), often detached. In contrast, in Asia (apart from China and Chinese communities located in other eastern Asia countries), new urban development is dominated by the construction of apartment buildings of up to 10 storeys. Dhaka and Khulna in Bangladesh, for example, which benefited from recent rapid economic development, experienced very fast suburban expansion and inner area redevelopment. Urban development here is characterised by sub-division of family-owned land into plots in cities and suburban areas and the construction of residential buildings of

Fig. 3.2 Private plot-based self-built urban expansion in Dodoma, Tanzania. Photo by Ya Ping Wang

Fig. 3.3 Selective housing redevelopment in Dhaka City. Photo by Ya Ping Wang

multi-storey apartments organised by either the original landlord (alone or in collaboration with a small developer), or small housing cooperation. If plot owners can accumulate enough funds to rebuild their traditional houses into modern apartment buildings, they will do it themselves and rent out the extra apartments and/or rooms to others to make a living. If they cannot afford to rebuild, they work with a developer, with the latter providing the funds and expertise for construction. The finished buildings will be shared between them according to an agreed percentage. For younger middle-class families without land of their own, they often work together as a small housing cooperative. They pool resources together to buy a plot in the city or the suburban areas and construct a multi-storey residential building. When completed, each participating family is allocated one flat. The extra apartments are either rented out to generate an income for the cooperative or sold to other families in similar conditions who will become members of the cooperative (Fig. 3.3).

Urban development in China exhibits many differences from other case study countries. Apart from the simultaneous development of the urban

Fig. 3.4 Garden and apartment buildings in a large commercial housing estate in Datong, China. Photo by Ya Ping Wang

villages, much of the recent city expansion in China was professionally planned and designed. Industries were grouped into large development zones, and residential areas were constructed into various commercial and social housing estates. While urbanisation is driven by rural to urban migration, urban development normally starts with the government (through politicians and planners) and property developers. The government acquires land from farmers and sells land use rights (for a fixed period) to commercial developers. Housing estates were planned without the involvement of the people, with properties being normally sold to individuals or institutions before construction started in order to secure funding from future residents. Home buyers normally pay the developer in instalments (one-third at the beginning of the construction, one-third when the building structure is completed, and one-third before moving in) (Fig. 3.4). Similar processes were followed for displacement and relocation housing schemes. This is very different from the other countries, as discussed above, where people occupied land before major development and were involved closely in the construction process.

3.5 Neighourhood Typology and Inequality

The SHLC project aimed to study the internal socio-economic and residential structure of these cities. Preliminary investigation identified some different types of residential areas in cities, such as a traditional/historic town (normally the city centre area), colonial settlement areas, post-war public housing areas, sites- and-services development project areas, areas of urban upgrading, recent commercial housing estates, gated luxury home communities, urban villages, and informal settlements. Research teams in each city were encouraged to be innovative in defining and identifying neighbourhoods, drawing on local physical and geographical features, road structures, historical style of development and expansion of the city, government/administrative division and governance, health and educational provision and catchment areas, etc. Reflecting on the cultural and local differences between the case study countries, the project teams adopted mixed and flexible approaches in defining neighbourhoods. The approaches include:

1. A systematic and statistical approach using cluster analysis and various indices to identify urban neighbourhoods, where research teams utilised highly disaggregated spatial data from the population census with variables covering a wide range of issues, including demographic, socio-economic, public services, housing, and other physical characteristics (for more details, see Lynge et al. 2022; SHLC Bangladesh 2020).
2. Administrative boundaries approach where cities have a local level management system) which were considered along with other geographical boundary features to identify neighbourhoods.
3. Participatory mapping, where research teams conducted local fieldwork and interviews with officials and community workers to identify and map the neighbourhoods in cities.

Neighbourhood typologies are identified below through the SHLC case studies.

3.5.1 Planned, Unplanned and Mixed Neighbourhoods

Rapid urban transformation and inadequate provision of affordable housing are causing spatial fragmentation in many cities. Unplanned settlements have become a common phenomenon in many urban settings. The UN estimated that in 2020, nearly 25% of the world's urban population confined to such areas (UNDESA 2020). Contextualising and understanding how such spatial growth patterns shape development dynamics remains a herculean task for city authorities, particularly in the developing world (Martinez et al. 2008). The SHLC study identified planned, unplanned, and mixed neighbourhoods based on their official designation. They were the dominant neighbourhood types in Kigali and Huye, Rwanda. In Rwanda, such neighbourhoods are classified based on how they emerged. While a planned neighbourhood has a physical spatial plan, unplanned neighbourhoods do not. The latter is usually spontaneous in nature. Planned neighbourhoods have both land leases and construction permits. Unplanned neighbourhoods also have a land lease and pay land taxes; however, they do not conform to construction guidelines since they have no planning and building permits (Jaganyi et al. 2020). As such, they are commonly referred to as 'informal'. The planned ones are classified as 'formal'. Mixed neighbourhoods (semi-formal) are in between the formal and the informal and are largely unplanned areas that have been given a facelift by way of infrastructural upgrading (Jaganyi et al. 2020). In practice, these typologies are not homogenous, as they are sub-classified based on the structure. Table 3.3 depicts the categories.

Table 3.3 Neighbourhood typology in Rwanda

Planned (formal)	Unplanned (informal)	Mixed (semi-formal)
Luxurious multi-storey houses	Old high density	Mixed planned-informal
Private estate	New high density	Upgraded informal
Modern villas	Old low density	Degraded formal
Old fashion planned	New low density	
Planned low-rise	Suburbs (urban periphery)	

Source Jaganyi et al. (2020)

Planning and regulation are also important factors in neighbourhood differentiation. In Delhi, the team identified 10 neighbourhood types: urban villages; rural villages; resettlement and rehabilitation; slums; planned colonies; walled cities; unauthorised regularised colonies; and unauthorised colonies (NIUA SHLC Team 2020). The walled city is the oldest part of Delhi, and its morphology is traced back to the era of Mughal rulers. Slums in Delhi are classified as either notified or non-notified, with the former emanating from section 3 of the Slum Areas (Improvement and Clearance) Act, 1956 (NIUA SHLC Team 2020). Although they both have poor ventilation, sunlight, and infrastructure facilities, those that are notified have semi-permanent structures. Unauthorised neighbourhoods are neighbourhoods that are situated on lands which are not included in the area Master Plan and are not zoned for residential purposes. Regularisation is a process by which a colony receives formal recognition. From time to time, the unauthorised colonies receive a facelift by way of regularisation. Resettlement colonies are those areas developed by the government to resettle evicted residents from the slums. Meanwhile, the rural villages exist within the boundary of the National Capital Territory of Delhi. These areas usually await notification to be classified as urban villages, and they are predominantly characterised by rural living lifestyles with mixed land-use patterns. Planned colonies are those built on lands earmarked in an area's development master plan. Unlike the slums, authorised and non-authorised colonies, such areas have better housing and service provision, including open spaces, road connectivity, water and sanitation facilities, and parks (NIUA SHLC Team 2020).

3.5.2 Rich and Poor Neighbourhoods

Income remains a key indicator that differentiates spaces in urban areas. With rapid urbanisation, a growing labour market, and different consumption patterns, cities have become places of self-actualisation. However, the reality is always undesirable. Growing income differences mean that spatial units are increasingly becoming unequal. The housing market is always designed to favour the rich. Affordable housing continues to be a dream that will never materialise for many urban residents in the developing world (Chan and Adabre 2019). While the rich move to the high-end neighbourhoods with all the latest facilities, the poor continue to live in poverty in slums, ghettos, and other low-income settlements. Housing is, therefore, polarised along the line of socio-economic status. In Tanzania, for instance, income remains the dominant variable that differentiates one neighbourhood from the other. The team identified six neighbourhoods that are classified based on the income level of residents: (1) Low-Income Neighbourhoods; (2) Mixed Low- and Medium-Income Neighbourhoods; (3) Medium-Income Neighbourhoods; (4) Mixed High- and Medium-Income Neighbourhoods; (5) High-Income Neighbourhoods, and (6) Super High-Income Neighbourhoods (Msuya et al. 2020). Dar es Salam was found to have a large diverse income group distribution, with most of the people residing and falling within the medium- and mixed high- and medium-income groups (Msuya et al. 2020). The minority high-income group, including diplomats, live in the best areas in the city, such as along the beach and islands.

A similar typology is found in Manila and Batangas in the Philippines. Following socio-economic, environmental, and physical attributes, the team identified diverse neighbourhoods that show the extent of social divisions in the city (SHLC Philippines 2020). The team identified 23 neighbourhoods based largely on housing and income levels. The middle-income areas were found to be dominant in both cities. However, most of the people who happened to be from low-income households live in urban slums and other poorly serviced areas, port facilities, and marketplaces. Other interesting, but unanticipated typologies include cemeteries and movie houses in residential areas. While the higher-income neighbourhoods are well-serviced areas, those of the lower-income are deprived, with rusting galvanised roofing, wooden structures, and dangerous and illegal electricity connections (SHLC Philippines 2020). Interestingly, although the conditions in low-income neighbourhoods, including slums, are inhuman, residents have a strong sense of belonging and are unwilling to relocate. There are strong social ties among residents that play a major role in securing economic opportunities (SHLC Philippines 2020). Social networking remains the major livelihood and resilience-building block for such neighbourhoods.

Employing cluster analysis, the Bangladesh team observed that socio-demographic and environmental characteristics, such as profession, religion, services accessibility, and housing are the key factors that differentiate residential areas in Dhaka and Khulna. Most unplanned neighbourhoods were found to be mixed-use areas (SHLC Bangladesh 2020). Utilising a similar approach, the South African team identified 10 neighbourhood types: informal settlements connected to electricity; informal settlements without electricity; institutional spaces; rented apartments or flats; predominantly whites living in larger houses; coloured and black Africans living in formal housing with backyards; coloured (Afrikaans-speaking); black Africans (isiXhosa) in planned settlements with backyard shacks; coloured (English speaking), middle-class; and black Africans (isiXhosa) living in planned settlements. Race, dwelling type, and income emerged as the dominant differentiating variables (Turok et al. 2020b).

3.5.3 Traditional and New Neighbourhoods

With the advent of technology, residential areas are rapidly being developed, redeveloped, renewed, and retrofitted. This transformation continues to shape the urban landscape and the built environment in many contexts. Given income differences, however, mixed development has become commonplace. Traditional or conventional housing types and modern developments coexist in many cities. As documented by Steuteville (2017) traditional neighbourhoods—built on historic neighbourhood concepts—started the new urbanism drive in the 1980s and 90s as alternatives to conventional master-planned communities. Such neighbourhoods act as laboratories to test housing ideas by creating small communities and spatial units that overcome legal and institutional barriers to holistic development. Today, many traditional neighbourhood developments are geared toward smaller infill projects, with occasional newly built neighbourhoods, particularly in areas where there is readily finance and a desire for large-scale transformation.

Traditional and newly built neighbourhoods are common typologies in China. In Chonqing and Datong, the Chinese team identified four neighbourhood types: (1) commercial housing communities; (2) affiliation-bonded residential communities (or work unit housing); (3) supportive housing; and (4) urban villages. Commercial housing in China is the product of housing market reform (SHLC China 2020). These are areas developed and sold to people by real estate developers for occupation. They form the larger share of new residential areas in Chinese cities. These neighbourhoods are further categorised into high-end and ordinary commercial housing units. A movement from an ordinary unit into a high-end unit is a key indicator of social mobility in China (SHLC China 2020). The affiliation-bonded neighbourhoods are special types of communities, where occupation is based on one's affiliation with a relevant institution. Many institutions have the power to develop residential housing units and lease them to employees according to their rank and length of service. Unlike other neighbourhood types, these residential units charge a nominal rate (both rent and purchase) that is lower than the market price, making it the better option for many workers (SHLC China 2020). The underlying motive behind the development of such neighbourhoods is to attract and retain top talent (SHLC China 2020).

Supportive housing neighbourhoods are meant for less privileged and low-income households. They include public rental housing, low-rental housing, affordable housing, and a few limited-price commercial apartments. The difference between this neighbourhood and the other two types is that supportive units are developed and maintained by government initiatives and financial funds. As such, they are social support properties and are usually affordable compared to the other types. Given their special nature, they are limited in number, particularly in Chongqing (SHLC China 2020). The urban village is a classic example of a traditional neighbourhood. In China, such residential areas are largely occupied by farmers who are transitioning to urban living. Importantly, they comprise simple housing and shed housing and they form a small proportion of the housing stock, particularly in Chongqing (SHLC China 2020). A key feature is that such neighbourhoods lack strict formal planning and spatial regulation; houses are randomly distributed, with poorly maintained structures, low disaster stability, poor lighting and ventilation, and crowded occupation. As a way of integration, however, municipal governments in recent times have invested hugely to redevelop such neighbourhoods (SHLC China 2020).

3.5.4 Urban Slums

In developing countries, the existence of large areas of slum settlements is a major manifestation of spatial inequality and poverty. They form a sharp contrast to the well-designed colonial settlements and the emerging middle-class colonies. Despite the major efforts made in poverty reduction during the first 15 years of the twenty-first century under the UN Millennium Development Goals (MDGs) programme, urban slums still house a significant proportion of the population. The UN *World Social Report 2020* estimated that in 2016 urban slums were home to more than 1 billion people. 'Slums—also known as bidonvilles, taudis, barrios marginales, tugurios, favelas and many other names—are urban areas characterised by substandard housing, overcrowding, unsanitary conditions, and lack of services' (UN *World Social Report 2020*: 119). As suggested in this definition people living in slums suffer various forms of deprivations, including lack of access to improved water sources, lack of access to improved sanitation facilities, lack of sufficient living area, lack of housing durability, and lack of tenure security (UN *World Social Report 2020*: 119). Slum dwellers also often suffer from chronic hunger and die younger than other urban dwellers in the same city (UN-Habitat 2006). Moreover, the unfavourable health and safety conditions found in slums affect women, children, and youth disproportionately. (UN *World Social Report 2020*).

The SHLC project countries for which data is available have all seen a decline in the slum population between 2000 and 2020, and Tanzania, Rwanda, and Philippines all made very significant progress, while the figure remains high for Bangladesh and India (Table 3.4). Overall, a large

Table 3.4 Proportion of urban population living in slum households (%) in SHLC case countries

Country	2000	2020	Reduction
Philippines	50.0	36.3	13.7
India	55.3	49.0	6.3
Bangladesh	58.3	51.9	6.4
Rwanda	71.4	38.3	33.1
Tanzania	81.3	40.9	40.4
South Africa	27.6	24.2	3.4

Source UN-Habitat (2022)

proportion, from a quarter to half of the urban population still lives in slums and many new residential developments are informal and spontaneous. In Johannesburg, for example, by 2018, informal land use had grown to the equivalent of almost one-third of the entire built-up area (Wang and Kintrea 2021). As the case study chapters will show, most slums lack formal legal recognition; they are often disconnected from the infrastructural grids for drinking water, gas, electricity, and sanitation. Public services, including basic services such as refuse collection, health care, and education in these informal settlements either do not exist or are of very poor quality. In Manila, for example, children in some slum areas can only attend school on a part-time basis due to limited classroom space. Where there are services available, the private sector often plays a major role and the cost for residents tends to be high. These slums tend to be located at marginal places such as riverbanks, along railway lines, beaches, wastelands left over by industrial facilities, under bridges, and other environmentally hazardous places. Slums and informal housing areas are always poorly connected by paved roads and transport, further limiting their job and employment opportunities. More recent data is not available, but it is widely accepted that Covid-19 has made the situation worse.

China does not officially recognise any slum settlements and there is no report of slum population. Rural to urban migrants were not given official urban citizenship; they were generally treated as the main low-income group in cities. Their residences are either found in urban villages or marginal land areas in temporary buildings not built for family housing purposes. The total number of migrants in China (defined as people who live in a place which is not theirs/ordinary resident registration place) was around 376 million, about 26.8% of the total population (National Statistics Bureau of China 2021). This is not a direct comparison with the slum population in other countries, but it does reflect the scale of inequality in Chinese cities. In recent years, most cities have introduced policies to integrate the migrant population with the normal urban community. Most migrant workers' living conditions are not comparable to those of established urban residents, but most of them are not in poverty.

3.6 Conclusion

This chapter sought to explore neighbourhoods and socio-spatial inequalities, trends in urbanisation, and the diverse urban development processes that shape urban and spatial development. In terms of the future trend of urbanisation it is interesting to know that both Asia and Africa are expected to be the main regions experiencing rapid urban expansion. The speed and style of urban growth however will be very different in the two continents. Major cities in Asia will grow much larger than cities in Africa and will show a much higher land use intensity and population density. Urbanisation in Asia will have a closer relationship with industrialisation, while urbanisation in Africa on the other hand will be driven mainly by population growth and migration.

This investigation identified very different ways of urbanisation. Cities in China benefited from the strong government control and public ownership of land, and consist of many professionally planned, designed, financed high-rise and high-density areas including various standard housing estates. Urban development in Bangladesh (and in India to some extent), driven by recent industrial development and migration, is dominated by small housing cooperatives and private housing renewal partnerships, and development mainly occurs on privately owned city or suburban land. Small pieces of land and traditional private ownership dramatically increased the urban land use intensity and population density and compromised the general living environment and communication/transportation system. Cities in the Philippines came in between the Chinese and South Asia development models. In places where there is a concentration of population of Chinese origin, larger apartment buildings and sizeable expensive middle-class new estates were built. Other places showed a similar development pattern to that observed in South Asia. Urbanisation in Africa follows very different routes. In South Africa, the emergence of large-scale informal settlements in its major cities made the government's effort to build affordable housing for the poor insignificant; the large new informal settlements further magnified the racial differences and enhanced the spatial and racial inequalities. In another model of urbanisation, in central Africa, countries like Tanzania and Rwanda, urban expansion is led mainly by family-based private housing development in new suburban neighbourhoods.

In the *World Cities Report* 2016, UN-Habitat states that 'the current urbanisation model is unsustainable in many respects, puts many people at risk, creates unnecessary costs, negatively affects the environment, and is intrinsically unfair' (UN-Habitat 2016: iv). Rapid urban expansion and sprawl in Asia and Africa have indeed resulted in new inequality and informality, which are demonstrated by the different types of neighbourhoods that were identified in the case study cities in this book. Living conditions and life quality vary substantially between the well-planned commercially constructed housing estates, gated or semi-gated communities for the rich, tenement and high-rise apartments for the middle-class and civil servants, and state-supported low-income settlements for the poor. In some cities, informal settlements such as urban villages, displacement and relocation settlements, as well as various slum settlements also grew at a much faster speed than the planned areas. Neighbourhood differentiation is an important indicator of urban inequality, especially spatial inequality. This is characterised by huge differences between rich and poor neighbourhoods, in income levels of residents, family wealth, access to public and social services, as well as housing conditions and the general living environment. There are also differences in the level of attention from urban planners, government officials, and investors. In all cities, the most striking spatial differences and inequality exist between the slum settlements and new middle-class areas.

At the micro level within each neighbourhood, there are significant inequalities between families and individuals. In most old and new neighbourhoods there is some degree of mixture of residence: rich families live side by side with the poor ones, and poor servant families live inside/or in the backyard of the large houses. Rich and poor areas are often superficially identifiable, but in reality there is no clear-cut distinction between them. In older areas of cities, there is more mixed living traditionally. In new areas, where land was bought individually and houses were built to various standards by each family. Many of the issues discussed in this chapter are relevant to the content of the case study chapters and will be explored further. The discussion here also helps to put these cases in a comparative framework.

References

Atkinson R, Kintrea K (2004) 'Opportunities and Despair, it's all in there': practitioner experiences and explanations of area effects and life chances. Sociology 38(3):437–455

Baffoe G (2019) Understanding the neighbourhood concept and its evolution: a review. Environ Urban ASIA 10(2):393–402. https://doi.org/10.1177/0975425319859115

Baffoe G, Malonza J, Manirakiza V, Mugabe L (2020) Understanding the concept of neighbourhood in Kigali City, Rwanda. Sustainability 12(4):1555. https://doi.org/10.3390/su12041555

Baker A, Bentley R, Lester L, Beer D (2016) Housing affordability and residential mobility as drivers of locational inequality. Appl Geogr 72(July):65–75

Campos MJZ, Kain J-K, Oloko M, Scheinsohn M, Stenberg J, Zapata P (2022) Residents' collective strategies of resistance in Global South cities' informal settlements: space, scale and knowledge. Cities 125:103663. https://doi.org/10.1016/j.cities.2022.103663

Cerf ME (2023) The social-education-economy-health nexus, development and sustainability: perspectives from low- and middle-income and African countries. Discov Sustain 4:37. https://doi.org/10.1007/s43621-023-00153-7

Chan APC, Adabre MA (2019) Bridging the gap between sustainable housing and affordable housing: the required critical success criteria (CSC). Build Environ 151:112–125

Delos Reyes MR, Gamboa MAM, Daguio KGL, Rivera RRB (2020) Manila: understanding neighbourhoods for a more sustainable city. SHLC Research Summary 05. University of the Philippines School of Urban and Regional Planning and Centre for Neighbourhood Studies (CeNS). https://www.centreforsustainablecities.ac.uk/research

DHHS (Department of Health and Human Services) (2021) Community health and economic prosperity: engaging businesses as stewards and stakeholders—a report of the surgeon general [Internet]. US Department of Health and Human Services, Washington, DC

Everatt D, Abrahams C, Lynge H (2020) Johannesburg: the growing importance of class in shaping neighbourhoods. SHLC Research Summary 10. Wits School of Governance, University of the Witwatersrand, Johannesburg, South Africa https://www.centreforsustainablecities.ac.uk/research

Fox S, Goodfellow T (2022) On the conditions of 'late urbanisation.' Urban Stud 59(10):1959–1980. https://doi.org/10.1177/00420980211032654

Goodman A (1988) An econometric model of housing price, permanent income, tenure choice, and housing demand. J Urban Econ 23(3):327–353

Hopkins E (2016) Understanding the different types of low-income neighborhoods. Community Invest 22(1):13–36

Jackson-Smith DB, Stoker PA, Buchert M, Joanna E-W, Carlos VL, Molly CS, Shujuan L, Zack B, Luke B (2016) Differentiating urban forms: a neighborhood typology for understanding urban water systems. Cities Environ (CATE) 9(1)

Jaganyi J, Njumwa K, Nzayirambaho M, Rutayisire C, Manirakiza V, Nsabimana A, Malonza J, Leon M, Nduwayezu G (2020) The emerging patterns of neighbourhood/community distribution: case studies of the cities of Kigali and Huye. SHLC RTP2 Report, University of Rwanda, Rwanda

Kigali City Council (2020) City of Kigali Master Plan 2050. https://masterplan2020.kigalicity.gov.rw/portal/apps/webappviewer/index.html?id=218a2e3088064fc6b13198b4304f3d35/

Kruize H, van der Vliet N, Staatsen B, Bell R, Chiabai A, Muiños G, Higgins S, Quiroga S, Martinez-Juarez P, Aberg Yngwe M, Tsichlas F, Karnaki P, Lima ML, García de Jalón S, Khan M, Morris G, Stegeman I (2019) Urban green space: creating a triple win for environmental sustainability, health, and health equity through behavior change. Int J Environ Res Public

Health 16(22):4403. https://doi.org/10.3390/ijerph16224403
Kundu A (2020) Urbanisation and Industrialisation in Africa and Asia in the Context of SDG Linked Issues of Sustainability, Inclusivity and Partnership. In: Chaturvedi S, Prakash A, Dash P (eds) Asia-Africa growth corridor. Springer, Singapore, pp 53–67
Kundu D, Debnath T, Chakravorty S, Sharma P, Kar B (2022) The urban century trends and patterns of urbanisation in Asia and Africa. Delhi: National Institute of Urban Affairs and SHLC. https://centreforsustainablecities.ac.uk/research/the-urban-century-trends-and-patterns-of-urbanisation-in-asia-andafrica/
Kundu D, Sharma P, Lahiri B, Sangtani C (2020) Delhi: a city of diversity and disparity. SHLC Research Summary 02. National Institute of Urban Affairs, India. https://www.centreforsustainablecities.ac.uk/research
Lombard M (2014) Constructing ordinary places: place-making in urban informal settlements in Mexico. Prog Plan 94:1–53. https://doi.org/10.1016/j.progress.2013.05.003
Lynge H, Visagie J, Scheba A, Turok I, Everatt D, Abrahams C (2022) Developing neighbourhood typologies and understanding urban inequality: a data-driven approach. Reg Stud Reg Sci 9(1):618–640. https://doi.org/10.1080/21681376.2022.2132180
Mace A (2018) Whiteness, class and place in two London suburbs. Ethn Racial Stud 42:1032–1049
Manirakiza V, Nduwayezu G, Malonza J, Mugabe L, Nsabimana A, Rutayisire PC, Nzayirambaho M, Njunwa JK, Jaganyi D (2020) Kigali: the rapid urban growth and neighbourhood dynamics of a prosperous city. SHLC Research Summary 06. University of Rwanda. https://www.centreforsustainablecities.ac.uk/research
Manzi T, Smith-Bowers B (2005) Gated communities as club goods: segregation or social cohesion? Hous Stud 20(2):345–359. https://doi.org/10.1080/0267303042000331817
Martinez J, Mboup G, Sliuzas RV, Stein A (2008) Trends in urban and slum indicators across developing world cities, 1990–2003. Habitat Int 32(1):86–108. https://doi.org/10.1016/j.habitatint.2007.08.018
Mouratidis K (2020) Neighborhood characteristics, neighborhood satisfaction, and well-being: the links with neighborhood deprivation. Land Use Policy 99:104886. https://doi.org/10.1016/j.landusepol.2020.104886
Msuya IR, Moshi I, Levira F (2020) RTP2 Report, Dar es Salam. SHLC—Ifakara Health Institute, Tanzania
Nassar DM, Elsayed HG (2018) From informal settlements to sustainable communities. Alex Eng J 57(4):2367–2376. https://doi.org/10.1016/j.aej.2017.09.004
National Statistics Bureau of China (2021) Communiqué of the Seventh National Population Census (No. 1)—the basic situation of the Seventh National Population Census. National Statistics Bureau of China. https://www.gov.cn/guoqing/2021-05/13/content_5606149.htm
Nijman J, Wei YD (2020) Urban inequalities in the 21st century economy. Appl Geogr 117(2020):1–8
NIUA SHLC Team (2020) Internal structure and differentiation of neighbourhoods in national capital city—Delhi. Unpublished SHLC Project Report. Centre for Sustainable, Healthy and Learning Cities and Neighbourhoods
Nutakor JA, Zhou L, Larnyo E, Addai-Danso S, Tripura D (2023) Socioeconomic status and quality of life: an assessment of the mediating effect of social capital. Healthcare 11(5):749. https://doi.org/10.3390/healthcare11050749
OECD (2020) The territorial impact of COVID-19: managing the crisis across levels of government. Available at: https://read.oecd-ilibrary.org/view/?ref=128_128287-5agkkojaaa&title=The-territorial-impact-of-covid-19-managing-the-crisis-across-levels-of-government
Randolph GF, Storper M (2023) Is urbanisation in the Global South fundamentally different? Comparative global urban analysis for the 21st century. Urban Studies 60(1):3–25
Roy S, Sowgat T (2020) Dhaka: diverse, dense, and damaged neighbourhoods and the impacts of unplanned urbanisation. SHLC Research Summary 08. Khulna University. https://www.centreforsustainablecities.ac.uk/research
Satterthwaite D, Archer D, Colenbrander S, Dodman D, Hardoy J, Mitlin D, Patel S (2020) Building resilience to climate change in informal settlements. One Earth 2(2):143–156. https://doi.org/10.1016/j.oneear.2020.02.002
Shadmi E, Chen Y, Dourado I et al (2020) Health equity and COVID-19: global perspectives. International Journal for Equity in Health 19:104. https://doi.org/10.1186/s12939-020-01218-z
SHLC Bangladesh (2020) Internal structure and differentiation of neighbourhoods in Dhaka. Khulna University, Khulna
SHLC China (2020) Social and spatial transformation in city and neighbourhood: case from Chongqing, China. Unpublished SHLC Project Report. Centre for Sustainable, Healthy and Learning Cities and Neighbourhoods
SHLC Philippines (2020) Manila: understanding neighbourhoods for a more sustainable city. SHLC Research Summary 05. https://www.centreforsustainablecities.ac.uk/research
Steuteville R (2017) Great idea: traditional neighborhood development. Public Sq: CNU J. Available at: https://www.cnu.org/publicsquare/2017/03/30/great-idea-traditional-neighborhood-development (accessed 15 January 2023)
Turok I, Borel-Saladin J (2014) Is urbanisation in South Africa on a sustainable trajectory? Dev South Afr 31(5):675–691. https://doi.org/10.1080/0376835X.2014.937524
Turok I, Scheba A, Visagie J (2020a) Cape Town: a city still divided by race and class. SHLC Research Summary 01. Human Sciences Research Council, South Africa. https://www.centreforsustainablecities.ac.uk/research
Turok I, Visagie J, Scheba A (2020b) RTP2 Report—Cape Town. Centre for Sustainable, Healthy and Learning

Cities and Neighbourhoods (SHLC), Human Sciences Research Council, Cape Town, South Africa

UN (Department of Economic and Social Affairs) (2020) UN World Social Report 2020: inequality in a rapidly changing world. United Nations

UN-Habitat (2006) State of the World's Cities 2008: the Millennium Development Goals and urban sustainability. UN-Habitat, Nairobi

UN-Habitat (2016) Urbanization and development: emerging futures World Cities Report 2016. Nairobi: UN-Habitat, Nairobi

UN-Habitat (2022) World Cities Report 2022: envisaging the future of cities. UN-Habitat, Nairobi https://unhabitat.org/wcr/

Van Ham M, Manley D, Bailey N, Simpson L, Maclennan D (eds) (2012) Neighbourhood effects research: new perspectives. Springer, Dordrecht

Voumik LC, Sultana T (2022) Impact of urbanization, industrialization, electrification and renewable energy on the environment in BRICS: fresh evidence from novel CS-ARDL model. Heliyon 8(11):e11457. https://doi.org/10.1016/j.heliyon.2022.e11457

Wang YP, Kintrea K (2021) Urban expansion and land use changes in Asia and Africa. Environ Urban ASIA 12(1S):13S–17S

Wei Y, Ewing R (2018) Urban expansion, sprawl and inequality. Landsc Urban Plan (September). https://doi.org/10.1016/J.LANDURBPLAN.2018.05.021

World Bank (2023) Harnessing urbanization to end poverty and boost prosperity in Africa. World Bank, Washington, DC

Zajacova A, Lawrence EM (2018) The relationship between education and health: reducing disparities through a contextual approach. Annu Rev Public Health 1(39):273–289. https://doi.org/10.1146/annurev-publhealth-031816-044628

4 Urban Education and Neighbourhood Inequality

Yulia Nesterova, Michael Osborne, and Michele Schweisfurth

Abstract

Education is viewed as a leveller of inequalities and key to socio-economic development. Yet, education is riven with inequalities that perpetuate broader socio-economic inequalities. Available research focuses on non-spatial markers of inequality (e.g. gender, minority status). Our study explores how participation in schooling and adult education differs between neighbourhoods of different socio-economic status. Using examples from the SHLC Household Survey in Africa and Asia, and city-level reports with secondary data, we show that the general advantage of urban education masks different levels of access, participation, and quality between neighbourhoods. While our findings show a predictable trend of educational advantage in wealthier neighbourhoods, there are outliers, including poorer neighbourhoods with high levels of parental satisfaction with schooling and high levels of participation in adult education.

Keywords

Urban education · Adult education · Lifelong learning · Educational inequalities · Educational disadvantage · Sustainable Development Goals (SDGs)

4.1 Introduction

Education is often hailed as a leveller of inequalities and a ticket to a better life for individuals and to economic and social development for nations. However, education is itself riven with inequalities of opportunity. Personal characteristics including gender, ethnicity, and socio-economic status help determine educational access, experiences, and outcomes. So, too, does where we live. In this chapter, we explore some of the ways that spatial inequalities in general and neighbourhood inequalities in particular shape educational opportunity in cities in the Global South. Using illustrative examples from the Sustainable Healthy Learning Cities (SHLC) Household Survey across 13 cities in Africa and Asia as well as city-level reports with secondary data compiled by our partners, we show how participation in schooling and adult education differs between neighbourhoods of different socio-economic status.

Y. Nesterova (✉) · M. Osborne · M. Schweisfurth
School of Education, University of Glasgow, Glasgow, Scotland, UK
e-mail: Yulia.Nesterova@glasgow.ac.uk

M. Osborne
e-mail: Michael.Osborne@glasgow.ac.uk

M. Schweisfurth
e-mail: Michele.Schweisfurth@glasgow.ac.uk

Y. P. Wang et al. (eds.), *New Drivers of Division*, Sustainable Development Goals Series,
https://doi.org/10.1007/978-981-96-5629-5_4

Putting the issue of spatial inequalities in education in a global context, the Education for All movement dates back to the 1990s when the inclusion of diverse groups of students in and their access to education became a dominant policy imperative. Key international agreements were forged and adopted, including the *World Declaration on Education for All* and *Framework for Action to Meet Basic Learning Needs* by the World Conference on Education for All in Jomtien, Thailand in 1990 (UNESCO 1999). It found further expression in the Millennium Development Goals (MDGs 2000–2015) with targets on increased enrolment at the basic/primary level and on greater parity of enrolment between boys and girls (United Nations General Assembly 2000). The United Nations Sustainable Development Goal 4 (SDG 4—2015–2030) extends the scope of these intentions and aims to ensure quality and equity of education and learning for all (UNESCO 2016).

Improving equality of access to and quality of education is understood to promote a range of benefits. According to the International Panel on Social Progress (IPSP, Spiel et al. 2018), education has economic, civic, humanistic, and equalising purposes. In terms of human capital, education prepares people for better jobs and a more productive work life, and promotes economic growth and labour market diversification for nations. The civic purpose points to education's role in developing local, national, and global citizenship, and fostering participation. The humanistic purpose celebrates the part that education plays in all forms of human flourishing, from spiritual growth to developing talents and appreciation for the arts. The equality dimension demonstrates how quality education can be a vehicle for justice and inclusion—or, where equality is lacking, injustice and exclusion. Above and beyond these instrumental aims, education is also a human right and primary level education should be free for all and compulsory (Article 26, UN Declaration of Human Rights). However, these purposes can only be met where education and its benefits are equally distributed. In reality, many learners are de facto excluded from the benefits of education as even when they have access to schooling and attend, they are not able to cope up with the curriculum due to the gap between their own learning background and official expectations (Lewin 2011).

Given the expectations attached to education and the consequences of limited access and poor outcomes, along with the investment in it as a global goal, it is not surprising that there is a vast literature surveying, analysing, and comparing educational inequalities. International organisations, including the World Bank and UNESCO (especially through its Global Education Monitoring Report that publishes annual thematic reports), maintain statistical databases that monitor the progress of countries towards Education for All and SDG 4, disaggregated along different parameters of inequality. However, the available evidence focuses largely on non-spatial markers of inequality. The research that shapes and influences policy and practice as well as interventions to reduce inequalities and provide quality services focuses overwhelmingly on differences determined by group belonging. Gender, for example, has been a significant issue, given that girls' participation in schooling has historically been lower than that of boys' for economic and cultural reasons, and also given the particular benefits to girls in terms of empowerment, and to societies through, for example, female economic participation and smaller and healthier families (e.g. UNESCO 2018, 2019a, 2020a, 2022). Other individual markers of disadvantage that affect education include ethnicity, class, religion, and migrant/refugee status (Nesterova 2023; UNESCO 2010, 2019b, b). When spatial inequalities are foregrounded, it is usually on a large scale, where progress toward global education goals is compared by nation or region, or between the Global North and Global South (UNESCO 2010, 2020b). It is rare to see sub-national units of analysis featuring in the discussion of inequalities, and the smaller the unit the less likely it is to be compared, generating a gap at the level of city and especially of neighbourhood.

There is also substantial literature on the so-called 'urban advantage' in education in relation to access and also outcomes. A sample of the evidence base shows that:

- In a sample survey of 52 low- and middle-income countries between 2010 and 2015 '5 children in rural areas attended for every 10 children in urban areas' (UNESCO 2017: 144).
- Between 2010 and 2015, for every 100 adolescents from urban areas who completed lower secondary, just 75 from rural areas did so (UNESCO 2017: 187).
- Average out-of-school rates stand at 16% in rural areas, and 8% in urban areas (UIS and UNICEF 2015 as cited in UNESCO 2015a: 8).
- In South and West Asia by the end of 2010 89% of urban adolescent boys from the richest families completed lower secondary school compared to just 13% of the poorest rural girls (UNESCO 2014: 8).
- '… even if primary attainment rates had remained constant in urban and rural areas in sub-Saharan Africa, the increase in the percentage of people living in high attainment urban areas would have been enough to raise the average primary attainment rate by 1.5 percentage points between 2000 and 2010' (UNESCO 2015b: 22).

Again, participation in education by adults is shaped by geography, alongside a range of other personal and situational factors, notably Indigeneity, gender, disability, and socio-economic class. The 4th Global Report on Adult Learning and Education (GRALE) highlights the challenges of rurality, reporting the stark contrast between literacy rates amongst adults in sub-Saharan Africa and South Asia (UNESCO-UIL 2019).

Indicators to monitor inequalities are crucial to redressing disadvantage and ensuring the provision of quality education and learning for all. However, the notion of urban advantage belies the heterogeneity of cities in countries that are becoming increasingly and uncontrollably urbanised in a context of ongoing rural-to-urban migration. While cities are viewed as offering comparatively more educational and learning opportunities to their residents, at the level of neighbourhood, these opportunities may play out very differently amid different socio-economic compositions and differing levels of demographic stability. The intersection of spatial segregation, socio-economic inequalities and schooling based on catchment area makes 'place' an important but complex marker of educational opportunity (Kintrea 2021).

In this study, we draw on SHLC research which gathered household-level data in five neighbourhoods in each of the 13 cities in seven countries of Africa and Asia. We focus specifically on data on educational participation. Although access and participation are the tip of the iceberg in terms of understanding the complexities of educational inequalities, they are the essential starting point, and these illustrative examples help to emphasise the need to understand and document educational inequalities through the neighbourhood as a unit of analysis.

4.2 Literature Review: Cities and Neighbourhoods as Shapers of Participation and Success in Education

4.2.1 Why the Type of Neighbourhoods Matter for Schooling

Despite the relative lack of data disaggregated to neighbourhood level, existing studies (mainly from the Global North) demonstrate that the type of neighbourhoods matter for schooling. Neighbourhood characteristics help to predict educational access and outcomes. In the United States, Catsambis and Beveridge (2001) discovered that disadvantages at the neighbourhood and school level may place students at risk, and a study by Andersson (2004) in Sweden showed that neighbourhoods have an enormous effect on students' futures as they affect their socio-economic careers. According to Gillock and Reyes' study in the United States (1999), this includes severe stressors that can be found in resource-poor neighbourhoods, such as worrying about being mugged or shot, experiencing changes in parental employment, and family finances. Moreover, as a study conducted in the Netherlands by de Vuijst et al. (2015) shows, when someone grows up in a deprived neighbourhood, his/her chances to remain in the deprived neighbourhood increase, perpetuating cycles of disadvantage.

The income of a neighbourhood may have an effect on a pupil's access, scores, achievement, and drop-out rates. From research in the Netherlands (see Sykes and Musterd 2011), we know that neighbourhoods differ in their provision of types of education and that an unequal distribution of schools might result in a relative advantage in educational opportunities for children residing in well-served neighbourhoods over children living in more poorly served areas, and this is especially evident in the context of the Global South. We also know that attendance in higher rated schools is associated with higher reading and math scores from research in the United States (e.g. Catsambis and Beveridge 2001; Eamon 2005), and that higher socio-economic and socio-cultural resources and neighbourhood stability are predictors of higher educational outcomes from research in Sweden (see Andersson and Subramanian 2006). A greater prevalence of high-status residents strongly predicts more time spent on homework and higher math/reading test scores which may rival the impact of family and school factors, whereas greater neighbourhood deprivation is a significant predictor of lower test scores (Ainsworth 2002; Hanson et al. 2011, both studies from the United States). In addition, as Kauppinen's work in Finland (2007) discovered, neighbourhoods with the highest educational level of the adult population increase the probability of finishing upper secondary school (the more academic option) rather than the vocational school. In contrast, an Australian study by Overman (2002) showed that drop-out rates are high in neighbourhoods with high concentrations of vocationally qualified adults, which are more often found in more disadvantaged neighbourhoods. In addition, schools in less affluent neighbourhoods provide less support, including from teachers who find themselves under situations of high chronic school stress, which leads to lower academic achievement, according to Gillock and Reyes's research (1999) in the United States.

At the same time, parents in low-income neighbourhoods may not want to send their children to better-resourced, higher rated schools outside their neighbourhoods to receive better education. For example, as Cameron's study in Bangladesh (2011) discovered, parents may not send children to schools outside their own low-income neighbourhood because they feel they do not belong or because of the entrance examinations. This phenomenon aligns with Thompson's (2002, as cited in Kintrea 2021: 211) explanation that schools are context-derived but also context-generative. As Kintrea explains this phenomenon, strong residential segregation where 'the default is for children to attend their local school' leads to 'a narrowing of the social mix in individual schools', ultimately reproducing broader socio-economic inequalities and disadvantage.

4.2.2 Why Neighbourhoods Matter for Adult Education and Lifelong Learning

There is little data available internationally on participation in adult education and lifelong learning disaggregated to neighbourhood level. There is however, one important tool to monitor participation at a national level as described in UNESCO's GRALE (Global Report on Adult Learning and Education) that uses a survey developed as part of the Belém Framework for Action (BFA), adopted at the sixth International Conference on Adult Education (CONFINTEA VI) in 2009. In 2019 the survey was completed by the ministries of 159 countries. However, despite these efforts, even reliable data at national and especially at disaggregated levels is patchy at best. Furthermore, although responses from ministries are important, the data that they present does not tell the full story since only a limited number of countries actually monitor participation in adult learning in a systematic fashion, and the countries that do are mainly from the Global North. Additionally, as has been reported by Osborne and Hernandez (2021), the quality of data in the field of adult and lifelong learning is low and the statistics are not comprehensive, primarily due to inconsistent definitions of participation as well as 'lack of methodologies and resources to produce robust and meaningful data' (57).

From UNESCO-UIL's (2019) GRALE IV we know that in general 'Progress in participation in adult learning and education is insufficient' (20). In particular, marginalised groups, such as adults with disabilities, older adults, minority groups, residents in remote and rural areas, as well as people living in poverty and/or those in menial work, do not participate in and do not believe they can benefit from adult learning and education (UNESCO 2019a, b).

This points to the skewing of opportunities for participation in adult education and learning towards those already advantaged and, as with schooling, those with particular socio-demographic backgrounds and personal characteristics are most in danger of exclusion.

The proxy of participation from the International Adult Literacy Survey (IALS), which was used from 1994 to 1998 in Organisation for Economic Co-operation and Development OECD countries, is typical of the broad-brush approach taken (if taken at all) by nations:

> During the past 12 months, that is, since …, did you receive any training or education including courses, private lessons, correspondence courses, workshops, on-the-job training, apprenticeship training, arts, crafts, recreation courses or any other training or education? (OECD 2000: 133)

Such definitions are so wide-ranging as to include provision within both formal, non-formal and informal settings. Formal education is 'institutionalised, intentional, and planned through public organisations and recognised private bodies' (UNESCO Institute for Statistics 2011: 11–12) that leads to recognised national qualifications. Non-formal education shares the characteristics of being institutionalised, intentional, and planned, but is a complement to formal provision that typically is short in duration, and does not have a continuous pathway of progression. It is offered as part of the lifelong learning offer and, as well as provision for adults, is also manifest in literacy, life skills, and social/cultural development programmes for all ages. Institutions of many types offer such programmes, including many that are not part of the formal system. Informal learning is offered as a result of an individual being involved in activities that are not overtly educational in purpose, and is often referred to as being experiential, incidental, or even accidental.[1]

Most surveys of participation in Adult Learning and Education (ALE) that have been developed for use internationally, in order to make national comparisons, tend to focus on skills, and in particular on forms of literacy. The most well known of these surveys are OECD's PIAAC (Programme for the International Assessment of Adult Literacy), International Adult Literacy Survey (IALS) and the Adult Literacy and Life Skills Survey (ALL) used from 2003 to 2007, after IALs and prior to PIAAC (OECD 2013). While there may be questions as what constitutes meaningful adult participation in learning (in terms of duration and qualifications achieved), these are well-validated instruments that map participation rates in all forms of adult learning against current skills levels, and can monitor these rates over time at national and supranational levels.

However, relatively little attention has been given to fine-grained analyses at urban and neighbourhood levels of difference in adult participation that reflect the impact and nuances of place, and of the geographical heterogeneity of opportunity. There are some exceptions at urban level, mostly linked to actualising the learning city concept that has been promoted most notably since 2013 by UNESCO in its Global Network of Learning Cities. This framework seeks to mobilise the resources of multiple actors to promote and facilitate learning that enhances the social and economic well-being of the locality (UNESCO-UIL 2013). Although largely applied in relatively wealthy parts of the world (Lido et al. 2019), the learning city foregrounds the idea that while national governments play major roles in assigning roles in setting policy agendas for education, it is city administrations that typically

[1] See https://www.coe.int/en/web/lang-migrants/formal-non-formal-and-informal-learning.

are the enablers of practice, working with a range of stakeholders, and those who are more likely to adopt a lifelong and life-wide approach. It also recognises that the successful building of a learning infrastructure within a city relies not only on top-down approaches, but the agency of people living in cities who are most likely connected by common concerns within their neighbourhoods (Ó Tuama 2020). However, while the framework encourages the gathering of nuanced data at the local level, and provides a consistent structure for operationalising this, the data is not systematically consolidated, aggregated, or compared across contexts.

There is thus a significant gap in our knowledge of how adult participation in education and learning plays out at smaller scales of geography, and the concomitant implications, particularly in the cities of the Global South. This is important in the light of the fact that, even more so than with schooling, it is often the city or region that has a predominant role in supporting the participation of adults in learning, both within the formal system of education (schools, colleges, and universities) and the non-formal system in community settings. This is because, typically, adult education is a lower priority for state support by comparison to schooling (Ireland and Spezia 2014). The most recent GRALE V Report (UNESCO 2022: 62) reports that only 15% (22 countries) out of 102 reporting spent more than 4% of total government educational expenditure on Adult Education.

Hence, SDG Target 4.6, which refers to 'ensuring inclusive and equitable quality education and promoting lifelong learning opportunities for all' is one that has become a particular responsibility of Local and Regional Governments (LRGs), and there exist many initiatives around the world, particularly in the Global South, of efforts that are led at this level. This is particularly the case with regard to ensuring adult literacy. Osborne (2022) has provided numerous examples of practices at the urban level from many parts of the world that seek to localise learning opportunities and subsequent participation for adults.

Our work with SHLC has added a further dimension to this work by for the first time assessing the extent to which adult participation relates to the nature of neighbourhoods in cities. We define adult participation for the purposes of our survey as the receipt of any adult education or training provided formally by either the formal or non-formal sector during the previous three years.

4.3 Research Design: The Survey and Secondary Data

The data we draw on in this chapter comes from two Research Task Packages (RTP) of the SHLC project: RTP2 and RTP3. RTP2 consisted of comprehensive reports prepared by each in-country team based on secondary data that included: the analysis of education data sets from national level down to the lowest level of aggregation such as a neighbourhood, where possible; informal 'go along' interviews with key stakeholders and residents in each neighbourhood; and auditing of built environment and service provision at neighbourhood level to evaluate availability and accessibility of educational institutions. RTP3 constituted the flagship work package of the SHLC project and involved a household survey conducted in five neighbourhoods in each city (two cities per country). The neighbourhoods were stratified partly by economic characteristics, but attention was also given to natural or administrative boundaries and the built environment. Over 14,000 survey questionnaires (around 2000 from each of the seven countries) were administered at the household level.

The questionnaire included 20 items on formal and non-formal or adult education. The themes explored on schooling included:

- Level of education completed by each household member

- Children in the household—who attends school, boys and girls, and reasons for non-attendance where applicable
- What kind of schooling children attend—public, private
- Mode and time taken for travel to school
- Levels of satisfaction with school
- Views on importance of schooling and the quality of schooling available nearby
- Educational support during the COVID-19 pandemic

The questions on adult education included themes of:

- Participation in adult education, mode and purpose
- Funding for adult education
- Benefits of courses taken
- Views on educational opportunities for adults.

COVID-19 delayed fieldwork for over a year in most of our partner cities but when it eventually took place it generated rich data with much potential for researchers who wish to explore how neighbourhood inequalities affect education across a range of themes, and how these intersect with other inequalities found across neighbourhoods.

Some illustrative examples based on a preliminary analysis of the data on participation in schooling and adult education are given below. The gaps in the literature on neighbourhood-level access and participation in schools and adult education and the slow and uneven progress toward SDG 4 cannot be addressed fully in one chapter, but the selective data demonstrates the importance of neighbourhood spaces in shaping educational and life opportunities for residents. These examples belie the richness of the data and the potential for further correlational and regression analyses, as also its potential for further qualitative focus group discussions.

4.4 Findings

4.4.1 Schooling: Access, Participation, and Drop-Out

Over the past decades, cities in the SHLC study have seen significant improvements in securing access to education (Nesterova and Young 2020); however, access continues to be a challenge across many neighbourhoods. As our data from two work packages shows, the general picture is that the higher the income level of a neighbourhood, the better access to, and higher quality of schools a neighbourhood is likely to have. To show the contrast, we provide a few illustrative examples between access to schools in high-income, higher middle-income, middle-income, lower middle-income, and low-income neighbourhoods.

Availability and coverage of schools: As data from RTP2 shows, low-income neighbourhoods tend to have insufficient availability and inadequate coverage of schools. This means that available schools are not enough to educate all children who require education in the neighbourhood and are not distributed well across the neighbourhood to allow easy access. This also means that classrooms in such neighbourhoods are likely to be overcrowded with few teachers and resources. Most schools in low-income neighbourhoods tend to be at primary level, with only a few exceptions at secondary and high school level. If a child needs to attend a better quality school and/or progress to secondary education, there is a high chance that he or she will have to travel to another neighbourhood, which may be a challenge, as discussed below. On the other hand, higher middle-income and high-income neighbourhoods have a wide variety of institutions from kindergartens to universities and a good spatial coverage to allow easy access. In the few places where high-income neighbourhoods do not have enough schools, children often attend renowned private schools around the city (e.g. in Delhi,

India). Lower middle- and middle-income neighbourhoods tend to have better school availability and coverage than low-income neighbourhoods but worse than in higher middle- and high-income neighbourhoods.

Physical access to schools within one's neighbourhood and in other neighbourhoods: As data from RTP2 shows, low-income neighbourhoods tend to have damaged, unsafe, and encroached roads and footpaths, heavy traffic, and long distances to reach public transport facilities. These conditions lead to children experiencing challenges with commuting between home and school within their neighbourhood and prevent them from attending schools outside their neighbourhood, or make their commutes lengthy and uncomfortable. While all children in such neighbourhoods are affected, this makes walking to school particularly challenging for children with special physical needs. It should be noted that, as an exception worthy of further exploration, low-income neighbourhoods in Madurai, India, have safe and good roads and are well-connected to the rest of the city. Higher middle-income and high-income neighbourhoods across all studied cities tend to have good connectivity, accessibility, and higher quality schools.

Another aspect regarding physical access to schools is how children travel to schools and how long it takes. Our survey showed that across the study children from higher income neighbourhoods tend to travel to school door-to-door by car, whereas low-income neighbourhood children travel mostly on foot. For example, in high-income neighbourhoods in Bangladesh, 40% of pupils travel by car versus 12% on foot, whereas, 5% travel by car and 54% travel on foot in middle-income neighbourhoods, and 1% and 69% respectively in low-income neighbourhoods. The trend is similar in the other countries studied, with a few exceptions such as in India and Rwanda where almost the same number of pupils use a car or walk to school—23% vs. 38% in India and 22% vs. 28% in Rwanda. This is a result of two factors: higher rates of car ownership and an ability to travel to further away schools (e.g. parent/driver to drive) in high-income neighbourhoods; both factors enable access to better schools.

As the survey shows, in the majority of cases, and in contrast to existing research evidence on rural schools, it does not take longer than 30 min for children in any type of neighbourhood to reach school, regardless of the method of transportation. For example, in Bangladesh, 68% of the pupils in high-income neighbourhoods, 80% in middle-income, and 75% in low-income reach school within 20 min. The figures are similar in Rwanda, where it takes up to 20 min for 80% of pupils in high-income neighbourhoods, 63% in middle-income, and 56% in low-income neighbourhoods. In India and Tanzania, for the majority it takes between 10 and 30 min. This applies to 70% of pupils in a high-income neighbourhood, 77% in middle-income, and 68% in low-income in India and 61%, 50%, and 51% respectively in Tanzania.

Quality and availability of resources in schools: As data from RTP2 shows, both public and private schools in higher middle-income and high-income neighbourhoods are relatively well-furnished, well-maintained, secure, and well-equipped, with clean water and toilets, multimedia facilities, computer labs, libraries, playgrounds, and sports facilities. They tend to have a lower student–teacher and student–classroom ratio and offer a range of extracurricular activities to their students. In comparison, public schools in low-income and lower middle-income neighbourhoods have less hygienic environments with limited access to clean water and few toilets, are poorly resourced (e.g. limited or no access to computers), offer no or few extracurricular activities, receive inadequate funding for maintenance, and face teacher shortages. As a result, as our survey showed, where a large number of children attend state schools where classrooms are overcrowded and there is shortage of teachers, low levels of satisfaction with schools are recorded.

Our research shows that exceptions exist, however. The RTP2 report from Bangladesh, for example, shows that there is an acceptable student–teacher ratio even in low-income neighbourhoods and the RTP2 report from India shows that one low-income neighbourhood in Madurai, India (exceptional again) has a public higher secondary school with good infrastructure, sufficient number of teachers and quality resources such as laboratories, aerobatics class, computers, and a library. As the RTP2 report on India notes, performance in this particular school is stated to be outstanding by stakeholders. Similarly, RTP2 reports from the other studied countries show that although schools in middle-income neighbourhoods tend to be adequately resourced, have relatively good student–teacher and student–classroom ratios, and offer extracurricular activities, schools within poverty clusters within middle-income areas have similar resources as schools in poorer neighbourhoods.

It is not surprising, therefore, that the survey showed that 72% of surveyed participants in low-income neighbourhoods in Bangladesh, 52.6% in China, 70.5% in India, 67.8% in the Philippines, and 50.3% in South Africa reported being satisfied with the schools their children attend. This can be explained by historically good schools, sustained support provided by non-governmental organisations, or access to neighbouring areas with better schools. It might also be explained by the perception of people in low-income neighbourhoods of the quality of their schools compared to others of a similar nature without personal knowledge of schools in more affluent neighbourhoods. In contrast, in high-income neighbourhoods there tend to be more people who are very satisfied with their schools. For example, while 28.1% were very satisfied with schools in high-income neighbourhoods in Bangladesh, only 6.4% were very satisfied in low-income neighbourhoods; in Rwanda, 60.6% in high-income neighbourhoods and 40.4% in higher middle-income neighbourhoods, whereas only 10% and 11.6% in lower middle and low-income neighbourhoods were very satisfied, respectively.

Drop-out and absenteeism rates tend to be considerably higher the less affluent the neighbourhood is: Rates for school drop-out and student absenteeism after primary school tend to be high in low-income and lower middle-income neighbourhoods. Our RTP2 data for Khulna (Bangladesh), for example, shows that as many as 30.6% of children do not attend school in low-income neighbourhoods, 42.8% in the lower middle-income, and 35.4% in middle-income neighbourhoods. It is important to note, however, that RTP2 data suggests that these neighbourhoods have clusters of areas poorer than the overall composition of the neighbourhood. While drop-out and student absenteeism rates decrease as the income levels of neighbourhoods increase, there are still concerns about these issues in some affluent neighbourhoods, particularly those that have poorer clusters. For example, in Khulna, again in a high-income neighbourhood, 31.8% of children do not go to school. Drop-out rates show that while 15.4% of children drop out of school in a mixed low-income neighbourhood, the proportion in a middle-income neighbourhood drops to 1.5% but then increases to 5.9% for a high-income neighbourhood that has clusters of poverty. Such statistics show how heterogenous neighbourhoods can be.

Our survey showed various reasons why children in different income neighbourhoods drop out of school, which included the need for a child to bring in additional income, unaffordability of schools, pregnancy and motherhood, disability, low value attached to education, and need to help with household chores. Interestingly, despite lower drop-out rates in higher income neighbourhoods, one of the primary reasons for dropping out is the need for a child to provide additional financial support (and the only reason in Bangladesh and Rwanda). For example, in Bangladesh, all households (100%) where children dropped out of schools stated this as the primary reason, although compared to 31% households in low-income neighbourhoods, other reasons also included lack of affordability (24.7%), little value attached to education (27.2%), and need to help around the house (13.6%). In the Philippines, 37.7% in high-income neighbourhoods need children to earn money and 19.7% cannot afford fees compared to 46.8% and 20.3% in low-income neighbourhoods, respectively. Such financial reasons may show that the respondents reside in poverty areas in high-income neighbourhoods. It is important to note that our research did not ask at what age pupils drop out of school, which is an important aspect as there is a difference between dropping out at 11 and dropping out at 17 years of age.

Next is the issue of access to nonstate schools where public sector schools tend to be overburdened and underfunded, as our RTP2 reports showed, and where private schools and those funded and run by faith-based institutions (e.g. madrassas and Christian churches) and NGOs have stepped in to improve access to education. In most countries under study, there has been a sharp increase in the number of private providers (Nesterova and Young 2020). While the picture is mixed, these providers tend to have better resources, offer better educational outcomes for pupils, and are associated with higher socioeconomic status (Nesterova and Young 2020). For example, in Bangladesh, private schools dominate in most neighbourhoods, while most private schools in India are located in middle- or higher income neighbourhoods. If there are private schools in low-income neighbourhoods, there tends to be financial support provided for poor children. However, while families in higher middle- and high-income neighbourhoods can more easily afford to send children to private schools, those in low-income and lower middle-income neighbourhoods, as well as less affluent families in richer neighbourhoods, are dependent on financial support for tuition and other school expenses provided by the state or NGOs. As a result, our RTP2 data shows that NGOs and faith-based schools are emerging as an important education provider for poorer populations. This raises concerns as the funding that NGOs receive is not sustainable and when it ends, so does the schooling of children from poorer families. In contrast, more financial stability and a better quality of education can be found in well-reputed schools in lower middle-income neighbourhoods that are run by missionaries (i.e. in Bangladesh). Missionary-run schools in India, on the other hand, are found in slums and charge fees.

Our survey showed that where there is a higher number of children attending private schools, there are higher levels of recorded satisfaction with schools. There are a few exceptions to this. For example, in Dar es Salaam (Tanzania) and Kigali (Rwanda), some participants were dissatisfied with private schools in low, lower middle, and even middle-income neighbourhoods, which can be explained by the fact that the more well-off attend higher quality private schools (see Cameron's 2011 study of Dhaka, and Gruijters, Alcott and Rose's 2021 study of countries in East Africa and South Asia). Interestingly, the survey showed some dissatisfaction among participants residing in Khulna's higher middle- and high-income neighbourhoods, which could be explained by them potentially residing and/or attending schools located in one of the poorer areas of a rich neighbourhood.

As with cities, neighbourhoods are heterogeneous. This means that lower middle-income neighbourhoods through to high-income neighbourhoods may contain areas of poverty or lower-income populations and schools compared with the overall neighbourhood profile. In such areas and schools, children's opportunities differ from those residing and attending schools in more affluent parts of the neighbourhood and their dropout and absenteeism rates are higher. These are the schools in more affluent areas that children from low-income neighbourhoods attend. Such areas, our RTP2 research shows, even in middle- and higher middle-income neighbourhoods may have unsafe buildings, insufficient number of classrooms and teachers, no playgrounds and sports facilities, unhygienic environment, and no clean water supply. As regards physical accessibility, while higher income neighbourhoods tend to have well-maintained footpaths and roads and are well-connected within the neighbourhood and with other neighbourhoods, there are exceptions here too. In some such neighbourhoods, footpaths, roads, and the transport system are not of a satisfactory condition or travel by car has taken priority.

It should also be noted that a lack of schools in a neighbourhood may not be an issue for wealthier residents. For example, our RTP2 report from Tanzania shows that in Dar es Salaam, a low-income neighbourhood with a cluster of university staff residents does not have good schools but those in nearby neighbourhoods are easily accessible and affordable. People in this neighbourhood are reported to have a very low illiteracy rate and a very high school completion rate, with children often attending good schools outside of the neighbourhood. In contrast, a middle-income neighbourhood in the same city that has no available local schools is in an inconvenient location which requires long commutes for children.

4.4.2 Adult Education and Lifelong Learning

While the SHLC RTP data goes some way in addressing the lack of neighbourhood-level data in cities in the Global South, patterns were occasionally difficult to discern. This may be because adult education in its broadest definition is actually distributed more evenly across neighbourhoods, or because there are issues with cross-cultural validity around the terms associated with adult education. It may also be reflective of the variability of funding for adult education, and meaningful accreditation for it. While the vocabulary of schooling is largely shared and robust, as noted above, adult education in its various forms can be more variably understood.

Our household questionnaire as it related to adult education and training began with asking whether individuals within the sample had received adult education or training of any kind in the last three years. The positive response rate across our countries varied from 2.5% of the sample (India) to 17% (Rwanda) (see Table 4.1).

Given the nature of our research, we were interested, as we were in schooling, in differences as they prevail across different types of neighbourhoods across the cities of the seven countries being studied. Levels of participation by adults in education and training tend to be highest within high-income and/or high- to middle-income neighbourhoods in some countries, with sometimes stark contrasts emerging especially between high-income and low-income neighbourhoods (see Table 4.2). This is particularly evident in Bangladesh and China where there are respectively 5.1 and 6.8 times as many participants proportionally in high-income with respect to low-income neighbourhoods. In the Philippines, however, the highest participation rates are in low- to medium-income neighbourhoods, and

Table 4.1 Participants reporting receipt of adult education or training and public expenditure by country

	Sample size	Percentage in receipt of education or training	Percentage of public education spending on ALE
Bangladesh	2586	13	0.5–0.9
China	860	12	1.0–1.9
India	2378	2.5	0–0.4
Philippines	2079	7	0.5–0.9
Rwanda	1485	17	Not available
South Africa	2001	13	Unknown
Tanzania	2841	6	No report
Overall	14,320	9	

Note Countries were asked to report rates in bands, starting with 0–0.4%
Source https://uil.unesco.org/adult-education/global-report/data;

in some countries the differences across types of neighbourhoods are not stark as is exemplified in India, South Africa, and Tanzania.

We have hypothesised that factors such as the perception of opportunity to participate in adult education and training, and of the availability of public funds might be accounting for participation, since these are long-standing and major factors in decision-making (Cross 1981). In Cross's Chain of Response Model, lack of information about education opportunity and attendant support mechanisms are key factors that interact with a series of dispositional (self-esteem and attitude) and institutional (e.g. entry requirements and scheduling of classes) barriers. The affordability of education is amongst a number of situational factors, which also includes personal responsibilities and availability of transport.

It is the case that there is a link overall between neighbourhood type and perception of opportunity with a decline in our sample from 34.4% in high-income to 14.5% in low-income zones. There are however differences across countries with levels of agreement with proportions varying from 42.9% in South Africa to 4.9% in Rwanda,

Table 4.2 Participation rates in adult education or training by type of neighbourhood and country

	High	High to medium	Medium	Low to medium	Low	Overall
Bangladesh	24	18	11	12	5	13
China	32	12	15	7	5	12
India	3	2	4	3	1	3
Philippines	9	5	6	11	6	7
Rwanda	29	27	15	12	14	17
South Africa	11	16	10	15	13	13
Tanzania	4	9	6	8	3	6
Overall	11	11	9	10	7	9

Note Figures are in per cent
Source Primary survey, 2021

Table 4.3 Sources of financial support for adult education and training reported by participants by country

	Union (%)	Employer (%)	Government (%)	NGO (%)	Self (%)	Family/friends (%)
Bangladesh	5.8	35.0	14.7	18.4	47.2	6.1
China	7.7	18.3	39.4	3.8	50.0	2.9
India	0	23.3	26.3	13.3	48.3	0
Philippines	26.8	36.3	59.5	4.6	19.4	29.2
Rwanda	6.9	46.2	16.9	28.2	10.5	1.6
South Africa	21.3	34.9	31.8	8.9	43.8	34.5
Tanzania	9.8	15.3	10.4	17.2	36.2	42.9
Overall	11.9	32.6	25.8	15.2	36.4	15.2

Note Rows add to more than 100% since some participants report more than one source of support
Source Primary survey, 2021

and neighbourhood differences are largely not evident in India and the Philippines. Neighbourhood type may therefore be a factor in some countries in perception of opportunity and in availing of that opportunity, but other factors are also in play.

It is however difficult to account for these differences based on reported proportions of public education spending that is directed to Adult Learning and Education because of the paucity of data, but the data from GRALE III (UNESCO 2016) is suggestive of a link for those countries who reported. Certainly in India where there is the lowest percentage of spending, there is also the lowest participation in the two cities of our study. However, in none of the countries for which GRALE provides data is there public expenditure that approaches UNESCO's recommended minimum of 4% of the total expenditure from government on education. This is reflected in responses that we received when we asked respondents how their participation had been funded (see Table 4.3).

Overall, self-funding (36.4% of participants) and employer-provided (32.6%) were the predominant sources of funding, followed by national or local government spending (25.8%), NGOs/charities (15.2%), family/friends-support (15.2%), and trade union/worker education provided (11.9%), with some participants supporting their learning from more than one of these sources. However, these figures disguise differences from country to country with, for example, self-funding at its highest in China (reported by 50.0% of participants) and at its lowest in Rwanda (10.5%). By contrast, employer support is at its highest in Rwanda (reported by 46.2% of participants) and lowest in China (18.3%). Support from government was highest in the Philippines (59.5%) and lowest in Tanzania (10.4%). Support from trade union/worker education was at its highest in the Philippines (26.8%) and South Africa (21.3%), and its lowest in India, where no participant reported this as a source of funding. NGO/charitable support was at its highest in Rwanda (28.2%) and lowest in China (3.8%). The importance of family and friends in supporting adult learners is reflected in the data from Tanzania, South Africa, and the Philippines, where respectively, 42.9%, 34.5%, and 29.2% provided funding, at least in part. This compares to no one reporting such support in India and only 1.6% in Rwanda.

There are also differences between sources of funding for adult education/training across neighbourhoods of different income levels. Residents in high-income neighbourhoods are likely to receive free adult education/training from their employers in much greater numbers that those in low-income neighbourhoods. The only exception is South Africa where 37% of residents in low-income neighbourhoods received free employer-provided adult education/training compared to 30% in the high-income group.

Access to union-provided education by neighbourhood shows inconsistency across countries. For example, 12% in the high-income compared to no participants in low-income neighbourhoods in Bangladesh and 25% compared to 14% in Tanzania access such funding. In the Philippines, however, it is 42% of residents in low-income and 32% in higher middle-income neighbourhoods that obtain such support, compared to 17% in high-income neighbourhoods. There is also inconsistency by neighbourhood across countries in government-funded adult education/training with similarities across all neighbourhoods in Bangladesh, China, and South Africa, but clear differences in the other countries. In India, for example, 56% of participants in lower middle-income and 60% in low-income neighbourhoods access such funding in contrast to 15% in high-income zones; in the Philippines the numbers are 75%, 59%, and 47% respectively; Rwanda, 23%, 28%, and 4% respectively; and in Tanzania, 29%, 11%, and 0% respectively.

More residents of high-income neighbourhoods access funding for adult education/training through NGOs in China, the Philippines, South Africa, and Tanzania. For example, in China, 6% of high-income residents reported accessing such funding in contrast to none in lower-income and low-income neighbourhoods. In contrast, in Bangladesh, 31% of low-income residents are funded through NGOs compared to 15% in high-income neighbourhoods (the trend is similar in India and Rwanda).

More people self-fund their adult education/training in high-income neighbourhoods in Bangladesh (51% compared to 42% in low-income), Rwanda (21% compared to 16% in low-income and 2% in lower-income), South Africa (53% compared to 33% in low-income). This is different to China where 88% of low-income residents self-fund compared to 44% in high-income, and in India, where 73% of middle-income residents self-fund compared to 46% in high-income and 40% in low-income. Few respondents reported receiving help from family or friends but those who did, reside primarily in lower middle-income neighbourhoods.

We were also interested in the importance of learning leading to a recognised qualification. On average this was the case for almost two-thirds of 1,312 participants, the only country that was an exception being Tanzania where this was the case for only 44.8% of the participants. There are no notable differences between neighbourhoods of different income levels in this regard, especially in Rwanda where numbers are very similar across the neighbourhoods. In Bangladesh, China, and South Africa, around 20% more residents in high-income neighbourhoods received formal qualifications than in low-income neighbourhoods, although in China, more lower middle-income neighbourhoods received a formal qualification (77%) than in the high-income category (68%). In India, a slightly higher number of residents in the low-income level received a formal qualification (60%) but significantly more in the lower middle-income (94%) than in high-income (54%) categories. This trend is similar in the Philippines.

Overall, in relation to how adult education and training is funded and whether participation leads to a qualification, there are some differences by neighbourhood in some countries, but this is not a universal phenomenon, suggesting, as might be expected, that neighbourhood type is a factor of importance, but not universally the case. Other factors specific to particular cities and countries may have greater influence in some sites.

We also asked a series of other questions concerning benefits of and reasons for participation, some inspired by the many instruments that exist in the literature related to motivation for engaging in adult learning (see Boeren 2017). Responses across neighbourhoods provide evidence of the self-perception that adult education/training made respondents better at their jobs. For example, in Bangladesh, between 73% (low-income) and 95% (high-income) answered this question positively. The figures in most other counties are similar. The exceptions are India where residents in lower middle-income neighbourhoods report perceived benefit in greatest numbers (81% compared to 40% for low-income and 39% for high-income counterparts), and the Philippines, where low-income residents reported benefiting more (92% compared to 86% in high-income). Residents across all neighbourhoods in most countries tended to express enjoyment at being able to learn something new. The exception is Tanzania where only 25% of participating residents in high-income neighbourhoods shared this view compared to 96% in high middle-income and 57% in low-income areas. Across all neighbourhoods there was an acknowledgement that adult education/training helped them contribute to their communities, with the highest figures at 93% in higher middle-income neighbourhoods in Tanzania. The lowest figures were in high-income neighbourhoods in Tanzania (30%) and low-income neighbourhoods in India (40%).

4.5 Conclusion

The so-called urban advantage in education masks vastly different levels of access, participation, retention, and quality in city neighbourhoods, particularly in relation to formal schooling. The motivation behind SHLC's comprehensive research agenda is the conviction that comparing neighbourhoods across cities reveals more locally relevant data that can lead to more strategic and targeted approaches in achieving SDG 4 and ensuring that no one is left behind. In this chapter, we have seen how sparse the pre-existing evidence on neighbourhood educational differences is, particularly from the Global South and especially regarding adult education.

The SHLC database will help to bridge this gap. A preliminary analysis shows predictable trends of educational advantage in wealthier neighbourhoods across countries, but also some outliers worthy of further exploration. Understanding how, for example, some poorer neighbourhoods secure high levels of parental satisfaction with schooling, or why in some contexts lower middle-income neighbourhood residents have high levels of participation in adult education, can provide lessons that could help to ameliorate disadvantages entrenched by spatial inequalities. Equally, the variations in response to the survey items on adult education warrant further investigation, for example, on how cultural expectations of adult education shape participation, and how structural differences such as funding levels and the usefulness of qualifications reflect human capital drivers in general and local uptake in particular.

Ultimately, urban neighbourhoods are profound shapers of educational advantage, and therefore of life chances. However, given the wider patterns and outliers, there is reason to believe they do not fully determine these outcomes, and there is much to be learned from a detailed comparison across scales and policy contexts. Differences across neighbourhoods are particularly prevalent in cities of some countries more than others, and with further analysis of a wider range of variables, we will be able to determine the extent of the neighbourhood effect.

Importantly, our research shows that although there is a satisfaction gradient between rich and poor areas, there is definitely satisfaction about education and other services in some poorer neighbourhoods. It would thus be useful to do more work on this aspect with a particular focus on what kinds of people are more satisfied, where they come from geographically and migrationally.

References

Ainsworth UW (2002) Why does it take a village? The mediation of neighbourhood effects on educational achievement. Soc Forces 81:117–152. https://doi.org/10.1353/sof.2002.0038

Andersson E (2004) From valley of sadness to hill of happiness: the significance of surroundings for socioeconomic career. Urban Stud 41(3):641–659. https://doi.org/10.1080/0042098042000178726

Andersson E, Subramanian SV (2006) Explorations of neighbourhood and educational outcomes for young swedes. Urban Stud 43(11):2013–2025. https://doi.org/10.1080/00420980600897834

Boeren E (2017) Understanding adult lifelong learning participation as a layered problem. Stud Contin Educ 39(2):161–175

Cameron S (2011) Whether and where to enrol? Choosing a primary school in the slums of urban Dhaka, Bangladesh. Int J Educ Dev 31(4). https://doi.org/10.1016/j.ijedudev.2011.01.004

Catsambis S, Beveridge AA (2001) Does neighborhood matter? Family, neighborhood, and school influences on eighth-grade mathematics achievement. Sociol Focus 34(4):435–457. https://doi.org/10.1080/00380237.2001.10571212

Cross P (1981) Adults as learners. Jossey-Bass, San Francisco

de Vuijst E, van Ham M, Kleinhans R (2015) The moderating effect of higher education on intergenerational spatial inequality. IZA Discussion Paper No. 9557. https://ssrn.com/abstract=2708370

Eamon MK (2005) Social-demographic, school, neighborhood, and parenting influences on the academic achievement of Latino young adolescents. J Youth Adolesc 34:163–174. https://doi.org/10.1007/s10964-005-3214-x

Gillock KL, Reyes O (1999) Stress, support, and academic performance of urban, low-Income, Mexican-American adolescents. J Youth Adolesc 28:259–282. https://doi.org/10.1023/A:1021657516275

Hanson MJ, Miller AD, Diamond K, Odom S, Lieber J, Butera G, Horn E, Palmer S, Fleming K (2011) Neighborhood community risk influences on preschool children's development and school readiness. Infants Young Child 24(1):87–100. https://doi.org/10.1097/IYC.0b013e3182008dd0

Ireland TD, Spezia CH (2014) Adult education in retrospect—60 years of CONFINTEA. UNESCO, Brasilia

Kauppinen TM (2007) Neighbourhood effects in a European city: secondary education of young people in Helsinki. Soc Sci Res 36:421–444. https://doi.org/10.1016/j.ssresearch.2006.04.003

Kintrea K (2021) Is there a place for place in educational attainment policy? Oxf Rev Educ 47(2):207–223

Lewin K (2011) Making rights realities: researching educational access, transitions and equity. CREATE: University of Sussex

Lido C et al (2019) Lifewide learning in the city: novel big data approaches to exploring learning with large-scale surveys, GPS, and social media. Oxf Rev Educ 45(2):279–295

Nesterova Y (2023) Towards social justice and inclusion in education systems. In: Postiglione G, Johnstone C, Teter W (eds) Handbook of education policy. Edward Elgar

Nesterova Y, Young G (2020) Education policies, systems, and progress in Africa and Asia: a comparative analysis of 12 cities in six countries. Centre for Sustainable, Healthy and Learning Cities and Neighbourhoods, Glasgow, UK. http://www.centreforsustainablecities.ac.uk/wp-content/uploads/2020/07/Education-Policies-Systems-and-Progress-in-Africa-and-Asia.pdf

OECD (2000) Education at a glance 2000. OECD Publishing, Paris

OECD (2013) The survey of adult skills: reader's companion. OECD Publishing, Paris. https://doi.org/10.1787/9789264204027-en

Osborne M (2022) Localizing SDG 4: ensure inclusive and equitable quality education and promote lifelong learning opportunities for all. Local and Regional Governments. Report to the 2022 UN High Level Political Forum. In: Global taskforce of Local and Regional Governments: towards the localization of the SDGs—Local and Regional Governments breaking through for a just and sustainable recovery. Series: Local and Regional Governments' Report to the 2022 UN High Level Political Forum. United Cities and Local Governments, Barcelona, pp 86–98

Osborne M, Hernandez S (2021) Sustainable learning cities: inclusion, equity and lifelong learning. In: Howells A, Valdés-Cotera R (eds) Inclusive lifelong learning in cities: policies and practices for vulnerable groups. UIL, Hamburg

Ó Tuama S (2020) Learning neighbourhoods: lifelong learning, community and sustainability in Cork Learning City. Andragoška Spoznanja 26(1):53–65. https://doi.org/10.4312/as.26.1.53-65

Overman HG (2002) Neighbourhood effects in large and small neighbourhoods. Urban Stud 39(1):117–130. https://doi.org/10.1080/00420980220099104

Spiel C, Schwartzman S, Busemeyer M, Cloete N, Drori G, Lassnigg L, Schober B, Schweisfurth M, Verma S

(2018) The contribution of education to social progress. In: Rethinking society for the 21st century: report of the International Panel for Social Progress. Cambridge University Press, Cambridge
Sykes B, Musterd S (2011) Examining neighbourhood and school effects simultaneously: what does the Dutch evidence show? Urban Stud 48(7):1307–1331. https://doi.org/10.1177/0042098010371393
UIS and UNICEF (2015) Fixing the broken promise of education for all: findings from the global initiative on out-of-school children. UNESCO Institute for Statistics, Montreal
UNESCO (1999) World Declaration on Education for All and Framework for action to meet basic learning needs. UNESCO, Paris
UNESCO (2010) EFA global monitoring report: reaching the marginalized. UNESCO, Paris
UNESCO (2014) EFA global monitoring report teaching and learning: achieving quality for all. UNESCO, Paris
UNESCO (2015a) Education for all 2000–2015: achievements and challenges; EFA global monitoring report, 2015. UNESCO, Paris
UNESCO (2015b) Education 2030 Framework for Action. UNESCO, Paris
UNESCO (2016) Incheon Declaration and Framework for Action for the implementation of Sustainable Development Goal 4. ED-2016/WS/28. UNESCO, Paris
UNESCO (2017) EFA global monitoring report accountability in education: meeting our commitments. UNESCO, Paris
UNESCO (2018) EFA global monitoring report: meeting our commitments to gender equality in education. UNESCO, Paris
UNESCO (2019a) EFA global monitoring report gender report: building bridges for gender equality. UNESCO, Paris
UNESCO (2019b) EFA global monitoring report: migration, displacement and education: building bridges, not walls. UNESCO, Paris
UNESCO (2020a) EFA global monitoring report: a new generation: 25 years of efforts for gender equality in education. UNESCO, Paris
UNESCO (2020b) EFA global monitoring report gender report: inclusion in education: all means all. UNESCO, Paris
UNESCO (2022) EFA global monitoring report gender report: deepening the debate on those still left behind. UNESCO, Paris
UNESCO Institute for Statistics (UIS) (2011) International Standard Classification of Education. UIS, Montreal
UNESCO-UIL (2013) Key features of learning cities—Introductory note. UIL, Hamburg
UNESCO-UIL (2019) Fourth global report on adult learning and education: leave no one behind: participation, equity and inclusion. UIL, Hamburg
UN General Assembly (2000) United Nations Millennium Declaration, resolution adopted by the General Assembly, A/RES/55/2. https://www.refworld.org/docid/3b00f4ea3.html

5 Neighbourhood Effects on Health in Urban Areas of Developing Countries

Ramjee Bhandari and Richard Mitchell

Abstract

This chapter examines the role of urban neighbourhoods in shaping health outcomes in low- and middle-income countries (LMICs) amidst rapid urbanisation and pronounced social inequalities. It explores how neighbourhood attributes contribute to health disparities and the potential for fostering healthier urban lifestyles. Drawing upon the SHLC Household Survey and relating the findings to established theories, this chapter reveals how physical, social, and economic neighbourhood factors intersect with broader socio-economic policies affecting well-being. The analysis is situated within the Sustainable Development Goals (SDGs), emphasising the way local actions can contribute to addressing global health commitments. The chapter focuses on both the challenges and opportunities in LMICs for achieving health equity at the neighbourhood level, providing insights for researchers, city planners, and policymakers.

R. Bhandari (✉)
Faculty of Education, Health and Human Sciences, School of Human Sciences, University of Greenwich, London, UK
e-mail: r.bhandari@greenwich.ac.uk

R. Mitchell
University of Glasgow, Glasgow, Scotland, UK
e-mail: Richard.Mitchell@glasgow.ac.uk

Keywords

Urban neighbourhoods · Health outcomes · Social inequalities · Socio-economic policies · Low- and middle-income countries (LMICs)

5.1 Introduction

This chapter explores how urban neighbourhoods influence overall health of their residents. It delves into the concept of 'neighbourhood' as a dynamic entity that extends beyond its physical borders. The social, economic, and physical environments within a neighbourhood greatly influence the health and well-being of residents and we take a close look at neighbourhoods as spaces where infrastructure, access to services, social connections, and environmental factors intersect to shape individual and collective well-being. This is particularly relevant in urban areas of low- and middle-income countries (LMICs). By adopting this perspective, we gain insight into how neighbourhood features contribute to health inequalities and how we can foster healthier urban lifestyles.

The chapter focuses on neighbourhoods within LMIC cities, recognising that while the social determinants of health (SDOH) (discussed in more detail later) are globally relevant, their manifestation in LMICs is distinct due to factors

Y. P. Wang et al. (eds.), *New Drivers of Division*, Sustainable Development Goals Series,
https://doi.org/10.1007/978-981-96-5629-5_5

like rapid urbanisation, informal settlements, and resource constraints. Central to the chapter's message is that neighbourhoods are not just physical spaces in which 'health happens' but are shaped by both the existing social structures and economic activities and, in turn, influence health outcomes. The interplay of socio-economic and environmental factors in LMIC urban neighbourhoods demands a focused examination to understand and address the health inequities they harbour. Amidst the backdrop of unprecedented urban growth in LMICs, we explore pressing questions: How do neighbourhood factors contribute to or mitigate health inequalities? What opportunities do these urban landscapes offer to pivot communities towards healthier lives? Our inquiry is not just an academic exercise; it is a critical step towards reimagining urban health in the context of LMICs, where the stakes for sustainable, equitable health outcomes have never been higher.

Why focus on cities? The global population has now reached eight billion and more than half of that population lives in cities. The world is experiencing unprecedented rates and levels of urbanisation. In 1950, only 30% of the world's population lived in cities. Today, that number is 55% and it is projected that 66% of the world will be urban dwellers by 2050 (Leeson 2018). This shift in human experience means that more people than ever before live near each other in an urban setting. While cities offer opportunities for residents to enjoy better health, they also present challenges, particularly so if they develop in an unplanned way. Poorly designed cities can lead to many problems, including air pollution, water contamination, and a lack of green spaces. In addition, unplanned urban development can result in overcrowding and a lack of basic amenities, which can be *pathogenic* (health-damaging) to city dwellers (Pearce et al. 2011). However, such vast gatherings of people can also lead to easier distribution of resources and services, a reduced carbon footprint compared to dispersed populations, increased social interaction, and the sharing of ideas. The key aim for researchers, planners, developers, and city managers is to find ways to balance the positive and negative influences of cities on health and well-being so that we can continue to thrive as a species.

Why focus on neighbourhoods within cities? We argue that thinking about neighbourhoods, rather than cities as a whole, offers the best way to make this balance. This, more granular perspective on the intricate interplay between urban environments and health outcomes enables an understanding of how diverse socio-economic and environmental conditions influence residents' well-being. A localised approach allows for tailored interventions addressing specific challenges within communities, such as access to health care, green spaces, or social support networks. Also, focusing on neighbourhoods may enable grassroots initiatives and community engagement, empowering residents to actively participate in improving their living conditions and fostering a sense of ownership over their health outcomes. If we build the city neighbourhood by neighbourhood, we can foster a more holistic and equitable approach to urban health, ensuring that interventions are responsive to the unique needs and assets of each neighbourhood.

What are we worried about in terms of cities and neighbourhoods? Unfortunately, a coherent 'bottom up' approach to building cities is not generally a reality in LMICs. The most rapid urbanisation the world has ever seen is occurring in developing countries where population growth and unplanned urbanisation are more likely than not to be uncontrolled. Developing countries are also home to the vast majority of the world's poor and vulnerable people (Riley et al. 2007) who, in general terms, experience much worse health than wealthier people. Uncontrolled urbanisation and lack of planning can exacerbate poverty and inequality, create slums, and increase vulnerability to climate change (Watson 2009). The implications of this are far-reaching; urban growth has the potential to harm, but given the gathering of so many people in such little space, cities also have the potential to help solve many of the world's most pressing problems. As urbanisation concentrates ever-larger portions of the world's population into relatively smaller areas, it eases the task of targeting and reaching those who need help most and helping people live sustainable lives.

The concept of a sustainable and healthy city is one that considers the environmental, social, and economic needs of current and future generations (Bibri and Krogstie 2017). A sustainable city is one that can meet the needs of its residents without compromising the ability of future generations to meet their own needs. A healthy city is one in which residents have both access to the resources they need, and an environment that enables a healthy and prosperous life. Sustainable cities and neighbourhoods should be better able to withstand and recover from shocks, even if these shocks are at a macro level, for example, due to climate change or disease outbreaks such as COVID-19 (Amirzadeh et al. 2023). From floods and droughts to hurricanes and heat waves, climate change is increasingly causing extreme weather events that can devastate entire communities. Since many city dwellers in developing countries live in informal settlements with little or no infrastructure, they are often ill-equipped to deal with these shocks (Wang and Kintrea 2019; Wilkinson et al. 2020). Furthermore, climate change is expected to exacerbate existing problems such as poverty, water scarcity, and food insecurity.

How can neighbourhoods help? As the management of stressors such as COVID and climate change becomes a global priority, the significance of neighbourhoods is increasingly recognised. These units or groups are pivotal in organising and ensuring effective responses and resilience, highlighting their significance in our collective efforts to address these challenges. Such areas provide a support system for residents and can offer a sense of stability during difficult times. Neighbourhoods and their communities can also serve as a resource for information and assistance.

What factors influence our health? Health is a complex system with influences that operate at scales from the molecular to the entire planet (Seltenrich 2018). Individual genetics play a role, as does lifestyle, but environmental factors, social factors, and economic circumstances are more important when we think about the health of whole populations. Researchers and planners use 'socio-ecological models of health' to provide a framework for understanding the complex interactions between different influences on health and their impact (Foster and Townsend 2013). Such models typically consider the influences on health which operate at different 'levels': on individuals; between individuals (interpersonal); on organisations; and then at wider geographical levels including the neighbourhood, town or city, nation, and region. The neighbourhood is special because it is where a lot of different levels of influence on health come together. It is where we live and work, where we meet and interact with others, but also is often the place where we encounter organisational influences (such as schools, health services, local government, housing agencies) and wider influences such as the labour market.

5.2 Health Inequalities in the Cities of LMICs

Evidence is growing for the existence and/or exacerbation of health disparities within cities of LMICs (Allen et al. 2017; Donkin et al. 2018; Miranda et al. 2019). These disparities refer to differences in health status that exist between different population groups. Disadvantaged population groups tend to have a worse health status when compared to other groups (Miranda et al. 2019). Crucially, the differences in health between these groups are reflected in geographical space—between neighbourhoods. See, for example Fig. 5.1, which shows the difference in people's general health by different neighbourhoods based on socio-economic classification in the cities within our study. The general health perceptions tend to vary by neighbourhood socio-economic status (SES), with a trend of higher proportions of 'very good' health ratings in higher SES areas and lower proportions in slums/informal settlements. Such differences are unfair but also place pressures and burdens on services. We need to better understand these differences in order to address them.

It's important to recognise that complex feedback and interaction loops are often implicated in these kinds of inequalities. Consider the rela-

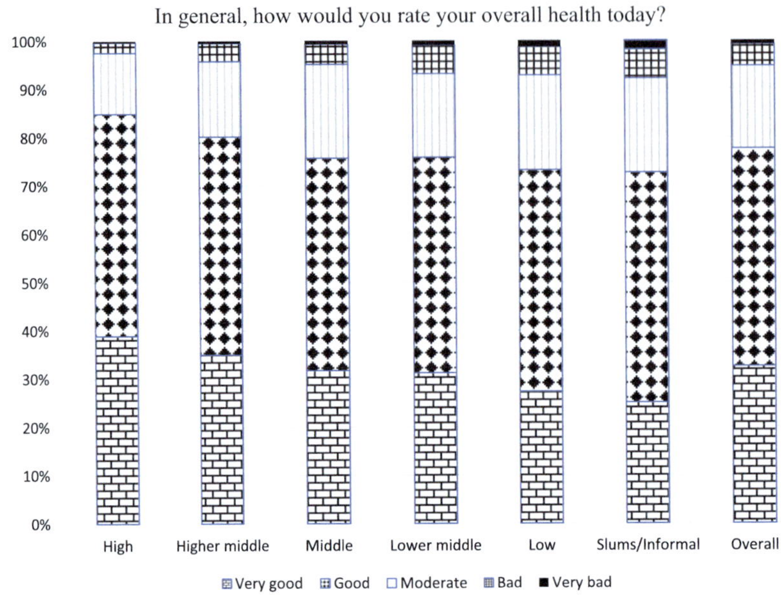

Fig. 5.1 General health by neighbourhood socio-economic status (SES) in the SHLC cities. *Note* SHLC—Sustainable Healthy and Learning Cities. *Source* Primary survey, 2021

tionships between poverty and air pollution as an example. Cities tend to have higher concentrations of poverty and pollution, which can lead to increased rates of disease and mortality (Kingham et al. 2007). They also often have large disparities in access to health care and other resources, which can exacerbate health risks and social inequalities (Vlahov et al. 2007). These inequalities and exposures interact with each other. Within cities, the problem of air pollution and associated health outcomes tends to be concentrated in the more economically deprived neighbourhoods (Ribeiro et al. 2019). Furthermore, those living on a lower income may be more affected by this exposure due to a higher likelihood of existing health problems and worse access to medical care. Their economic situation also means they are less able to move away from the pollution. Finally, concentrations of adverse health outcomes in particular places then creates a significant and uneven burden on health systems, further exacerbating the inequalities in health outcomes (Asada 2005; Hansson 2003).

In the next section, we will consider what factors influence health in cities in general and how these lead to inequalities in health within the cities themselves. Although we set these out, step by step, it is important to understand that these factors interact with each other in a system.

5.3 Determinants of Urban Health

It is increasingly recognised that neighbourhood context has an impact on the health and well-being of the citizens. Cummins et al. (2007: 1835) argue that there exists a 'mutually reinforcing and reciprocal relationship between people and place'. Although the variation in health between neighbourhoods is mostly explained by the characteristics of the individual residence (i.e. who lives here?), there is a significant contribution of the contextual factors (i.e. what is the place like?) (Bhandari et al. 2017; Pickett and Pearl 2001). Contextual factors not only make direct contribution to health (e.g. exposing people to polluted air); they influence health indirectly by, for example, affecting people's interaction with each other or with services (Cummins et al. 2007). However, the impacts of these neighbourhood factors may depend upon the context, extent of their exposure, and the individuals themselves. Some evidence suggests that people's vulnerability to their neighbourhood environment is a function of their own individual characteristics (Cassarino et al. 2020).

A common assumption by planners, politicians, and the public is that health care systems (e.g. doctors, hospitals, and clinics) are the most important determinants of population health. In fact, sectors and interventions other than these health care services are far more influential in shaping population health outcomes (Ehrenberg and Ault 2005). The overall role of health care services in shaping population health and well-being is relatively small compared to other determinants. It has been quantified at around 10% (Bibby 2017). Assuming that all health problems can be solved with better access to, and quality of health care services, is a problem referred to as 'medicalisation' (Tanner 2014). Access to good health care is, of course, important, but if we *only* think about population health as driven by, and the responsibility of, health care services we erroneously ignore other vital determinants such as employment, economy, sanitation, and urban transport.

The social determinants of a health model (SDOH) acknowledge the role of individual, neighbourhood, city, country, and planetary level factors in shaping health outcomes. Cities are a complex system, composed of natural, societal, and economic components, with multiple nodes of interactive relationships (Su et al. 2010). Furthermore, these elements are dynamic in nature: they are constantly changing (rapidly in developing countries). Consequently, the people living there are 'constantly adapting to new opportunities and threats' (Kearns et al. 2007: 48). The health problems of modern cities are different from the past; in many locations they have shifted from communicable diseases to non-communicable diseases. This shift is partly the result of our way of life in urban spaces—the physical, social, economic, and political environment of the cities—functioning as an 'ecosystem' within which residents exist. A framework is required that acknowledges health and well-being as the result of interaction of factors at different levels. In this context, the SDOH model highlights the interplay between individuals and their environments, with key factors across various sectors influencing health outcomes. These include individual behaviours, social and community networks, and broader socio-economic policies. For example, housing quality, access to transportation, and urban infrastructure directly affect physical health, while employment and education opportunities impact economic stability and well-being. Governance and policy decisions further shape these determinants, illustrating the multi-layered nature of health gaps (Bentley, 2014; Bronfenbrenner 1994; Shareck et al. 2013).

The SDOH model (see Fig. 5.2) acknowledges that there is an ongoing interaction between these levels and their relationship is mutually reinforcing (Golden and Earp 2012). SDOH signifies the role of one's exposure to day-to-day attributes at home, neighbourhoods, or at work for long-term health and well-being. Moreover, because of the comprehensive nature of SDOH, it offers more routes to dealing with complex urban health problems in the developing countries.

Many people who are new to thinking about health in cities arrive at the topic believing that individual lifestyle choices explain a huge proportion of any variation in health between people and places. If some people seem less healthy than others, the typical assumption is that it must be down to the way they behave; perhaps they smoke, drink, eat poorly, and do not take enough care of themselves. The SDOH model understands that individual lifestyle choices are important causes of poor health, however, it gets us to ask, 'what are the causes of those causes?'. In fact, evidence demonstrates that an individual's lifestyle choices are strongly determined by his/her wider social, cultural, and economic environments in the first place. For example, if healthy food is expensive or hard to find in a neighbourhood, and alcohol is cheap, so the culture is to consume the latter in large quantities—a healthy lifestyle 'choice' in this case becomes very much harder at an individual level.

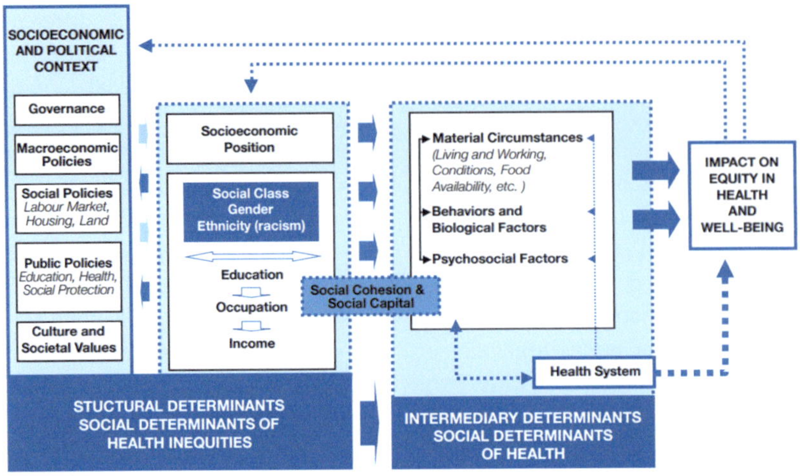

Fig. 5.2 Conceptual framework from commission of SDOH. *Source* Solar and Irwin (2010: 6)

5.4 Neighbourhood Effects on Health

There is a growing body of evidence which suggests that the differences in health between areas or neighbourhoods which we observe are partly created by the nature of the place itself, in terms of its economic, social, cultural, and physical environment (Agampodi et al. 2015; Bilger and Carrieri 2013; Pickett and Pearl 2001). Natural and human-made factors present in neighbourhoods have an impact on the physical and/or mental health of the population. Neighbourhoods hold and disperse the environmental '*goods*' and '*bads*', the mechanism of which, according to Pearce (2015: 26) can be understood as 'environmental (in)justice and health'. The proximity to and concentration of health-damaging factors ('*pathogenic*' such as pollution, crime) and health-promoting factor ('*salutogenic*' such as public parks, water bodies and healing places) in the neighbourhood often follow a distinctive pattern. While higher concentrations of *pathogenic* attributes are more commonly found in and around deprived neighbourhoods, *salutogens* are usually more accessible and common in the least deprived neighbourhoods (Pearce 2015). The neighbourhood-level average health therefore often follows a gradient; an increase in deprivation is correlated with a decline in health (Bhandari et al. 2017).

The socio-ecological model put forward by Bernard et al. (2007) provides and extends the idea of SDOH and proposes the identification and grouping of neighbourhood characteristics into physical and social influences (see Fig. 5.3).

Neighbourhoods in LMIC cities often face particularly stark adversities across many or all of the domains identified in this model, most particularly in those neighbourhoods that comprise informal settlements or slums (Riley et al. 2007). Poor housing conditions, lack of access to clean water and sanitation facilities, and exposure to environmental pollutants are all major risk factors for poor health and these exposures tend to be particularly high in some neighbourhoods in LMICs. Furthermore, macro-scale economic and development issues such as poverty and limited access to (quality) education also tend to be more prevalent in these settings. Such combinations of social and physical environmental factors create a daunting set of challenges for addressing health inequities in LMIC cities.

Fig. 5.3 Neighbourhood environments and rules of access. *Source* Bernard et al. (2007: 1843)

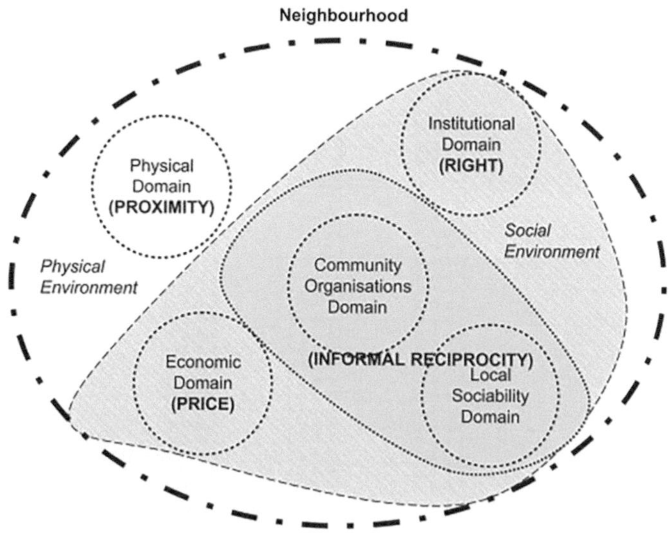

5.4.1 The Neighbourhood Physical Environment

The physical environment can be natural or human-made/built. The influence of physical environment on health and well-being is often determined by the proximity and/or extent of exposure to these environmental attributes. Growing evidence from LMIC cities has highlighted the importance of the physical environment in creating or addressing health inequalities (Helbich et al. 2019; Jiao et al. 2018; Liu et al. 2019; Qiu et al. 2019; Tao et al. 2019; Zeng et al. 2010). Such studies have been able to identify a variety of physical environment attributes and measure their contribution to the health and well-being of the population. For example, while the problem of infectious diseases is generally in decline in LMIC cities, the neighbourhood physical environment still plays a role in their incidence. Housing quality and proximity of open water bodies are causal factors in the endemicity of diseases such as malaria in some LMIC cities (Ngom and Siegmund 2015). Lack of proper sanitation and safe water supply is another important influence, with poor sanitation linked both to infectious and non-communicable disease prevalence. Residents are also often aware of the toxicity of their environment. Perceptions of pollution or proximity to main roads has, for example, been shown to contribute to poor mental health and well-being (Ma et al. 2018).

On the other hand, evidence also suggests protective roles for other neighbourhood characteristics such as ambient greenness (Helbich et al. 2019; Liu et al. 2019; Qiu et al. 2019; Tao et al. 2019), proximity to parks, vehicular restriction, and street layouts (Tao et al. 2019) on both physical and mental health and well-being. For example, our study at SHLC has shown that there is a 20% difference in the access to green spaces between the high and the low socio-economic status (SES) neighbourhoods (see Fig. 5.4). Access to nature is crucial for mental and physical health; however, the benefits are maximised only when these spaces are both high-quality and safe (Slater et al. 2020). Poorly maintained or unsafe natural areas can discourage their health-promoting use, underlining the need for well-kept and secure environments (Austin 2021).

The level of physical activity is another important risk or protective factor for several Non-communicable diseases (NCDs). From the geographical perspective, engagement in physical activity is influenced by the supportive neighbourhood attributes (Gomez et al. 2010). Our study has found that almost half (47%) of the people did not engage in any form of physical activity in the week prior to survey and more than half (52%) of these residents reported not having access to

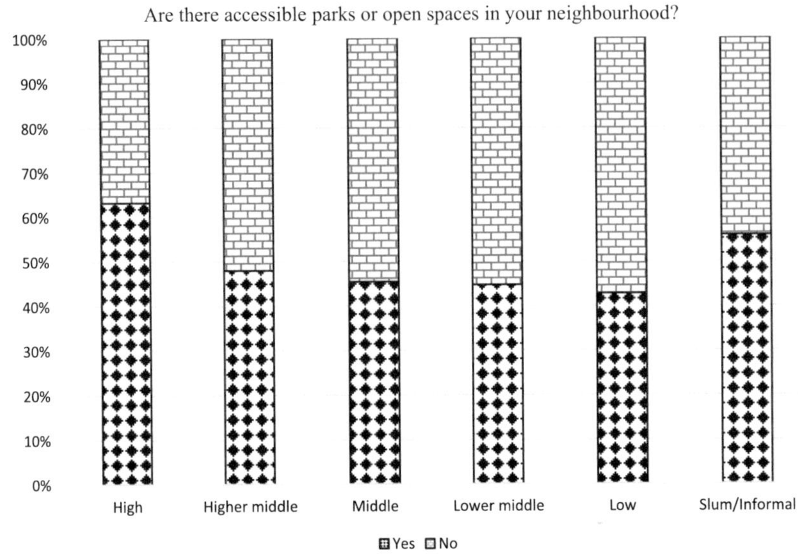

Fig. 5.4 Accessible parks and open spaces by neighbourhood types in the SHLC cities. *Source* Primary survey, 2021

a park or green space in their neighbourhood. From the built environment perspective, the presence of active commute/cycle routes (Gomez et al. 2015), protective streetscapes (traffic calming, safe street crossing) (Lopes et al. 2018) and walkable, connected streetscapes have been found to promote engagement in physical activity and to be associated with lower obesity rates (Araujo et al. 2018).

5.4.2 The Neighbourhood Social Environment

A strong and close-knit neighbourhood can make an important contribution to the health and well-being of its residents. When people feel connected to their neighbours and have a sense of community, they are more likely to take care of their surroundings and look out for one another—the idea of informal reciprocity (Bernard et al. 2007). This creates a safer and more pleasant environment for everyone. Relationships with service providers, local managers and politicians are also an important component of the social environment. If residents do not trust these people, they will not engage with them. If there is no engagement, how will those service providers know what the residents need?

Not surprisingly then, evidence suggests that social inequalities in health not only reflect the distribution of material (dis)advantage, but also the social environment (Uphoff et al. 2013; Wilkinson 1999). This is particularly relevant in low- and middle-income countries (LMICs), where social inequality is pronounced, and institutional capacities are limited. Evidence suggests that the four sub-domains of social environment—institutions, community organisations, local sociability, and economic domain (as proposed by Bernard et al. 2007) (see Fig. 5.3) are relevant in understanding the roles of neighbourhood attributes in health and well-being in LMICs. Effective local governance structured around the institutions and community organisation is an important social-environmental factor that determines the service landscape within the cities and plays a significant role in shaping health outcomes (Thomson et al. 2019). The economic domain also plays a role in shaping health in LMICs, as constraints of urban poverty (individual-level) as well as area-level resources tend to be higher (Abascal et al. 2022). Another important factor is social cohesion—this is when people feel a sense of belonging to their community and have support networks in place. For example, studies have found that a high level of neighbourhood social capital, that is, membership of community organisations, have protective benefits on mental health (Qiu et al. 2019; Wang et al. 2018a). Conversely, the presence of socio-economic inequalities has a detrimental effect on the health and well-being of neighbourhood residents (Chiavegatto Filho et al. 2013; Liu et al. 2019; Wang et al. 2018b). An unfavourable social environment is key to the existence of risk factors for NCDs, as well as the onset and prevalence of these long-term health conditions. The higher neighbourhood SESs are found to be associated with better general health and well-being (Chen et al. 2017), lower rates of obesity (Araujo et al. 2018), healthy diet consumption (Pessoa et al. 2015), and lower substance (mis)use (Borges et al. 2019).

5.5 Inequities Between Neighbourhoods in Health-Related Services

Uneven or inequitable distribution of access to and quality of health services contributes to health inequalities between neighbourhoods, but other kinds of services also matter. The presence of institutions such as schools, finance, transport, and police in the neighbourhood makes a difference to the health and well-being of the people (Blas et al. 2008). Availability is not the only characteristic that matters; quality and reliability are significant too. Where demand is particularly prevalent, for example, services can be overwhelmed leading to poorer quality. The unequal geographical distribution of limited resources is a particular problem when it comes to access to health care services (Rathod et al. 2017; Shadmi et al. 2020). Disadvantaged groups often have less access to quality health care than other groups. In addition to there being fewer service outlets in poor neighbourhoods in the cities of LMICs, those available often are not of standard quality and do not satisfy the needs of the communities (Arroyave et al. 2021). See Fig. 5.5, which shows the level of satisfaction of health services available in the neighbourhoods of SHLC cities. The services are of more acceptable quality in the richer neighbourhoods compared to those in poor neighbourhoods.

5.6 Neighbourhoods as Foundations for Achieving Urban Health SDGs

How can we bring together such a complex story, and focus action? The Sustainable Development Goals (SDGs) adopted by all United Nations member states in 2015 provide a shared blueprint for peace and prosperity for people and the planet, now and into the future. At its heart are the 17 SDGs, which are an interconnected set of goals designed to be a 'plan of action for people, planet and prosperity' (UN Department of Economic and Social Affairs (DESA) 2022). Though health and well-being are clearly addressed in SDG 3, health is also a determinant of SDG 11, which focuses on the creation of inclusive, safe, resilient, and sustainable cities (Ramirez-Rubio et al. 2019).

City governments have been assigned a crucial role to play in the implementation of the SDGs and the New Urban Agenda (NUA). In addition to being home to more than half of the world's population, cities are responsible for an unsustainable consumption pattern with more than 70% of global greenhouse gas emissions (Satterthwaite 2008). The New Urban Agenda, adopted in 2016, provides a comprehensive and integrated framework for sustainable urban development. It recognises that cities play a key role in achieving the Sustainable Development Goals (Caprotti et al. 2017). While both initiatives are ambitious and progressive in nature, success relies on the effective involvement and engagement of cities and local authorities (Zinkernagel et al. 2018). In doing so, neighbourhoods offer an appropriate entry point for actions to address disparities in health, and finally to achieve urban sustainability (Valencia

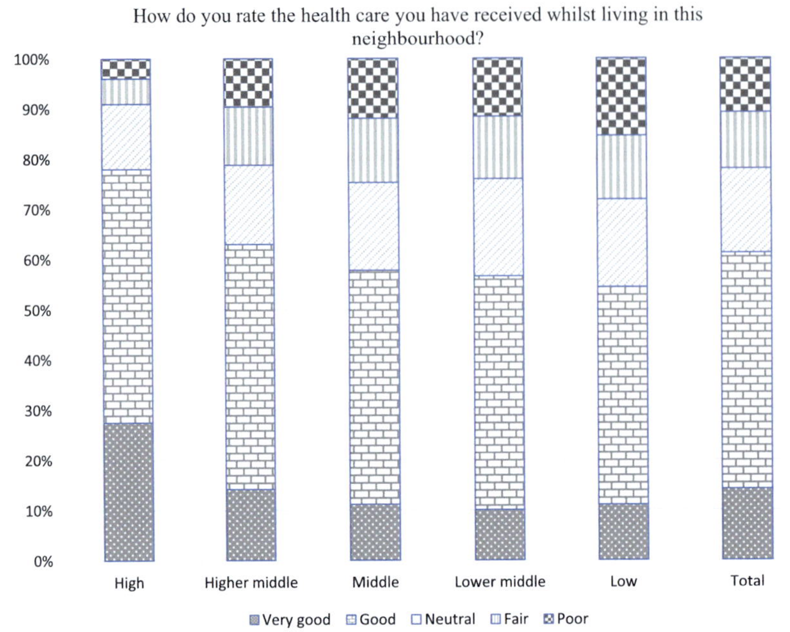

Fig. 5.5 Level of satisfaction of health care provision in the neighbourhoods of SHLC cities. *Source* Primary survey, 2021

et al. 2019). This, therefore, opens an avenue for wider research to explore the role of neighbourhoods and their attributes as the primary determinants of health and well-being.

However, while the importance of the social and environmental determinants of health is relatively uncontested, the rapid unplanned urbanisation in developing countries is a challenge because even if planners and politicians know what kinds of neighbourhoods are required, they might not be in control of the neighbourhoods actually created. Poor or absent urban planning can create neighbourhoods that promote and foster poor health behaviours, as well as threaten health directly via toxic or unsafe environments and hampering access to services (Maciejewska and Ulanicka-Raczyńska 2023; Siri and Geddes 2022).

5.7 Conclusion

There are many opportunities to improve urban health by investing in infrastructure and services that promote health and well-being. For example, investing in public transport, green spaces, and walkable neighbourhoods can create healthy environments that encourage physical activity and social interaction (Nawrath et al. 2022; Salvo et al. 2023). These investments also have the potential to create jobs and support economic development. In addition, cities need to create inclusive and resilient societies by reducing inequality, providing access to essential services, and promoting social cohesion. By implementing these measures, cities can make a major contribution to sustainable development.

This chapter articulates the multifaceted influences of neighbourhood characteristics on health outcomes in LMICs, a subject critical in the era of rapid urbanisation. It underscores the influence of physical, social, and economic environments within neighbourhoods on the well-being of urban residents. By leveraging insights from the 14-city household survey, this chapter highlights the intricate ways in which neighbourhood factors can exacerbate or mitigate health disparities. It emphasises the potential of localised, neighbourhood-level interventions in advancing the health-related Sustainable Development Goals (SDGs).

We conclude by calling for targeted policy actions that harness the unique dynamics of LMIC urban neighbourhoods to foster healthier, more equitable urban communities. An approach rooted in understanding the SDOH model, and connecting it to the SDG framework could not only address health inequities but also contribute to the broader vision of sustainable urban development. Neighbourhoods are a scale that matters for health and sustainability. They are where we observe health inequalities, but they might also be where we can begin to address them.

Funding Statement Rich Mitchell is funded by the UK Medical Research Council (MRC) Places and Health Programme (MC_UU_00022/4) and the Chief Scientist Office (CSO) (SPHSU19) at the MRC/ CSO Social and Public Health Sciences Unit, University of Glasgow.

References

Abascal A, Rothwell N, Shonowo A, Thomson DR, Elias P, Elsey H, Yeboah G, Kuffer M (2022) "Domains of deprivation framework" for mapping slums, informal settlements, and other deprived areas in LMICs to improve urban planning and policy: a scoping review. Comput Environ Urban Syst 93:101770. https://doi.org/10.1016/j.compenvurbsys.2022.101770

Agampodi TC, Agampodi SB, Glozier N, Siribaddana S (2015) Measurement of social capital in relation to health in low and middle income countries (LMIC): a systematic review. Soc Sci Med 128:95–104. https://doi.org/10.1016/j.socscimed.2015.01.005

Allen L, Williams J, Townsend N, Mikkelsen B, Roberts N, Foster C, Wickramasinghe K (2017) Socioeconomic status and non-communicable disease behavioural risk factors in low-income and lower-middle-income countries: a systematic review. Lancet Glob Health 5(3):e277–e289. https://doi.org/10.1016/S2214-109X(17)30058-X

Amirzadeh M, Sobhaninia S, Buckman ST, Sharifi A (2023) Towards building resilient cities to pandemics: a review of COVID-19 literature. Sustain Cities Soc 89:104326. https://doi.org/10.1016/j.scs.2022.104326

Araujo CAH, Giehl MWC, Danielewicz AL, Araujo PG, d'Orsi E, Boing AF (2018) Built environment, contextual income, and obesity in older adults: evidence from a population-based study. Cad Saude Publica 34(5):e00060217. https://doi.org/10.1590/0102-311X00060217 (Ambiente construido, renda contextual e obesidade em idosos: evidencias de um estudo de base populacional)

Arroyave L, Saad GE, Victora CG, Barros AJD (2021) Inequalities in antenatal care coverage and quality: an analysis from 63 low and middle-income countries using the ANCq content-qualified coverage indicator. Int J Equity Health 20(1):102. https://doi.org/10.1186/s12939-021-01440-3

Asada Y (2005) A framework for measuring health inequity. J Epidemiol Community Health 59(8):700–705. https://doi.org/10.1136/jech.2004.031054

Austin C (2021) Evaluating the health and wellbeing benefits associated with outdoor interventions: informing the development of natural health services. Doctoral, Liverpool John Moores University. http://researchonline.ljmu.ac.uk/id/eprint/14651/

Bentley M (2014) An ecological public health approach to understanding the relationships between sustainable urban environments, public health and social equity. Health Promot Int 29(3):528–537. https://doi.org/10.1093/heapro/dat028

Bernard P, Charafeddine R, Frohlich KL, Daniel M, Kestens Y, Potvin L (2007) Health inequalities and place: a theoretical conception of neighbourhood. Soc Sci Med 65(9):1839–1852. https://www.sciencedirect.com/science/article/pii/S0277953607003085?via%3Dihub

Bhandari R, Kasim A, Warren J, Akhter N, Bambra C (2017) Geographical inequalities in health in a time of austerity: baseline findings from the Stockton-on-Tees cohort study. Health Place 48:111–122. https://doi.org/10.1016/j.healthplace.2017.10.002

Bibby J (2017) Health care only accounts for 10% of a population's health. https://www.health.org.uk/blogs/health-care-only-accounts-for-10-of-a-population%E2%80%99s-health

Bibri SE, Krogstie J (2017) Smart sustainable cities of the future: an extensive interdisciplinary literature review. Sustain Cities Soc 31:183–212. https://doi.org/10.1016/j.scs.2017.02.016

Bilger M, Carrieri V (2013) Health in the cities: when the neighborhood matters more than income. J Health Econ 32(1):1–11. https://doi.org/10.1016/j.jhealeco.2012.09.010

Blas E, Gilson L, Kelly MP, Labonté R, Lapitan J, Muntaner C, Östlin P, Popay J, Sadana R, Sen G, Schrecker T, Vaghri Z (2008) Addressing social determinants of health inequities: what can the state and civil society do? The Lancet 372(9650):1684–1689. https://doi.org/10.1016/S0140-6736(08)61693-1

Borges G, Lown EA, Orozco R, Cherpitel CJ (2019) The relationship between social inequalities, substance use and violence in border and non-border cities of northern Mexico [Research Support, N.I.H., Extramural]. Drug Alcohol Depend 200:1–5. https://doi.org/10.1016/j.drugalcdep.2019.02.029

Bronfenbrenner U (1994) Ecological models of human development. Read Dev Child 2:37–43

Caprotti F, Cowley R, Datta A, Broto VC, Gao E, Georgeson L, Herrick C, Odendaal N, Joss S (2017) The New Urban Agenda: key opportunities and challenges

for policy and practice. Urban Res Pract 10(3):367–378. https://doi.org/10.1080/17535069.2016.1275618
Cassarino M, Bantry-White E, Setti A (2020) Neighbourhood environment and cognitive vulnerability—a survey investigation of variations across the lifespan and urbanity levels. Sustainability 12(19)
Chen H, Liu Y, Li Z, Xue D (2017) Urbanization, economic development and health: evidence from China's laborforce dynamic survey. Int J Equity Health 16(1):207. https://doi.org/10.1186/s12939-017-0705-9
Chiavegatto Filho AD, Kawachi I, Wang YP, Viana MC, Andrade LH (2013) Does income inequality get under the skin? A multilevel analysis of depression, anxiety and mental disorders in Sao Paulo, Brazil [Research Support, Non-U.S. Gov't]. J Epidemiol Community Health 67(11):966–972. https://doi.org/10.1136/jech-2013-202626
Cummins S, Curtis S, Diez-Roux AV, Macintyre S (2007) Understanding and representing 'place' in health research: a relational approach. Soc Sci Med 65(9):1825–1838. https://doi.org/10.1016/j.socscimed.2007.05.036
Donkin A, Goldblatt P, Allen J, Nathanson V, Marmot M (2018) Global action on the social determinants of health. BMJ Glob Health 3(Suppl 1):e000603. https://doi.org/10.1136/bmjgh-2017-000603
Ehrenberg JP, Ault SK (2005) Neglected diseases of neglected populations: thinking to reshape the determinants of health in Latin America and the Caribbean. BMC Public Health 5:119. https://doi.org/10.1186/1471-2458-5-119
Foster C, Townsend N (2013) Developing and applying a socio-ecological model to the promotion of healthy eating in the school. Public Health Nutr 16(6):1101–1108. https://doi.org/10.1017/S1368980011002655
Golden SD, Earp JAL (2012) Social ecological approaches to individuals and their contexts: twenty years of health education & behavior health promotion interventions. Health Educ Behav 39(3):364–372. https://doi.org/10.1177/1090198111418634
Gomez LF, Sarmiento OL, Parra DC, Schmid TL, Pratt M, Jacoby E, Neiman A, Cervero R, Mosquera J, Rutt C, Ardila M, Pinzon JD (2010) Characteristics of the built environment associated with leisure-time physical activity among adults in Bogota, Colombia: a multilevel study. J Phys Act Health 7(Suppl 2):S196-203. https://doi.org/10.1123/jpah.7.s2.s196
Gomez LF, Mosquera J, Gomez OL, Moreno J, Pinzon JD, Jacoby E, Cepeda M, Parra DC (2015) Social conditions and urban environment associated with participation in the Ciclovia program among adults from Cali, Colombia [Research Support, Non-U.S. Gov't]. Cad Saude Publica 31(Suppl 1):257–266. https://doi.org/10.1590/0102-311X00086814
Hansson L (2003) Inequality and inequity in use of mental health services. Acta Psychiatr Scand 107:161–162. https://doi.org/10.1034/j.1600-0447.2003.00076.x
Helbich M, Yao Y, Liu Y, Zhang J, Liu P, Wang R (2019) Using deep learning to examine street view green and blue spaces and their associations with geriatric depression in Beijing, China [Research Support, Non-U.S. Gov't]. Environ Int 126:107–117. https://doi.org/10.1016/j.envint.2019.02.013
Jiao K, Xu M, Liu M (2018) Health status and air pollution related socioeconomic concerns in urban China [Research Support, Non-U.S. Gov't]. Int J Equity Health 17(1):18. https://doi.org/10.1186/s12939-018-0719-y
Kearns A, Beaty M, Barnett G (2007) A social-ecological perspective on health in urban environments. N S W Public Health Bull 18(3–4):48–50. https://www.ncbi.nlm.nih.gov/pubmed/17601404
Kingham S, Pearce J, Zawar-Reza P (2007) Driven to injustice? Environmental justice and vehicle pollution in Christchurch, New Zealand. Transp Res Part D: Transp Environ 12(4):254–263. https://doi.org/10.1016/j.trd.2007.02.004
Leeson GW (2018) The growth, ageing and urbanisation of our world. J Popul Ageing 11(2):107–115. https://doi.org/10.1007/s12062-018-9225-7
Liu Y, Wang R, Xiao Y, Huang B, Chen H, Li Z (2019) Exploring the linkage between greenness exposure and depression among Chinese people: mediating roles of physical activity, stress and social cohesion and moderating role of urbanicity. Health Place 58:1. https://doi.org/10.1016/j.healthplace.2019.102168
Lopes A, Kienteka M, Fermino RC, Reis RS (2018) Characteristics of the environmental microscale and walking and bicycling for transportation among adults in Curitiba, Parana State, Brazil [Observational Study]. Cad Saude Publica 34(1):e00203116. https://doi.org/10.1590/0102-311X00203116
Ma J, Li C, Kwan MP, Chai Y (2018) A multilevel analysis of perceived noise pollution, geographic contexts and mental health in Beijing. Int J Environ Res Public Health 15(7):Article 1479. https://doi.org/10.3390/ijerph15071479
Maciejewska A, Ulanicka-Raczyńska M (2023) Lack of spatial planning as a cause of environmental injustice in the context of the provision of health safety to urban residents based on the example of Warsaw. Sustainability 15(3)
Miranda JJ, Barrientos-Gutiérrez T, Corvalan C, Hyder AA, Lazo-Porras M, Oni T, Wells JCK (2019) Understanding the rise of cardiometabolic diseases in low- and middle-income countries. Nat Med 25(11):1667–1679. https://doi.org/10.1038/s41591-019-0644-7
Nawrath M, Elsey H, Rijal ML, Dallimer M (2022) Greenspaces and human well-being: perspectives from a rapidly urbanising low-income country. Environments 9(12)
Ngom R, Siegmund A (2015) The key role of socio-demographic and socio-environmental factors in urban malaria occurrence and control—an illustration using the city of Yaounde. Soc Sci Med 133:269–279. https://doi.org/10.1016/j.socscimed.2014.12.001
Pearce J (2015) Invited commentary: history of place, life course, and health inequalities-historical geographic information systems and epidemiologic research. Am

J Epidemiol 181(1):26–29. https://doi.org/10.1093/aje/kwu312

Pearce JR, Richardson EA, Mitchell RJ, Shortt NK (2011) Environmental justice and health: a study of multiple environmental deprivation and geographical inequalities in health in New Zealand. Soc Sci Med 73(3):410–420

Pessoa MC, Mendes LL, Gomes CS, Martins PA, Velasquez-Melendez G (2015) Food environment and fruit and vegetable intake in a urban population: a multilevel analysis [Research Support, Non-U.S. Gov't]. BMC Public Health 15:1012. https://doi.org/10.1186/s12889-015-2277-1

Pickett KE, Pearl M (2001) Multilevel analyses of neighbourhood socioeconomic context and health outcomes: a critical review. J Epidemiol Community Health 55(2):111–122. https://www.ncbi.nlm.nih.gov/pubmed/11154250; http://jech.bmj.com/content/55/2/111.long

Qiu Y, Liu Y, Liu Y, Li Z (2019) Exploring the linkage between the neighborhood environment and mental health in Guangzhou, China [Research Support, Non-U.S. Gov't]. Int J Environ Res Public Health [Electronic Resource] 16(17):02. https://doi.org/10.3390/ijerph16173206

Ramirez-Rubio O, Daher C, Fanjul G, Gascon M, Mueller N, Pajín L, Plasencia A, Rojas-Rueda D, Thondoo M, Nieuwenhuijsen MJ (2019) Urban health: an example of a "health in all policies" approach in the context of SDGs implementation. Glob Health 15(1):87. https://doi.org/10.1186/s12992-019-0529-z

Rathod S, Pinninti N, Irfan M, Gorczynski P, Rathod P, Gega L, Naeem F (2017) Mental health service provision in low- and middle-income countries. Health Serv Insights 10:1178632917694350. https://doi.org/10.1177/1178632917694350

Ribeiro AG, Downward GS, Freitas CUd, Chiaravalloti Neto F, Cardoso MRA, Latorre MdRDdO, Hystad P, Vermeulen R, Nardocci AC (2019) Incidence and mortality for respiratory cancer and traffic-related air pollution in São Paulo, Brazil. Environ Res 170:243–251. https://doi.org/10.1016/j.envres.2018.12.034

Riley LW, Ko AI, Unger A, Reis MG (2007) Slum health: diseases of neglected populations. BMC Int Health Hum Rights 7(1):2. https://doi.org/10.1186/1472-698X-7-2

Salvo D, Jáuregui A, Adlakha D, Sarmiento OL, Reis RS (2023) When moving is the only option: the role of necessity versus choice for understanding and promoting physical activity in low- and middle-income countries. Annu Rev Public Health 44(1):151–169. https://doi.org/10.1146/annurev-publhealth-071321-042211

Satterthwaite D (2008) Cities' contribution to global warming: notes on the allocation of greenhouse gas emissions. Environ Urban 20(2):539–549. https://doi.org/10.1177/0956247808096127

Seltenrich N (2018) Down to earth: the emerging field of planetary health. Environ Health Perspect 126(7):072001. https://doi.org/10.1289/EHP2374

Shadmi E, Chen Y, Dourado I, Faran-Perach I, Furler J, Hangoma P, Hanvoravongchai P, Obando C, Petrosyan V, Rao KD, Ruano AL, Shi L, de Souza LE, Spitzer-Shohat S, Sturgiss E, Suphanchaimat R, Uribe MV, Willems S (2020) Health equity and COVID-19: global perspectives. Int J Equity Health 19(1):104. https://doi.org/10.1186/s12939-020-01218-z

Shareck M, Frohlich KL, Poland B (2013) Reducing social inequities in health through settings-related interventions—a conceptual framework. Glob Health Promot 20(2):39–52. https://doi.org/10.1177/1757975913486686

Siri JG, Geddes I (2022) Mainstreaming health in urban design and planning: advances in theory and practice. Cities & Health 6(5):853–857. https://doi.org/10.1080/23748834.2022.2148844

Slater SJ, Christiana RW, Gustat J (2020) Recommendations for keeping parks and green space accessible for mental and physical health during COVID-19 and other pandemics. Prev Chronic Dis 17:E59. https://doi.org/10.5888/pcd17.200204

Solar O, Irwin A (2010) A conceptual framework for action on the social determinants of health. Social Determinants of Health Discussion Paper 2 (Policy and Practice), Issue. http://apps.who.int/iris/bitstream/10665/44489/1/9789241500852_eng.pdf

Su M, Fath BD, Yang Z (2010) Urban ecosystem health assessment: a review. Sci Total Environ 408(12):2425–2434. https://doi.org/10.1016/j.scitotenv.2010.03.009

Tanner M (2014) Urban health in developing countries: progress and prospects. Routledge

Tao Y, Yang J, Chai Y (2019) The anatomy of health-supportive neighborhoods: a multilevel analysis of built environment, perceived disorder, social interaction and mental health in Beijing. Int J Environ Res Public Health 17(1):Article 13. https://doi.org/10.3390/ijerph17010013

Thomson DR, Linard C, Vanhuysse S, Steele JE, Shimoni M, Siri J, Caiaffa WT, Rosenberg M, Wolff E, Grippa T, Georganos S, Elsey H (2019) Extending data for urban health decision-making: a menu of new and potential neighborhood-level health determinants datasets in LMICs. J Urban Health 96(4):514–536. https://doi.org/10.1007/s11524-019-00363-3

UN DESA (2022) Transforming our world: the 2030 agenda for sustainable development. Retrieved 20 November 22 from https://sdgs.un.org/2030agenda

Uphoff EP, Pickett KE, Cabieses B, Small N, Wright J (2013) A systematic review of the relationships between social capital and socioeconomic inequalities in health: a contribution to understanding the psychosocial pathway of health inequalities. Int J Equity Health 12(1):54. https://doi.org/10.1186/1475-9276-12-54

Valencia SC, Simon D, Croese S, Nordqvist J, Oloko M, Sharma T, Taylor Buck N, Versace I (2019) Adapting the Sustainable Development Goals and the New Urban Agenda to the city level: Initial reflections from a comparative research project. Int J Urban Sustain Dev 11(1):4–23. https://doi.org/10.1080/19463138.2019.1573172

Vlahov D, Freudenberg N, Proietti F, Ompad D, Quinn A, Nandi V, Galea S (2007) Urban as a determinant of health. J Urban Health 84(1):16–26. https://doi.org/10.1007/s11524-007-9169-3

Wang YP, Kintrea K (2019) Sustainable, healthy and learning cities and neighbourhoods. Environ Urban ASIA 10(2):146–150. https://doi.org/10.1177/0975425319859129

Wang R, Xue D, Liu Y, Chen H, Qiu Y (2018a) The relationship between urbanization and depression in China: the mediating role of neighborhood social capital [Research Support, Non-U.S. Gov't]. Int J Equity Health 17(1):105. https://doi.org/10.1186/s12939-018-0825-x

Wang R, Xue D, Liu Y, Liu P, Chen H (2018b) The relationship between air pollution and depression in China: is neighbourhood social capital protective? Int J Environ Res Public Health 15(6):Article 1160. https://doi.org/10.3390/ijerph15061160

Watson V (2009) 'The planned city sweeps the poor away…': Urban planning and 21st century urbanisation. Prog Plan 72(3):151–193. https://doi.org/10.1016/j.progress.2009.06.002

Wilkinson RG (1999) Income inequality, social cohesion, and health: clarifying the theory—a reply to Muntaner and Lynch. Int J Health Serv 29(3):525–543. https://doi.org/10.2190/3QXP-4N6T-N0QG-ECXP

Wilkinson A, Ali H, Bedford J, Boonyabancha S, Connolly C, Conteh A, Dean L, Decorte F, Dercon B, Dias S, Dodman D, Duijsens R, D'Urzo S, Eamer G, Earle L, Gupte J, Frediani AA, Hasan A, Hawkins K et al (2020) Local response in health emergencies: key considerations for addressing the COVID-19 pandemic in informal urban settlements. Environ Urban 32(2):503–522. https://doi.org/10.1177/0956247820922843

Zeng Y, Gu D, Purser J, Hoenig H, Christakis N (2010) Associations of environmental factors with elderly health and mortality in China. Am J Public Health 100(2):298–305. https://doi.org/10.2105/AJPH.2008.154971

Zinkernagel R, Evans J, Neij L (2018) Applying the SDGs to cities: business as usual or a new dawn? Sustainability 10(9):3201. https://doi.org/10.3390/su10093201

6 Continuity or Change? Racial Segregation in Cape Town

Justin Visagie, Ivan Turok, and Andreas Scheba

Abstract

Cape Town's racial segregation has declined since 1994, but demographic shifts complicate comparisons over time. We decompose post-apartheid segregation trends and examine neighbourhood-level patterns to distinguish between city demographic effects and deeper segregation changes. Most of the decline stems from citywide population growth among Africans, with limited evidence of deeper social integration. Yet there are some notable neighbourhood-level patterns. Middle-income corridors, such as along Voortrekker Road, show notable desegregation through black upward mobility, while former-white suburbs are changing slowly. Historic townships remain predominantly black, reflecting persistent housing market inequalities. Urbanisation is leading to densification in poorer areas away from the core rather than socio-spatial integration. Thus, while Cape Town appears less segregated, apartheid's spatial legacy endures, as most black households remain priced out of former-white areas.

Keywords

Segregation · Racial integration · Spatial inequality · Neighbourhood inequality · Cape Town · South Africa

6.1 Introduction

South Africa has among the highest levels of income and wealth inequality in the world, and Cape Town is no exception. A recent book examining socio-economic inequalities ranked Cape Town as one of the most unequal cities among more than two dozen prominent metropolitan areas across the globe (Van Ham et al. 2021). While there is a large literature measuring national levels of inequality in South Africa (Francis and Webster 2019; Bhorat et al. 2021; Leibbrandt and Pabón 2021), there has been far less attention given to how such inequalities are generated and manifest at the local level, including in large cities such as Cape Town.

J. Visagie (✉) · I. Turok · A. Scheba
Human Sciences Research Council, Pretoria, South Africa
e-mail: JPVisagie@hsrc.ac.za

I. Turok
e-mail: turokin@ufs.ac.za

A. Scheba
e-mail: AScheba@hsrc.ac.za

J. Visagie · I. Turok
Department of Economics and Finance, University of the Free State, Bloemfontein, South Africa

A. Scheba
Centre for Development Support, University of the Free State, Bloemfontein, South Africa

Y. P. Wang et al. (eds.), *New Drivers of Division*, Sustainable Development Goals Series,
https://doi.org/10.1007/978-981-96-5629-5_6

An important issue is the extent to which South African cities have managed to break with the past in building more racially integrated and inclusive neighbourhoods. While the transition to democracy in 1994 signalled an end to deliberately engineered racial divisions, there are still stark differences in the quality of life at the neighbourhood level which correspond closely with the divides of the old apartheid city (Harrison et al 2014; McLennan et al 2016; Turok et al 2021; Lynge et al 2022; Maharaj 2020). For instance, former black townships continue to house the bulk of the population but are poorly located some distance away from economic opportunities concentrated in the city centre and suburban nodes.[1] As a consequence, poorer urban residents can spend as much as 40% of their income on transport alone (Kerr 2017). The same poorer communities have much lower levels of education and struggle to compete in a slack labour market with extreme levels of unemployment (Crankshaw 2012). The ten wards in Cape Town with the highest rates of unemployment (over 50%) are all on the Cape Flats (Turok et al. 2020). This is the low-lying, flat area situated to the southeast of the city centre. During apartheid, approximately 150,000 black residents were resettled to the Cape Flats after being forcefully removed from inner-city areas (IJR 2019).

The quality of public infrastructure provision is also very uneven across the city. The former white urban core continues to offer superior amenities, partly due to historic investments and planning for open spaces and greenery, but also because subsequent urban expansion has continued to be governed by formal town planning processes, which maintain high standards of urban design and service delivery. Affluent communities can also afford to subsidise or supplement essential services, such as through fee-paying schools, hospitals and security. By contrast, historically disadvantaged communities face serious backlogs in service provision, with a legacy of poor town planning for a growing population. The built environment has expanded through a mix of state-subsidised public housing alongside the unregulated infill of rudimentary backyard structures and the growth of informal settlements that lack most services and facilities (Harrison et al 2014; Turok et al. 2020; Massey and Gunter 2020).

There is little doubt that communities in different parts of Cape Town face very different standards of living and opportunity structures. Therefore, one might expect to see a significant reshuffling of individuals from less to more desirable neighbourhoods as their financial circumstances improve. However, while apartheid-era controls on the physical movement of individuals have long been abolished, people are limited in where they live based on what they can afford, irrespective of personal preferences or aspirations. This is exacerbated by a very steep house price gradient which could be getting worse (City of Cape Town 2021). In this way, the sorting of individuals by market forces, despite equal rights set out in the constitution, continues to drive segregation. The emergence of the black middle class may have allowed for some reshuffling but only for a limited number of households who can afford to move into better areas. Quantitative estimates of the black middle class differ widely, depending on the definition, but are usually below 5 million (Schlemmer 2005; Visagie 2015; Southall 2016).

The primary question addressed in this chapter is whether or not Cape Town is still characterised by racially divided neighbourhoods in light of entrenched social inequalities? Furthermore, we ask the following sub-questions: Has the transition to democracy translated into more racially integrated neighbourhoods? Or does the city largely reflect continuity with the past? In particular, which parts of the city have the highest levels of racial segregation and which have been the most resistant to change? Are there positive examples

[1] Statistics South Africa, which administers the national census, uses the following terms with regard to population groups: black African, coloured, white, Indian/Asian, other/unspecified. We make use of the same terminology in classifying race groups in this paper with the exception of distinguishing between 'African' and 'black'. We use the label 'African' to refer to individuals of African native descent while 'black' is exclusively used to refer to non-Whites (i.e. African, coloured and Indian). This language and labels are widespread in academic and policy documents and should in no way be interpreted as pejorative. At the same time, we recognise that like any racial categorisation, these labels are socially constructed and inherently flawed.

of neighbourhoods which have experienced racial desegregation and why? The focus of the analysis is on measuring changes in racial segregation, although the underlying class dynamics are an important part of the interpretation.

The contribution of the chapter to the literature is twofold. First, while there is a lot of evidence about socio-economic and race-based patterns of inequality at a national level, the geographic spread and changes in inequality at the neighbourhood level, including the degree of segregation, is understudied (Lynge et al. 2022). The analysis of country-level inequality is blind to the extent of local inequalities or segregation and can even be misleading at the local level. The geographic separation between different groups is not wired into standard tools for measuring inequality such as the Gini coefficient. In fact, if Gini inequality measures are crudely applied at a localised scale, disaggregated inequality scores can improve with *greater* levels of spatial inequality, because localities would be more homogenous in themselves. It is instructive to visually map patterns of racial segregation in Cape Town and to assess which areas have experienced most and least change. We make use of a variety of maps as well as a segregation plot tool to show how the full sample of neighbourhoods in Cape Town are ranked from the most to least segregated.

Second, we present a more thorough and comprehensive assessment of changes in racial segregation than prior studies. Related analyses of the pace and scale of racial desegregation in South African cities are hampered by methodological concerns that have led to mixed findings (Schensul and Heller 2011; Hamann and Horn 2015; Crankshaw 2022). A key challenge is that citywide demographic changes undermine common statistical indicators of segregation, such as the dissimilarity index or entropy index, because of the mathematical properties of the formulae. This is the case in Cape Town, where the white and coloured population shares have fallen significantly, with the African share rising. Our solution is to employ a type of segregation index (known as the M-index) which allows us to explicitly decompose the influence of demographic changes and allow for a robust comparison of 'pure' segregation scores over time (i.e. holding constant the influence of wider demographic changes).

6.2 Literature: Studies of Racial Segregation in South Africa

There is mixed evidence about the extent of racial desegregation in post-apartheid cities including Cape Town. The earliest studies by Christopher (2005a, 2005b), based on the 1996 and 2001 census for a full sample of South African cities and towns, report that desegregation is 'progressing at a very slow pace'. Schensul and Heller (2011: 26) agree that there is 'little change beyond perhaps a slow and small erosion' when examining the index of dissimilarity for eThekwini. Hamann and Horn (2015) find almost no change in a multi-group segregation index for Tshwane between 1991 and 2011. Finally, Katumba (2019) reports that 'patterns of racial integration across Gauteng have remained similar from 1996 to 2018'.

By contrast, a number of other studies, particularly when including data from Census 2011, tend to paint a more positive view. For instance, Parry and Eeden (2014: 47) find evidence that 'the level of segregation has exhibited a consistent downward trend in both Johannesburg and Cape Town'. Solomon (2023: 23) produces a wide variety of dissimilarity scores for both race and class (by occupation) for Cape Town and notes that most scores improve over the period although racial segregation is 'still the more dominant of the two'.

One reason for mixed evidence about the pace of desegregation in South African cities is purely methodological. While citywide segregation scores, such as the index of dissimilarity, are widespread in the literature because of their simplicity and elegance, a direct comparison between different cities or different groups can be misleading in cases where the baseline proportions are fundamentally different. This is also true when comparing changes for a particular city over a long period of time if there has been significant demographic change—which is the case in most South African cities. This is a point raised by Crankshaw (2022) in discussing why there are sometimes conflicting findings about the rate of desegregation in Johannesburg. Crankshaw (2022: 168) warns that '…the dissimilarity index should never be used to compare segregation levels using data for cities with different total racial proportions. Nor should the index be used to compare segregation levels in the same city if the total racial proportions of the population change over time'. Notwithstanding these methodological concerns, the above studies tend to agree that South African cities display extreme levels of social or racial segregation, even if the pace of change, and rankings between cities is a moot point.

A complementary approach to estimating a single summary measure of segregation is to examine changes at the neighbourhood level. This includes descriptive changes on maps, but also calculating localised segregation indices for each neighbourhood.

For instance, Parry and Eeden (2014) examine localised segregation scores at a variety of spatial scales for Cape Town and Johannesburg. They find general improvements in racial diversity across both cities with the exception of former African and coloured townships which remain highly segregated. This is the view of Geyer and Mohammed (2016) who examine neighbourhood-level changes in Cape Town with a focus on 'hyper segregation'. They argue that poorer communities get trapped in a negative cycle of disinvestment and decay as upwardly mobile residents exit as soon as they get a chance. The result is the perpetuation of relatively homogenous and impoverished 'ethnic enclaves'.

Ballard et al. (2021) examine ward-level demographic data for the Gauteng city-region. They show that it is affluent neighbourhoods in particular which have experienced the greatest racial desegregation. Crankshaw (2022) presents similar evidence for Johannesburg. However, he adds that racial integration is contingent on affordability and shows that transformation has been most rapid in middle-class neighbourhoods (note: these are probably closer to working class neighbourhoods compared with income levels in Northern cities) where property prices are more affordable and there is greater availability of rental housing stock.

Crankshaw (2022) backs up this view of class-driven racial integration by showing that black residents living in affluent neighbourhoods tend to be employed in high-earning occupations. Similarly, Turok et al. (2021) and Ballard et al. (2021) examine occupational or class-based segregation in Cape Town and Johannesburg respectively. Interestingly, they both report that aggregated class-based segregation scores diminished noticeably over the period, especially in Cape Town. At the neighbourhood level, this change appears to have been driven by a slight dilution of top occupations in affluent neighbourhoods through spreading out into adjacent neighbourhoods.

In summary, neighbourhood-level studies of segregation are better suited and more consistent in their findings than citywide single summary measures. A key message is that an increase in racial integration is often noticeable in affluent neighbourhoods but most so in middle income neighbourhoods which are more affordable (these might be interpreted as lower 'middle class' compared with the average South African). By contrast, many poorer communities, particularly former townships, have remained highly segregated. Stronger population growth in poorer communities amplifies this effect. This also explains why some studies disagree about the overall extent of racial desegregation at the city level because different neighbourhoods seem to have experienced quite different demographic and social processes.

6.3 Data and Methods

The data for this chapter comes from the South African censuses for 1996, 2001 and 2011. The data is analysed at a granular 'sub-place' or census tract level. While ward data is more commonly utilised by metropolitan municipalities for regular reporting, ward boundaries sometimes include neighbouring communities of different social

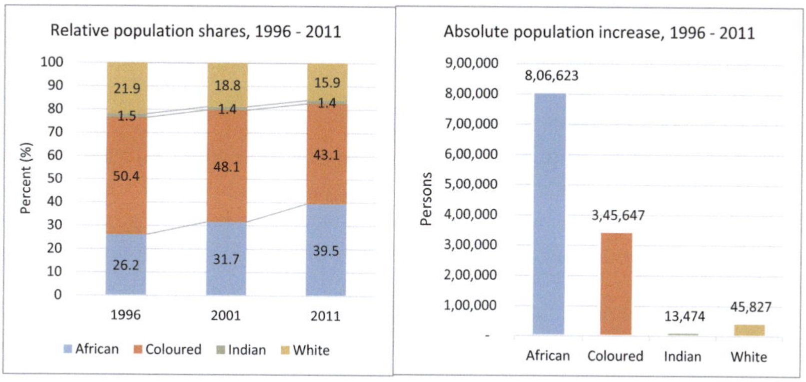

Fig. 6.1 Aggregate demographic change in Cape Town, 1996–2011. *Source* Census 1996, 2001 and 2011

standing which are in fact physically separate. Hence, the more fine-grained sub-place spatial units are preferable. The sub-place administrative boundaries have also been harmonised to match exactly between census years due to a few cases where there were changes in the boundaries over time, as released by Quantec (2016). There are a total of 889 sub-places in the Cape Town sample which each include less than 10,000 persons, notwithstanding a few exceptions. Sub-places may vary widely in geographic size but tend to keep total population numbers reasonably consistent.

For the purposes of this study, we are interested in segregation as the extent of spatial 'clustering' (or the converse of integration as the extent of 'evenness') (Massey and Denton 1988; Feitosa et al. 2007). Hence segregation is defined as the extent to which a particular group is represented within a sub-unit to the same degree that they are for the city as a whole. For example, in the case of Cape Town in 2011, a perfectly racially integrated census tract of 100 persons would match the city-wide demographic shares exactly of 43 individuals who are coloured, 40 Africans, 16 whites and 1 Indian (see Fig. 6.1 for demographic shares). In other words, segregation as defined here (as in most other studies) should always be interpreted and understood in reference to the city's overall demographic shares. This is our definition of pure segregation.

A related, but distinct, concept is exposure (or its converse, isolation) which is the extent of mixing between groups. Exposure/isolation is independent of the overall population shares.[2] In our example of a 100-person census tract in Cape Town, perfect exposure would mean an equal split between African, white, coloured and Indian individuals, or in other words, 25 persons from each demographic group. Hence it is neither realistic nor feasible to expect perfect mixing or exposure in each and every Census tract (and mathematically impossible) although it is still an important concept to bear in mind.

Many studies of segregation make use of a statistical index to provide a one-number summary of the degree of segregation (Reardon and O'Sullivan 2004; Feitosa et al. 2007; Mora and Ruiz-Castillo 2011). This is also the case for studies of segregation for South African cities as discussed above.

The most common measure of segregation is the index of dissimilarity or D-Index (see Box 6.1 for mathematical specification). A large attraction of the D-index is its simplicity and ease of interpretation. The index score, which ranges from 0 (perfect integration) to 1 (perfect segregation), can be interpreted as the proportion of the population that would need to move in order to achieve perfect

[2] Massey and Denton (1988) identify five distinct dimensions of segregation: (i) evenness (ii) exposure/isolation (iii) concentration or density (iv) centralization (v) clustering. However, Reardon and O'Sullivan (2004) clarify that these categories can be simplified into just two dimensions on the continuum between (i) 'evenness/clustering' and (ii) 'isolation/exposure'.

integration. A limitation of the D-index is that it is constructed as a comparison between only two groups (e.g. African and white, or white and non-white). This is an obvious drawback when most large cities have representation from more than two race or ethnic groups. The D-index also does not have a local level application, or in other words a score for each neighbourhood.

Box 6.1:

Mathematical formula for the Index of Dissimilarity (D-index) [two groups]

The index of dissimilarity for two groups in a particular location is expressed as:

$$\text{D index} = \frac{1}{2}\sum_{i=1}^{N}\left|\frac{a_i}{A} - \frac{b_i}{B}\right|$$

where i is a census tract, N is the total number of tracts in the area of study, a is race group 1 in tract i, A is the total population of race group 1, b is race group 2 in tract i, and B is the total population of race group 2.

Another well-known measure of segregation is Theil's entropy index or H-index (also commonly calculated in studies of inequality) (see Box 6.2 for the mathematical specification of H-index). Like the D-index, the H-index also varies between 0 (perfect integration) and 1 (perfect segregation) although the interpretation is less intuitive. The H-index can be understood as the average difference between a unit's group proportions and that of the system as a whole. In studies of income inequality, it measures the statistical 'distance' that the population is away from the state of everyone having the same income. In studies of segregation, it measures the distance that the population is away from a state in which every neighbourhood is perfectly integrated. The main attraction of the H-index is that it can be used to measure multi-group segregation. It can also be easily adapted to give a local census tract segregation score. We make use of a segregation plot tool to visually show the rankings of neighbourhoods sorted from least to most segregation based on the local H score (Elbers and Gruijters 2022).

A major problem with one-number summary indicators of segregation, like the D and H index, is that they struggle to disentangle different components of the segregation score which are fundamental to comparing levels of segregation over time, across countries or cities, or between ethnic or race groups (Grusky and Charles 1998; Elbers 2021; Crankshaw 2022). For example, changes in the demographic shares over time, or distributions between different census tracts, also impact on the calculation of final segregation scores. The literature sometimes refers to this as the issue of 'margin-dependency' which is referencing the mathematical properties of the formula when summing up the baselines or margins (for instance, across racial population shares, or across census tract population shares). This is distinct from what is sometimes referred to as 'pure' or structural segregation which is a comparison between two scenarios where the margins (such as the racial shares and total population size) are identical (i.e. apples with apples) although this seldom happens in practice (see Box 6.2 for a further discussion).

The solution applied in this chapter is to decompose the source of the change to segregation into each of these components (Elbers 2021; Mora and Ruiz-Castillo 2011). Rather than attempting a margin-free measurement of segregation at each point in time, we instead acknowledge and measure the contribution from each. This is done by calculating and decomposing the Mutual Information or M-index (which is a closely related component of the H-index) (see Box 6.3 for mathematical specification of M-index).

Box 6.2:

Illustrative example of margin-dependency in the calculation of segregation scores

Segregation by gender in the workplace

The issue of margin dependency can be illustrated by the following example:

For the baseline scenario, imagine a region or city with a total workforce of 50 males and 50 females working in either manufacturing or agriculture. Each sector is the same size (50 persons in each sector) but manufacturing (agriculture) is heavily skewed towards males (females). There are only 10 females (40 males) working in manufacturing and 40 females (10 males) working in agriculture. Let us consider the impact on the measurement of segregation by gender in the workforce in the following two scenarios.

In scenario 1, the number of females in the workforce is increased by tenfold from 50 to 500 while the number of males does not change [we also assume industry demand changes accordingly so there is no unemployment]. If we also hold constant the share of females allocated between sectors (so that among females a 20% share work in manufacturing, and 80% in agriculture– as previously in the baseline) our manufacturing sector now consists of 40 males (unchanged) and 100 females, and the agricultural sector consists of 10 males (unchanged) and 400 females.

In this new scenario, the manufacturing sector appears to be much less segregated because females outnumber males in manufacturing. However, upon closer inspection, the sector is still heavily biased towards males. The degree of 'pure' segregation has not changed. Instead it is the change in 'gender margins' which is responsible for the increase in females within manufacturing. In other words, the increase in labour force participation for females, rather than a deeper shift in segregation, is responsible for the change.

In scenario 2, the number of workers in manufacturing increases tenfold from 50 to 500 while demand for agriculture is unchanged [assuming the workforce grows to match the change in industry demand]. If we hold constant the share of female workers within the manufacturing sector, than there would be 100 females and 400 males working in manufacturing.

Notice that in this scenario, the number of females working in manufacturing is greater in absolute number than females working in agriculture (100 female workers in manufacturing, and 40 female workers in agriculture) but there is still a heavy bias towards males in manufacturing. Once again, the degree of 'pure' segregation has not changed. Instead, it is the change in 'sector margins' which is responsible for the increase in female workers in manufacturing. In other words, it is industrialisation, rather than an improvement in 'pure' segregation, which explains the change.

A salient point from these scenarios is that changes in the margins do reflect a change in segregation on the ground. However, the underlying source of this change is distinct from what might be identified as a pure or structural changes in segregation. This is why different statistical segregation scores can sometimes diverge depending on how they approach the issue of margin dependency.

(Note: for our study of racial segregation, 'gender' is analogous to race and the 'sector' is analogous to neighbourhood or census tract.)

Box 6.3:

Mathematical formula for Mutual Information Index (M-index) and Theil's Entropy Index (H-index) [multiple groups]

The Mutual Information Index and Theil's Entropy Index are closely related statistical measures and are derived through the following estimation procedure:

First, the entropy score (h) is calculated for each census tract i:

$$h_i = -\sum_{j=1}^{k} p_{ij}\ln(p_{ij}) \quad (6.1)$$

where i is a census tract, k is the total number of race groups j, p_{ij} is the population share of the jth race group in census tract i.

In a similar manner, the citywide entropy score (E) is calculated as:

$$E = -\sum_{j=1}^{k} p_j\ln(p_j) \quad (6.2)$$

The **Mutual Information Index** (M) is estimated by taking the difference between each census tract's entropy score and the citywide entropy score. The census tract scores are added together and weighted by their population share into a single aggregated indicator:

$$\text{M index} = \sum_{i=1}^{N}[t_i(E - h_i)] \quad (6.3)$$

where t_i is the proportion of the total population in census tract i, N is the total number of census tracts

Theil's Entropy Index (H) is estimated through the same procedure, but in addition, the scores are standardised to a range of 0 and 1 by dividing by the citywide entropy score:

$$\text{H index} = \sum_{i=1}^{N}\left[\frac{t_i(E - h_i)}{E}\right] \quad (6.4)$$

Therefore the H-index and M-index are interrelated in that:

$$H = \frac{M}{E} \quad (6.5)$$

where H is the H-index (4), M is the M-index (3) and E is the citywide entropy score (2).

It should also be noted that both the H and M index are produced by summing together the weighted local scores for each tract. Hence, local M and H scores for each neighbourhood can be reported although neither of these local scores is constrained to fall between 0 and 1.

Local H-index score:

$$\text{Local H score} = \frac{(E - h_i)}{E} \quad (6.5)$$

Local M-index score:

$$\text{Local M score} = (E - h_i) \quad (6.6)$$

For the purposes of this study, the decomposition of the M-index allows us to be specific about the sources of change in Cape Town's segregation score over the period. The four components of the decomposition include:

i. **Racial group margins**: this component captures the impact of changing demographic shares on the calculation of the M-index. The intuition is that faster population growth for a particular race group makes it easier for there to be greater exposure to that group which can lower the overall segregation score.[3] For example, in Cape Town, Africans experience far higher population growth over the period which means that we would expect the share of Africans to rise across all neighbourhoods—because the share of Africans is rising at the city level. This means that former white wealthy neighbourhoods can experience a rise in their share of Africans but this does not reflect pure racial desegregation (as a statistical measure—see also Box 6.2) if it is only in line with citywide demographic change.

[3] If all race groups experience exactly the same rate of population growth, then the racial shares would be constant and there would be no impact from 'racial margins' over time.

ii. **Census tract margins:** this component captures the impact of changing population shares between neighbourhoods on the calculation of the M-index. The intuition is that some neighbourhoods have higher levels of segregation than others. Hence, if there is faster population growth in those neighbourhoods, it can lead to greater segregation overall. For example, in Cape Town, population growth is often highest on the Cape Flats in racially homogenous black townships which means more people live in highly segregated neighbourhoods. This means a rise in the citywide segregation score over time, even if the extent of racial mixing (i.e. pure segregation in each neighbourhood) is unchanged.
iii. **Additions/removals:** this component captures the impact on the calculation of the M-index when adding or deleting particular sub-places (as defined by the census tracts). The impact on the citywide level of segregation could be positive or negative depending on the degree of segregation in each sub-place appearing or disappearing. For instance, in Cape Town new neighbourhoods can form through illegal occupations or approved formal developments. In practice, sub-places do not tend to appear or disappear because new neighbourhoods tend to occur on parcels of land which are already included as part of existing sub-places.
iv. **Pure segregation**: this component captures the impact of the distribution of racial shares on the calculation of the M-index. In this instance, the influence of other components (racial group margins, census tract margins, additions/subtracts) are already accounted for. So any change to the level of segregation on the M-index is purely due to structural changes in racial mixing i.e. pure segregation.

As a final step, we also make use of the M-index to calculate and illustrate local segregation scores (which we adjust to reflect 'pure segregation' when comparing changes over time) (see Box 6.3 for the mathematical specification of local M scores). Local segregation scores help draw attention to which parts of Cape Town are most segregated and which have experienced the most, or least, change. This is critical in offering more depth and nuance to the aggregated citywide segregation measure which fails to highlight how different neighbourhoods often experience different trends.

6.4 Results

6.4.1 A Descriptive Review of Racial Desegregation in Cape Town

A striking feature of population data for Cape Town over the period 1996–2011 is a process of fairly rapid demographic change. Figure 6.1 shows the absolute and relative change in the number of individuals represented by each demographic group in the city. The demographic profile of Cape Town in 2011 is fairly distinct with a majority share for coloureds (43%), followed by Africans (40%) and whites (16%). This is different from country-level demographics in 2011 which was majority African (79%) with much smaller shares for coloureds (9%) and whites (9%). In fact, only 20% of South African municipalities in 2011 registered a minority share for Africans (which includes the case of Cape Town).

Figure 6.1 shows how the population shares across demographic groups changed significantly between 1996 and 2011 in Cape Town. This is because of a large increase for Africans from just 26% of the population in 1996 to 40% of the population in 2011. These changes are even more pronounced when looking at absolute numbers. Africans increased in size by more than 800,000 persons which is almost double that of all other groups combined. While none of the population groups experienced a population decline, the relative growth rates work out very differently with a 126% increase in the number of Africans, 28% for coloureds and 8% for whites over the 15-year period. This means that the African population grew in size 4.5 times faster than the growth among coloureds and more than 15 times faster than the growth among whites.

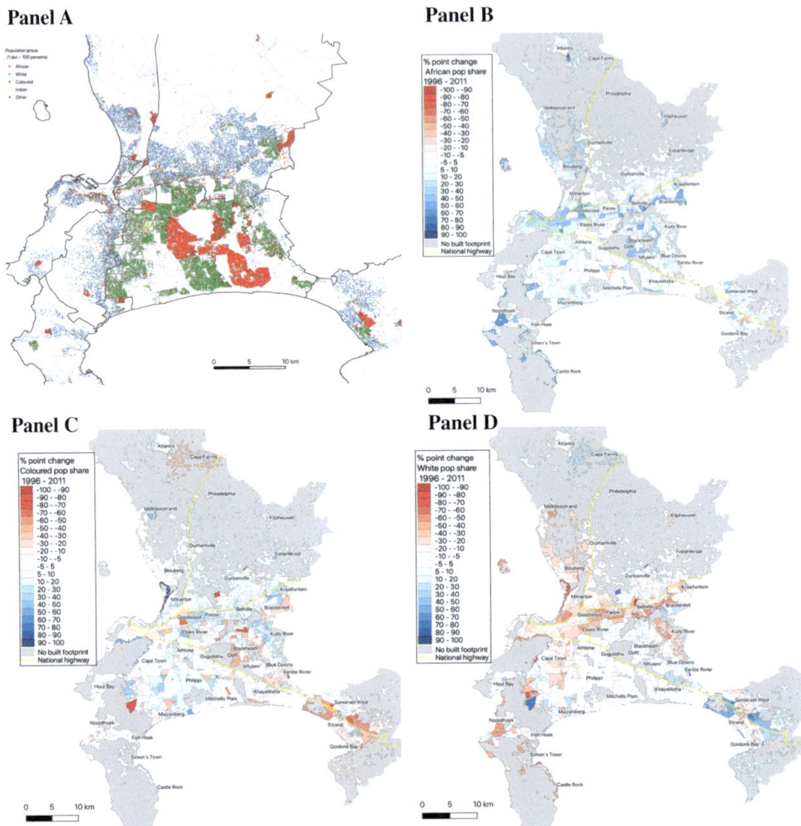

Fig. 6.2 Demographic change at the neighbourhood level in Cape Town, 1996–2011. *Source* Census 1996 and 2011, harmonised sub-place tracts (Quantec 2016)

These contrasting demographic trends have clear implications for the extent of racial mixing across the city over time. Figure 6.2 (Panels A, B, C and D) shows descriptive changes to the demography of the city at the neighbourhood level.

Panel A shows the raw population numbers by race in 2011 (where 1 dot represents 100 persons). The visualisation is striking in the extent of similarity with apartheid-era settlement patterns and the obvious degree of racial segregation in many parts of the city. The most intense concentration of Africans is on the Cape Flats in large townships such as Khayelitsha, Langa and Mfuleni. These are clearly distinct from coloured townships on the Cape Flats such as Athlone, Philippi and Mitchells Plain. You can also make out Rylands, a former Indian township on the Cape Flats. Many former white areas are still dominated by whites, such as Durbanville, Bellville and Milnerton in the Northern Suburbs, Camps Bay and Llandudno on the Atlantic Seaboard and Constantia in the Southern suburbs.

There are obvious divides between historically white neighbourhoods in the Southern and Northern suburbs, moving outwards to coloured communities on the margins of the Cape Flats and finally African communities at the centre of the Cape Flats. There are also other noticeable racial enclaves for African and coloured communities such as Masiphumelele near Noordhoek, Dunoon near Bloubergstrand and Nomzamo in Somerset West. Informal settlements are also identified as racially segregated spaces with dense concentrations of Africans in settlements such as Imizamo Yethu in Hout Bay, Joe Slovo Park near Milnerton, Wallacedene in Kraaifontein and many parts of the Cape Flats. A few informal settlements are located in relative close proximity to neighbouring White (and affluent) neighbourhoods, such as Joe Slovo Park, but it is hard to argue that this represents racial desegregation even if the distance between communities is closer.

Yet despite continuity with historic patterns of segregation in Cape Town, there is also evidence of change. Panels B, C and D of Fig. 6.2 show the percentage point change in the share of African, coloured and white race groups within each sub-place respectively between 1996 and 2011. The picture is not static. For instance, many neighbourhoods in and around the City Bowl have noticeably changed in character to become fairly diverse and even majority black. This starts in the CBD and moves up towards the Southern suburbs including Woodstock, Salt River, Observatory and Rosebank. Another example is along the Voortrekker corridor with large increases in both coloured and African populations. This is the area stretching along Voortrekker Road below the N1 highway starting from the Salt River Circle to Stikland Bridge, and includes parts of Maitland, Goodwood, Parrow and Belville. There are also signs of racial change along the border of the Southern Suburbs and the Cape Flats such as in Wynberg, Kenilworth, Plumstead and Diep River. Even within the Cape Flats, there are some places which appear to have become more integrated (between coloureds and Africans), such as parts of Delft and Philippi.

The areas of greatest change tend to be middle income neighbourhoods (or lower 'middle class'), and usually with higher densities, which points to the role of affordability and class mobility in enabling racial integration. Wealthy neighbourhoods do show rising numbers of black households, but at a slower rate compared to better priced areas.

Overall, the descriptive evidence shown in Figs. 6.1 and 6.2 presents a mixed picture of racial desegregation in the city of Cape Town. On the one hand, there are still many racial enclaves such as townships within the Cape Flats, informal settlements and affluent suburbs. Continuity with the apartheid city is unmistakable. On the other, there are signs of change, particularly in and around the City bowl as well as in some former white affluent neighbourhoods. Middle-income neighbourhoods appear to be most conducive in supporting racial integration arguably because of greater affordability.

6.4.2 Statistical Measures of Desegregation in Cape Town

What is the net impact in terms of the level of segregation in Cape Town between 1996 and 2011? What is the contribution from densification in poorer parts of the city? What is the contribution to racial mixing from faster population growth among Africans? Is there a deeper structural change in the spread of race groups in neighbourhoods or, in other words, the level of 'pure' desegregation?

These questions are better answered by statistical segregation measures as discussed in the data and methods section. Figure 6.3 presents the results for the Dissimilarity Index (D-index) (for Africans/whites and blacks/whites), Theil's Multigroup Index (H-index) and the Mutual Information Index (M-index). They all represent perfect integration as 0 with perfect segregation as 1.[4]

South African cities reflect extreme scores compared to studies of segregation in other countries (Van Ham et al. 2021). The dissimilarity indices are much closer to the upper bound of 1 than to the lower bound of 0. The implication is that the city of Cape Town is still highly segregated. For instance, an African/white D-index of 0.88 in 2011 means that 88% of whites (or Africans) would need to relocate in order to achieve perfect integration between these two groups. The black/white D-index is very similar at 0.87 in 2011.

Figure 6.3 shows more variation when comparing the trends over time for each of the segregation measures. This is in terms of the

[4] The exception is the M-index which does not have a uniform upper bound although a higher number would represent greater segregation. In fact, the H-index is a transformation of the M-index into a standardised form from 0 to 1.

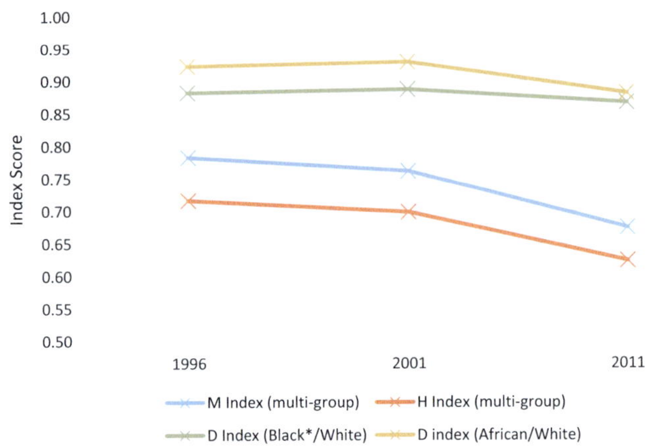

Fig. 6.3 Segregation score indicators, 1996, 2001 and 2011. *Note* *Black here refers to non-white (i.e. grouping together coloured, African and Indian population groups). *Source* Census 1996, 2011 and 2011, harmonised sub-place tracts (Quantec 2016)

magnitude of the change rather than the direction. The indices suggest that segregation score for Cape Town declined very little between 1996 and 2001. The reduction was more noticeable over the period 2001 and 2011, particularly for both multi-group indicators (H-index and M-index). Overall, the multi-group indictors fell by 9 to 11 index points over the period compared to only 2 to 5 index points for the D-indices.

Table 6.1 goes further to decompose the specific sources of the reduction calculated in the M-index between 1996 and 2011. The M-index declined by a significant, but modest, 10.9 index points (or a 13.9% decline) over the period. The observed decline is split between changes in (i) racial group margins (i.e. demographic change) (ii) census tract margins (i.e. changes in population intensity across neighbourhoods) (iii) additions/removals of Census tract units; and (iv) pure segregation (i.e. racial mixing after controlling for other components) (see: data and methods section above). An apparent reduction in segregation attributable to citywide demographic changes is quite different from a deeper structural shift in the racial shares of neighbourhoods i.e. pure desegregation.

Table 6.1 Decomposition of the change in segregation score (M-index)

	Estimate	Std error	% share
M-index			
1996	0.783	0.000	
2011	0.674	0.000	
Decomposition			
Index change	**−0.108**	**0.001**	**100.0**
Racial group margins	−0.110	0.001	101.7
Census tract margins	0.016	0.001	−15.0
Additions	0.000	0.000	−0.2
Removals	0.001	0.000	−1.0
Pure segregation	*−0.016*	*0.001*	*14.5*

Source Census 1996 and 2011, harmonised sub-place tracts (Quantec 2016)

The first component, changes to racial group margins, makes the greatest contribution to reducing levels of segregation in Cape Town, and by a long way. What this component means is that the stronger population growth for Africans (and to a lesser extent coloureds) in comparison to whites naturally causes desegregation in Cape Town. This is because, even if we assume the relative share of Africans *across all neighbourhoods* does not change (i.e. the absolute number of Africans increases by the same percentage in all neighbourhoods), stronger population growth for Africans relative to other groups means that the share of Africans *within each neighbourhood* is also rising. The implication is that we expect many former white middle-class neighbourhoods to now have a higher share of Africans. However, it is important to realise that this change is simply in line with demographic shifts for the whole city. It is a form of desegregation—because more Africans are entering former white neighbourhoods—however strictly speaking, this is not structural, because we would expect as much because of macro level changes in the citywide population shares (i.e. statistically, this is not pure desegregation).

The second component, changes in census tract margins, is responsible for a slight increase in the segregation score. What this means is that population growth tended to be stronger in neighbourhoods with higher segregation scores. This is because densification occurred in poorer communities, such as townships and informal settlements, which remain highly segregated spaces. In practice, these populations expanded through delivery of state-subsidised housing as well as informal processes of densification such as backyarding, subdivisions and informal settlements. This had the effect of adding to the level of segregation.

The third component, additions or removals in census tracts, had almost no impact on the segregation score. This makes sense because the number of populated sub-places is very consistent over the period.In other words, only a tiny share of Census sub-places either fall away or appear for the first time in containing people. Hence the contribution from this component can effectively be ignored.

The final component, changes in pure segregation, is of prime interest in understanding the reduction in segregation in Cape Town over the period. Pure segregation represents the extent to which the racial shares in each neighbourhood align with or differ from the citywide shares. It is disentangled from the influence of the other components (as listed above), and is the 'strict' meaning behind segregation. While the extent of pure segregation in Cape Town has reduced, its contribution to the overall decline is small. In fact, its contribution to the decline is only 15% in comparison to the contribution of the racial group margins score. Hence, the impact from pure desegregation is small.

In summary, Cape Town experienced mild desegregation over the fifteen-year period as registered by a fall in the M-index of nearly 14% between 1996 and 2011. Interestingly a decomposition of the components of change suggests that the observed decline was in line with a higher rate of population growth among Africans (and to a lesser extent for coloureds) (known as racial group margins). This implies that we should expect to find more blacks in former white neighbourhoods because of citywide demographic change. The structural pattern of racial mixing (i.e. pure segregation) for Cape Town, after controlling for changing demographic shares, contributed only slightly to desegregation over time.

As a final step in the statistical measurement of segregation, we can also calculate local segregation scores for different parts of the city, and also calculate a change in 'pure segregation' for each neighbourhood over time (i.e. adjusted net of race and census group margins). The results of this exercise are captured in Fig. 6.4.

First, it is interesting to see that the places with the highest segregation scores are often in affluent neighbourhoods in the northern and southern suburbs (see Fig. 6.4: Panel A). Even though many former white neighbourhoods contain a mix of racial groups in practice, the fact that whites have a low share at the city level (only 16% in 2011) means that they are grossly over-represented in these areas. For instance, suburbs such as Bishopscourt, Newlands, Bantry Bay, Bothasig and Brackenfell North have high segregation scores even though they are at least 25% black. That said, there are still affluent suburbs that are over 90% white, such as Sonstraal in Durbanville, Dennendal in Tokai, and Kommetjie. By comparison, most neighbourhoods in the Cape Flats tend to have lower segregation scores even though they are extremely homogenous. For example, Gugulethu and Khayelitsha are 98% African while Hanover Park and Parkwood are 96% coloured. The local segregation scores are more flattering in townships than in suburbs because African and coloured populations are more numerous at the citywide level. This recognises the fact that it is arithmetically 'easier' for these neighbourhoods to be more homogenous.

Panel B of Fig. 6.4 shows how local segregation changed over the period 1996 and 2011, after adjusting for margins in order to reflect the pure segregation effect. The figure reveals how different parts of the city experience opposite trends, with some neighbourhoods becoming more segregated while others become less segregated. This is why the net impact is small in

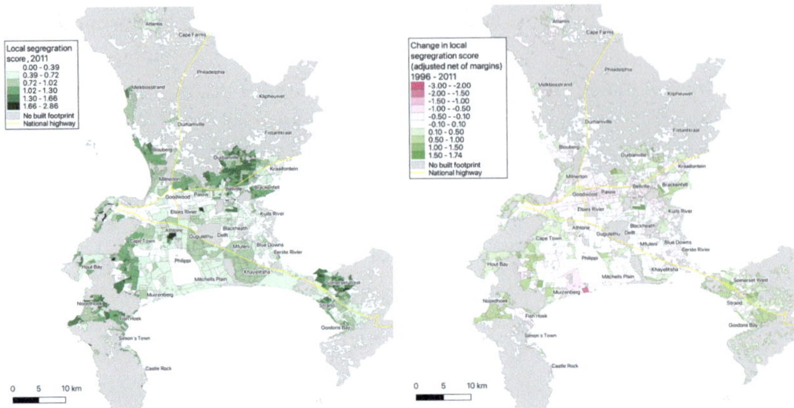

Figure. 6.4 Local segregation scores, 1996–2011. *Notes* Local segregation scores are derived from the M-index. *Source* Census 1996 and 2011, harmonised sub-place tracts (Quantec 2016)

Table 6.1. It is important to note that many wealthy neighbourhoods actually become more segregated over time even though they experience an inflow of black residents. This apparent contradiction is explained by the fact that one would expect some growth of black residents because of their stronger population growth. In practice, these neighbourhoods have been slow to change. Yet there are also positive cases of transformation. This is most noticeable for a lengthy stretch of middle-income (or lower middle class) neighbourhoods starting in the inner city in Woodstock and moving along the Voortrekker corridor to Kraaifontein in the North-East. There are also other clusters which became less racially segregated, such as around Milnerton, Blouberg and some parts of the Cape Flats.

6.5 Conclusion

The chapter has attempted to answer the question: has Cape Town experienced racial desegregation since the transition to democracy? The answer is not straightforward. Even though apartheid-era restrictions on place of residence have long been abolished, people are constrained in where they can live by affordability. Indeed, the literature on racial desegregation in South African cities presents a mixed picture of progress.

On the surface, we confirm that there has been progress in the extent of racial desegregation in Cape Town. Several indices of segregation show decline over the period 1996, 2001 and 2011, although from a very high level. This reflects an increase in the share of black individuals in some wealthy neighbourhoods. However, the biggest changes have been in middle-income neighbourhoods and around the inner city and along the Voortrekker corridor. Meanwhile, many large townships on the Cape Flats, such as Khayelitsha and Mitchells Plain, have retained their profiles as predominately African or coloured settlements. These places have also experienced higher population growth.

A deeper interrogation of the sources of desegregation suggest that qualifications need to be added. A key finding is that the main reason behind desegregation in Cape Town is closely related to demographic change. Growth in the population of Africans is some 15 times faster than whites and 4.5 times faster than coloureds. This requires that former white neighbourhoods absorb more black households at a faster rate over time just to maintain the same level of pure segregation. This also implies that the distribution of Africans between neighbourhoods does not change. In contrast, the impact of pure desegregation (i.e. holding constant the contribution of demographic change) is positive but very small at only 15% of the size of demographic change. In fact, many former white suburbs regressed (despite some inflow of black households) because the pace of change was too slow. In addition, strong population growth in highly segregated places, such as the townships, had the net effect of cancelling out the contribution from a slight improvement in the degree of pure segregation. In the end, Cape Town has become slightly less segregated, but mainly because of its skewed demographic growth.

The main implication for policy is that reducing racial segregation in South African cities demands a more proactive stance from the state. The sorting of households based on income and affordability continues to limit the options available to poorer groups who remain mostly black. It has been middle income (or lower middle class) neighbourhoods which have been most conducive to racial integration but these standards are still unaffordable for the average black South African. There is a limit to growth of the black middle class, with a poor economic outlook for the country in the foreseeable future. The state cannot rely on the market forces alone to achieve racial desegregation. Nor is it desirable for South African cities to simply replace race with class -based forms of segregation.

In the light of stark contrasts in opportunities and amenities between neighbourhoods, creating more affordable housing options in well-located areas is one of several crucial mechanisms for economic inclusion and social integration. A second mechanism is to improve the education and vocational skills of black households to enable them to compete more effectively in the labour market. Third, it is important to increase the employment opportunities in and around poorer communities to raise household incomes and facilitate social and spatial mobility.

A critical area for further research is to understand how changes to the built environment, including formal and informal processes of housing formation and densification, are related to demographic outcomes. Supporting the creation, a variety of private and public housing options is crucial for promoting affordability and in breaking down race and class barriers. In addition, it is interesting to explicitly consider how class-based measures of segregation correlate and compare with race.

References

Ballard R, Hamann C, Mosiane N (2021) Spatial trends in Gauteng. Gauteng City Region Observatory: Occassional Paper #19. GCRO, Johannesburg. https://doi.org/10.36634/lmwn5165

Bhorat H, Stanwix B, Thornton A (2021) Changing dynamics in the South African labour market. In: Oqubay A, Tregenna F, Valodia I (eds) The Oxford handbook of the South African economy. Oxford University Press, Oxford. https://doi.org/10.1093/oxfordhb/9780192894199.013.29

Christopher A (2005a) Further progress in the desegregation of South African towns and cities, 1996–2001. Dev South Afr 22(2):267–276. https://doi.org/10.1080/03768350500163006

Christopher AJ (2005b) The slow pace of desegregation in South African cities, 1996–2001. Urban Stud 42(12):2305–2320. https://doi.org/10.1080/00420980500332122

City of Cape Town (CoCT) (2021) Human settlements strategy. Cape Town.

Crankshaw O (2012) Deindustrialization, professionalization and racial inequality in Cape Town. Urban Aff Review 48(6):836–862. https://doi.org/10.1177/1078087412451427

Crankshaw, O (2022) Urban inequality: theory, evidence and method in Johannesburg, Bloomsbury Academic & Professional. ProQuest Ebook Central. http://ebookcentral.proquest.com/lib/uoct/detail.action?docID=6819016.

Elbers B (2021) A method for studying differences in segregation across time and space. Sociol Methods Res. https://doi.org/10.1177/0049124121986204

Elbers B, Gruijters RJ (2022) Segplot: a new tool for visualizing patterns of segregation. https://doi.org/10.31235/osf.io/ruw4g

Feitosa FF, Câmara G, Monteiro AMV, Koschitzki T, Silva MPS (2007) Global and local spatial indices of urban segregation. Int J Geogr Inf Sci 21(3):299–323. https://doi.org/10.1080/13658810600911903

Francis D, Webster E (2019) Poverty and inequality in South Africa: critical reflections. Dev South Afr 36(6):788–802. https://doi.org/10.1080/0376835x.2019.1666703

Geyer HS, Mohammed F (2016) Hypersegregation and class-based segregation processes in Cape Town 2001–2011. Springer 27:35–38. https://doi.org/10.1007/s12132-015-9264-6

Grusky DB, Charles M (1998) The past, present, and future of sex segregation methodology. Demography 35(4):497–504

Hamann C, Horn AC (2015) Continuity or discontinuity? Evaluating the changing socio-spatial structure of the city of Tshwane, South Africa. Urban Forum 26(1):39–57. https://doi.org/10.1007/s12132-014-9231-7

Harrison P, Gotz G, Todes A, Wray C (2014) Changing space, changing city. Johannesburg after apartheid. Wits University Press, Johannesburg

Institute for Justice and Reconciliation (IJR) (2019) Forced removals: a case study on Constantia. An oral history resource guide for teachers. IJR, Cape Town

Kerr A (2017) Tax(i)ing the poor? Commuting costs in South African cities. S Afr J Econ 85(3):321–340. https://doi.org/10.1111/saje.12161

Katumba S (2019) Gauteng's ward level racial diversity: 2018. GCRO map of the month. https://doi.org/10.36634/IUCM9765

Leibbrandt M, Pabón FAD (2021) Inequality in South Africa. In: Oqubay A, Tregenna F, Valodia I (eds) The Oxford handbook of the South African economy. Oxford University Press, Oxford. https://doi.org/10.1093/oxfordhb/9780192894199.013.9

Lynge H, Visagie J, Scheba A, Turok I, Everatt D, Abrahams C (2022) Developing neighbourhood typologies and understanding urban inequality: a data-driven approach. Reg Stud Reg Sci 9(1):618–640

Massey DS, Denton NA (1988) The dimensions of residential segregation. Soc Forces 67:281–315

Massey R, Gunter A (2020) Urban geography in South Africa: perspectives and theory. Springer, Cham. https://doi.org/10.1007/978-3-030-25369-1

Maharaj B (2020) The apartheid city. In Massey R, Gunter A (eds) Urban geography in South Africa. Springer, Cham, pp 39–54

McLennan D, Noble M, Wright G (2016) Developing a spatial measure of exposure to socio-economic inequality in South Africa. S Afr Geogr J 98(2):254–274. https://doi.org/10.1080/03736245.2015.1028980

Mora R, Ruiz-Castillo J (2011) Entropy-based segregation indices. Sociol Methodol 41(1):159–194

Parry K, van Eeden A (2014) Measuring racial residential segregation at different geographic scales in Cape Town and Johannesburg. S Afr Geogr J 97(1):31–49. https://doi.org/10.1080/03736245.2014.924868

Quantec (2016) 1996, 2001 and 2011 harmonised Census sub-place community profiles. https://www.easydata.co.za/

Reardon S, O'Sullivan D (2004) Measures of spatial segregation. Sociol Methodol 34:121–162

Schensul D, Heller P (2011) Legacies, change and transformation in the post-apartheid city: towards an urban sociological cartography. Int J Urban Reg Res 35(1):78-109. https://doi.org/10.1111/j.1468-2427.2010.00980.x

Schlemmer L (2005) Lost in transformation? South Africa's emerging middle class. Centre for Development and Enterprise Focus Papers 8, Centre for Development and Enterprise, Johannesburg

Solomon J-P (2023) A neo-apartheid city? Labour market inequalities and residential segregation in Cape Town, South Africa. https://doi.org/10.31124/advance.24124851.v2

Southall R (2016) The new black middle class in South Africa. Jacana, Auckland Park

Turok I, Scheba A, Visagie J (2020) Cape Town: a city still divided by race and class. SHLC Research Summary 01. GCRF Centre for Sustainable, Healthy and Learning Cities and Neighbourhoods (SHLC) University of Glasgow

Turok I, Visagie J, Scheba A (2021) Social inequality and spatial segregation in Cape Town. In: van Ham M, Tammaru T, Ubarevičienė R, Janssen H (eds) Urban socio-economic segregation and income inequality. The Urban Book Series. Springer, Cham. https://doi.org/10.1007/978-3-030-64569-4_4

Van Ham M, Tammaru T, Ubarevičienė R, Janssen H (2021) Rising inequalities and a changing social geography of cities. An introduction to the global segregation book. In: Van Ham M, Tammaru T, Ubarevičienė R, Janssen H (eds) Urban socio-economic segregation and income inequality: A global perspective. Springer, Cham

Visagie J (2015) Growth of the middle class: two perspectives that matter for policy. Dev South Afr 32(1):3–24. https://doi.org/10.1080/0376835x.2014.965387

7 Measuring Local Exposure to Economic Inequality in the City of Johannesburg

Halfdan Lynge, David Everatt, and Samy Katumba

Abstract

This chapter introduces a novel methodology for measuring local exposure to economic inequality in South Africa, specifically focusing on the City of Johannesburg. Using census data, the study calculates Gini-like coefficients for 5800 neighbourhoods, considering both within-neighbourhood inequality and inequality within specified radii around each neighbourhood. The research addresses a significant gap in understanding how directly experienced inequality differs from awareness of inequality, particularly relevant in South Africa's context of extreme economic disparity and apartheid-legacy spatial planning. While primarily methodological, the study yields empirical insights into Johannesburg's inequality patterns. By analysing census tracts as neighbourhood proxies, the research contributes to the literature on 'neighbourhood effects' while acknowledging the limitations of using administrative constructs for spatial analysis. The findings lay groundwork for future research on local inequality exposure.

Keywords

South Africa · Johannesburg · Space · Apartheid spatial planning · Economic inequality

7.1 Introduction

South Africa is among the most unequal countries in the world (Sulla et al. 2022). However, at local level, these inequalities are often hidden. Due to the legacies of apartheid spatial planning, most South Africans live in neighbourhoods that are relatively homogeneous. Poorer (predominantly black) people live among other poorer (predominantly black) South African communities in former townships. Similarly, wealthier (predominantly white) South Africans live among other wealthier (predominantly white) South Africans, in the suburbs.[1]

[1] It is recognised from evidence that suburbia is beginning to deracialise, certainly in the City of Johannesburg, as upwardly mobile non-white South Africans move to the more affluent northern neighbourhoods (Everatt et al. 2020).

H. Lynge (✉) · D. Everatt
Wits School of Governance, University of the Witwatersrand, Johannesburg, South Africa
e-mail: halfdan.lynge@wits.ac.za

D. Everatt
e-mail: David.Everatt@wits.ac.za

S. Katumba
Department of Geography, Geoinformatics and Meteorology, University of Pretoria, Pretoria, South Africa
e-mail: samy.katumba@up.ac.za

Y. P. Wang et al. (eds.), *New Drivers of Division*, Sustainable Development Goals Series,
https://doi.org/10.1007/978-981-96-5629-5_7

Economic inequality affects human behaviour. However, it does so in different ways. Knowing about inequality is one thing. Experiencing it on a daily basis is something else. While the effects of the former have been examined, in surveys and controlled experiments, the effects of the latter remain largely unexplored.[2] The main reason for this, the authors believe, is methodological. It is easy to manipulate survey respondents and participants in laboratory experiments by providing them with information, images, or staged actions representing inequality, and measuring their responses. It is far more complex to measure their exposure to economic inequality as they go about their daily lives.

This chapter takes a first stab at measuring local exposure to economic inequality. It does so by calculating and mapping Gini-like coefficients for 5800 neighbourhoods across the City of Johannesburg. The coefficients draw on household income data extracted from census data and calculated for concentric circles around each neighbourhood. This means they measure the degree of inequality not just within each neighbourhood but also in a specified radius around them. The main objective is methodological: to create a measure of local exposure to economic inequality and lay the groundwork for future research which would contribute to the literature on 'neighbourhood effects' (see Galster 2012 for an overview). However, by calculating and mapping the measure across the City of Johannesburg, the chapter also generates interesting empirical findings in their own right.

In this chapter, 'neighbourhoods' are defined as *bundles of spatially based attributes associated with clusters of residences* (Galster 2001: 2112) and official census tracts are used as proxies. While the limitations of using census tracts as proxies are recognised—as 'legal–administrative constructs that [...] embody specific political and administrative objectives of the government and the national statistics office' (Lynge et al. 2022: 620)—they nevertheless have value in neighbourhood research. First, they are statistically convenient units (Delmelle 2015) that have already been used in other studies (Delmelle 2015, 2017; Hincks 2015, 2017), including in South Africa (Lynge et al. 2022). Second, they are the smallest spatial unit for which census data are available in South Africa.

The chapter is structured as follows. First, the background is provided and the chapter is situated in the literature on space and inequality. Second, the relevant literature is reviewed, focusing on two central concepts or ideas: exposure to economic inequality and residential segregation. People are exposed to economic inequality in different ways and at different levels. The focus of this chapter is primary exposure at local level. This means the measure of exposure to economic inequality adopted in this study must capture economic inequality at neighbourhood level and take into consideration the degree of residential segregation. Third, the literature on economic inequality and residential segregation in South Africa is reviewed in order to contextualise the empirical findings of the chapter. Fourth, the measure of exposure to economic inequality is calculated and mapped across the City of Johannesburg, and the patterns that emerge are discussed. Finally, conclusions are drawn by reflecting on the limitations of the measure, summarising the key empirical findings, and discussing future research.

7.2 Space and Inequality

Inequality is inherently relational, as the distribution of goods and services, needs and opportunities, is influenced by power. If poverty is 'the condition of basic lack' (Soudien et al. 2019: 3), analysing inequality can help understand power dynamics and the socio-political formations they create in society (here, at neighbourhood level), such as clearly defined areas of poverty or of affluence and so on. This also explains why tackling poverty—changing the conditions under which poor people live—is far easier, politically, than tackling inequality, which requires redistribution

[2] For rare exceptions, see McLennan et al. (2016) and Sands and de Kadt (2020).

from those with to those without. The challenge comes in seeking to understand the *impact of* inequality. Poverty and its pernicious effects are frequently measured using the multiple global and local indicators and methodologies that exist. This is not true for inequality.

In the South African literature, Soudien et al. (2019: 5) argue that the impacts of inequality—the differential and unequal states in which different people and groups find themselves—are clear:

> In the common spaces in which human beings find themselves, geographically of course, but critically, also socially, they occupy very different positions of power and possibility. Some have access to material resources. Others do not. Some are able to experience, beyond the material, a bounteous sense of providence. They have the confidence to be what they want to be, to do things. Others, by contrast, feel only debilitating affliction—affliction which impacts upon their sense of self, of self-worth and of possibility. They doubt themselves immensely [...] Enablement and disempowerment are distributed very differentially. It is precisely here, at the conjunction of these disparate material and psychological states that the difficulties arise. (Soudien et al. 2019: 5)

These observations operate well on a large canvas, though what kind of research informed them is not made clear. In the literature, spatial inequality rapidly morphs into debates about spatial justice. At national, provincial or even city scales, spatial justice arguments are measurable. Israel and Frenkel's (2018) comment—drawing on Sen (1995, 2008)—that the 'extent to which these capabilities are equally distributed in space will define whether a given spatial arrangement is (un)just' (Israel & Frenkel 2018: 648), is more easily said than measured or analysed when the focus is on the individual located within their own very local space.

The focus of this study is on the micro-scale, looking to see if the impact of income inequality on people in neighbourhoods can be measured, and then (if it can) to measure what impact it has on them. It may well be that at that micro-scale, people are not marked in their everyday lives by inequality per se, unlike the immediacy of poverty, which is not relational. Do people in deprived or under-served spaces think about those living elsewhere, in well-served areas, transforming spatial injustice into a process of thought, reflection and action? As it is, according to one commentator, 'current discussions on spatial justice are moralising' (Philippopoulos-Mihalopoulos 2010: 205).

The starting point is to see if and with what level of accuracy inequality can be measured, using a Gini-like coefficient, at suburb or neighbourhood levels. If the City of Johannesburg is profoundly unequal, what can be and should be done about it? South Africa's National Development Plan, and its accompanying Vision 2030, say very little about inequality at all; and even less about how to reduce it. Primarily, reducing inequality is seen as a natural result of functional policies and institutions. If education works as planned, inequality will diminish; if the economy grows and is inclusive, inequality will drop, and so on.

The study also tries to locate and analyse inequality in space. For some this is a requirement, that 'one cannot understand social life without understanding the arrangements of particular social actors in particular social times and places [...] Social facts are *located*' (Abbott 1997: 1152). Part of the challenge, at this stage of the research, is to focus on the possibilities of measurement of exposure to economic inequality at low level, rather than immediately engage in the social justice debates that accompany spatial inequality. There is a dialectic at work: inequality is relational, not absolute; and spatial justice is concerned with 'the social justice of distributions in space or within and across territories that is being evaluated' (Pirie 1983: 470). As Pirie (1983: 471) noted, 'space has been treated in the familiar way as some sort of container, or as an existing or physical expression made up of individual locations and their distance relations', but if space is 'a social creation [...] a structure created by society and not merely as a context for society' it requires more nuanced attention. As Dikeç (2001: 1797) noted,

> If one holds the hypothesis that space is socially produced—and is not simply a container—there are important implications to be drawn from the 'segregation problem': segregation is not a question of distribution in space, but of spatialization itself. It is the very structural dynamics of the spatial organising processes in the city—and not simply some

> 'distributive force' external to space, distributing each to his or her 'proper' place. (Dikeç 2001: 1797)

These dynamics create spatially discrete pockets of poverty and wealth. Space itself is implicated, rather than providing simply a container or frame. Dikeç (2001: 1797) goes on to note that spatial dynamics (deriving from the way power is exercised) 'contribute to the formation of such segregated areas with a concentration of poverty' and oblige a group of people to live there, often replicated over generations. Soja (2009: 3) argued that class, race, and gender were the 'three most familiar forces shaping locational and spatial discrimination'; but the political organisation of space—particularly acute in post-apartheid South Africa—bears 'the imprint of colonial and/or military geographies of social control' (Soja 2009: 3).

7.3 Exposure to Economic Inequality

Economic inequality is associated with a range of negative outcomes. Research in economics has found that countries with higher levels of economic inequality have lower growth rates (Alesina and Rodrik 1994; Easterly 2007; Persson and Tabellini 1994),[3] are less likely to sustain growth over time (Berg and Ostry 2011), and tend to translate growth into wealth accumulation among the already wealthy (Piketty and Saez 2014). The negative outcomes are not limited to economic growth. Research across the social sciences has demonstrated that economic inequality is associated also with poorer physical health, both among the poor (Wilkinson 1992) and also the wealthy (Subramanian and Kawachi 2006); higher levels of obesity and obesity-related deaths (Pickett 2005); poorer mental health, measured as mental illness (Burns et al. 2014) and depression (Messias et al. 2011); lower levels of general well-being (Oishi et al. 2011; Posel and Casale 2011; Schneider 2016); higher crime rates, especially for murder and assault (Hsieh and Pugh 1993) higher levels of social distrust (Delhey and Dragolov 2014) lower levels of political participation (Solt 2008, 2010) and a range of other negative outcomes, including drug abuse, teenage pregnancy, racism, and incarceration (reviewed in Wilkinson and Pickett 2009).

The associations have been attributed—at least, partially—to psychological and behavioural processes at the individual level (Easterly 2007; Kondo et al. 2009; Pickett and Wilkinson 2015; Uslaner and Brown 2005). Experimental research has studied these processes by 'treating' individuals with information about inequality or about their relative position in an inequality distribution. This research suggests that information about inequality lowers levels of generosity (Côté et al. 2015) and cooperation (Nishi et al. 2015) increases status anxiety and risk-taking (Mishra et al. 2015; Payne et al. 2017); shapes beliefs about inequality and economic opportunity (Davidai 2018; McCall et al. 2017) suppresses trust in the government (Kuziemko et al. 2015); increases support for organised labour (Kane and Newman 2019) increases beliefs in the legitimacy of inequality (Bénabou and Tirole 2006; Trump 2018), and—under certain conditions—affects support for redistribution (Trump 2021).

People are exposed to economic inequality in different ways and at different levels. At the international level it is observed between countries or what is commonly referred to as the Global North and Global South. At national level, it typically manifests itself as economic inequality between regions or cities and across different population groups (e.g. gender, race, age, etc.). At local level, economic inequality is visible within and between neighbourhoods, in the workplace, among friends, or even within families. The extent to which individuals are exposed to economic inequality is determined by their encounters with people from different economic classes (*primary exposure*), and by the information they receive about economic inequality through friends, families, or the media (*secondary exposure*). Exposure to economic inequality is thus determined by space (where people live, work, and socialise), individual characteristics (e.g. friend and family

[3] The authors recognise that this has been challenged (Banerjee and Duflo 2003; Barro 2000).

relationships and interest in obtaining information), and societal characteristics (e.g. supply of information provided through the media).

This study focuses on exposure to economic inequality at the local level; more specifically, how the neighbourhood in which people reside affects their primary exposure to economic inequality. This requires a measure that captures economic inequality at neighbourhood-level and takes into consideration the degree of residential segregation. As already noted, neighbourhoods in South Africa are relatively homogeneous: poorer South Africans live among other poorer South Africans. What matters therefore is the relative location of neighbourhoods: whether poorer South Africans are exposed to wealthier South Africans in nearby neighbourhoods, and vice versa. In the following section, some of the problems with measuring residential segregation are discussed.

7.4 Residential Segregation

A lot of the existing research on residential segregation has been limited by its reliance on measures that do not adequately capture space. Several papers have pointed out that the most widely used measures—the dissimilarity index, the exposure index, the variance ratio index, and the entropy-based information index—are essentially aspatial (Grannis 2002; Massey and Denton 1988; Morrill 1991; Reardon and O'Sullivan 2004; Wong 1993, 2002). Concretely, the papers point out two problems. The first is the checkerboard problem (Morrill 1991; White 1983), which arises when measures of residential segregation do not take into consideration the relative location of neighbourhoods. To illustrate, consider a checkerboard where each square represents a neighbourhood. The neighbourhoods are all homogeneous, but they are surrounded by others with different characteristics, which means they are exposed to variation. Now, imagine rearranging the checkerboard with all of the black squares on one side and all of the white squares on the other. Still the neighbourhoods are homogeneous but, in addition, most of them are now surrounded by neighbourhoods with similar characteristics. A meaningful measure of residential segregation should change when we rearrange the checkerboard. Aspatial measures do not.

The second problem is the modifiable areal unit problem (MAUP). The MAUP arises because the data that is often used in neighbourhood research is collected, aggregated, and reported for spatial units. In the case of census data, this would be the census tracts. This implies that two households at opposite ends of the same spatial unit are presumed to be closer than two households across the road from each other but located in different spatial units. In South Africa, this is a pronounced problem, as census tracts and other spatial units used in the census are often demarcated by roads. Again, a meaningful measure of residential segregation should capture this dimension. Aspatial measures do not do this.

Reviewing the most widely used measures of residential segregation, Reardon and O'Sullivan (2004: 124–125) suggest that two primary dimensions of segregation are identified: (1) spatial exposure (or isolation), and (2) spatial evenness (or clustering). Spatial exposure, they state, refers to 'the extent that members of one group encounter members of another group (or their own group, in the case of spatial isolation) in their local spatial environments'; while spatial evenness, or clustering, refers to 'the extent to which groups are similarly distributed in residential space'.

To demonstrate the difference between spatial exposure (or isolation) and spatial evenness (or clustering), Reardon and O'Sullivan present a diagram, recreated in Fig. 7.1. Imagine the white dots represent poorer neighbourhoods, while the black dots represent wealthier neighbourhoods. The horizontal axis captures the degree of spatial exposure (or isolation), while the vertical axis captures the degree of spatial evenness (or clustering). The result is four distinct patterns capturing four distinct types of residential segregation. In the upper half, the wealthier neighbourhoods are evenly distributed, while in the lower half, they are concentrated in larger or

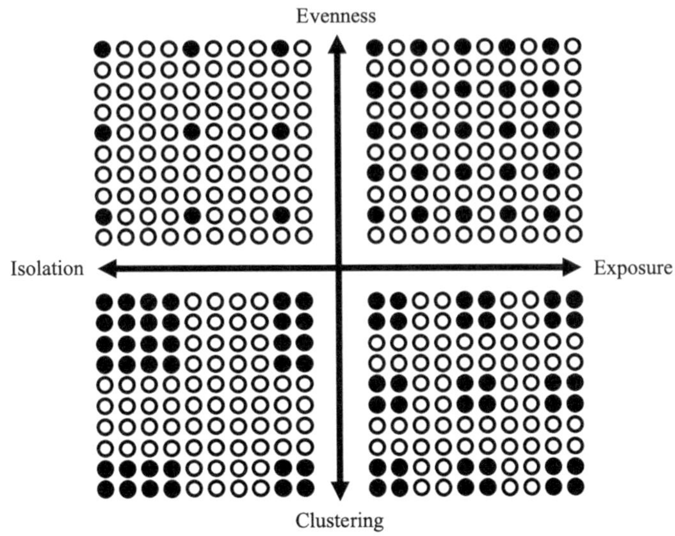

Fig. 7.1 Dimensions of segregation. *Source* Adapted from Reardon and O'Sullivan (2004)

smaller clusters. On the left side, the wealthier neighbourhoods are generally isolated, while on the right side they are exposed to poorer neighbourhoods.

Again, the focus of this study is on local exposure to economic inequality. This means the measures adopted must take into consideration both dimensions of residential segregation and minimise both the checkerboard problem and the MAUP. Reardon and O'Sullivan (2004: 124) suggest that 'a segregation measure that used information on the exact locations of individuals and their proximities to one another in residential space could eliminate the checkerboard problem and MAUP issues entirely'. Unfortunately, such fine-grained, geo-referenced data is unavailable in South Africa. Instead, information about the income distribution in neighbourhoods has been used, which is represented by census tracts; the locations of neighbourhoods; and their proximities to one another. This, however, does not fully address the checkerboard problem and the MAUP, as discussed later. However, it does allow an estimation of primary exposure: the likelihood that the residents in a neighbourhood will encounter residents from another neighbourhood belonging to a different economic class.

7.5 Economic Inequality and Residential Segregation in South Africa

In this chapter, a general measure of local exposure to economic inequality is created. It is calculated and mapped specifically for the City of Johannesburg. This is done to demonstrate the measure. However, calculating and mapping the measure for the City of Johannesburg generate interesting empirical findings in their own right. To contextualise these findings, in the following, economic inequality and residential segregation in South Africa is briefly discussed, drawing on literature and descriptive statistics.

South Africa had a Gini coefficient of 0.67 in 2018 (World Bank 2023), one of the highest in the world. The wealthiest 10% of the population, mainly white, earn about 50% of the national income, while the poorest 10%, mainly black, earn less than 0.1%. Inequality in assets is even higher, and intergenerational mobility is low, meaning inequalities are passed down from generation to generation with little change over time. A recent paper from the Southern Centre for Inequality Studies estimates that the wealthiest 10% of the population own 86% of the national wealth and that approximately one-third of this is owned by the wealthiest 0.1% (Chatterjee et al. 2020).

Economic inequality is starkest in South Africa's larger cities (metropolitan municipalities). Images of affluent, leafy neighbourhoods separated from destitute, informal settlements only by a fence or a wall have popularised the idea that South Africans from different ends of the income distribution scale live right next door to each other. While the images certainly capture the inequalities of the country, and while there are some places where economic inequality is this visible, the reality is that most South Africans live in racially, socially, and economically homogeneous neighbourhoods, relatively speaking. Moreover, both South Africa's racist past and ongoing racialised inequality have ensured that, for elites, in particular, there is little need to be anywhere near the poverty that undergirds the city.

Intra-neighbourhood homogeneity and inter-neighbourhood inequality are legacies of South Africa's apartheid past. People of colour (black African, coloured, and Indian)[4] were removed from designated 'white areas' and forced into townships, usually on the margins of the cities. Rivers, mine dumps, highways, railways, and industrial zones were used as barriers, ensuring that movement in and out of the townships, and into 'white' suburbs and cities, was kept under control. As urbanisation increased, more townships were created to contain especially the Black African population, who were prohibited from owning property and starting most types of businesses and were required to carry an internal passport (the hated 'dompas') to move around.

The transition to democracy in 1994 paved the way for a reform of cities. The 1996 Constitution gave the state powers to expropriate land. However, in line with the political settlement, little was subsequently done to address the spatial inequalities and injustices of the past and reform South African cities. The government adopted a market-oriented approach to land reform. Poverty was addressed through large social welfare programmes and investments in public services and infrastructure. However, there was no integrated approach to spatial transformation in cities, with government and the ruling African National Congress (ANC) seemingly more concerned with developing the country's rural hinterlands (Ebrahim and Everatt 2023).

Figure 7.2 plots annual household income across the City of Johannesburg. Each dot on the map represents 20 households. Race (whose patterns are known, and which are sadly predictable) has been deliberately ignored and this paper exclusively looks at annual household income. Households have been trichotomised into the 'poor' (ranging from no income to ZAR 76,400/USD 3960 per annum), a 'middle-income' group (between ZAR 76,401/USD 3961 and ZAR 614,400/USD 31,850), and the 'wealthy' (ZAR 614,001/USD 31,851 or greater annual income). These are rough and ready divisions, for the purpose of showing general distribution. Light grey dots represent poor households, darker grey are the middle-income households, and black dots are the wealthy households.

Immediately visible (and reflecting the City of Johannesburg's apartheid past and current situation) are the wealthier north and the poorer south. The two large concentrations of poor households that are in the north are found in an apartheid-era township, Ivory Park, and a more recent informal settlement-turned-township, Diepsloot. Much of the light grey scattering within the otherwise darker grey and black northern suburbs represents poorly paid (but locally resident) cleaners, gardeners, and others who attend to the wealthy. The remainder are either the middle-income or the wealthy, with Sandton containing some of the most expensive real estate in Africa and housing some of Johannesburg's wealthiest residents.

The very large light grey concentration that bisects the lower portion of the map (i.e. a concentration of poorer households) represents Soweto, the largest township in South Africa, with Kagiso further to the west. The light grey, however, continues further south to include large informal settlements, such as Orange Farm. The west to east line of lighter grey that bisects Johannesburg (just above Soweto) is the old gold mining belt, which gave rise to Johannesburg's existence, with many townships and informal settlements running along the belt (these would have been controlled labour pools under apartheid, many are now re-mining old mine dumps and mines).

---

[4] Race classifications currently used by Statistics South Africa (Stats SA) have been used.

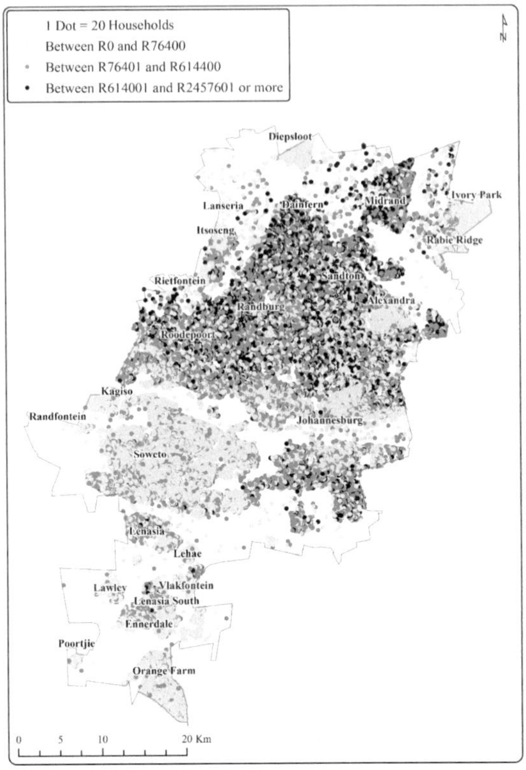

Fig. 7.2 Annual household income distribution. *Source* Authors 2021

Using a variable such as annual household income reveals how a homogenous city can be pictured, viewed at city scale; the heterogeneity in the north sees the middle income and wealthy classes living cheek-by-jowl, as do the poor in the south. But beneath the broad sweep, the appearance of the poor, middle-income, and wealthy households in all parts of the city beg for analysis at a smaller scale than the city alone.

7.6 Creating a Measure of Local Exposure to Economic Inequality

A description of the measure of local exposure to economic inequality is given here. To create the measure, the following steps were taken. First, a data set containing information about household income in every neighbourhood in the city of Johannesburg was compiled. Data was extracted from the 2011 census.[5] The small area layer (SAL), which is the smallest geographical unit for which census data is available was used. The SAL uses census tracts, known in South Africa as enumeration areas (EA). Where EAs have a population below 500, they are merged with neighbouring EAs to protect the identity of the individuals. The result is a data set with information about household income in 5800 SAL polygons in the City of Johannesburg.

Second, concentric circles were drawn around each of 5800 SAL polygons (represented by their centroids) at a radius of 500 m, 1000 m, and 2000 m, and aggregated the data within each circle. As explained above, the SAL polygons are census tracts (or merged tracts), demarcated on the basis of population density and administrative and logistical considerations. This means they vary in size. Polygons with higher population densities tend to be smaller. By drawing concentric circles

[5] A census was completed in 2022 but the data is not yet available.

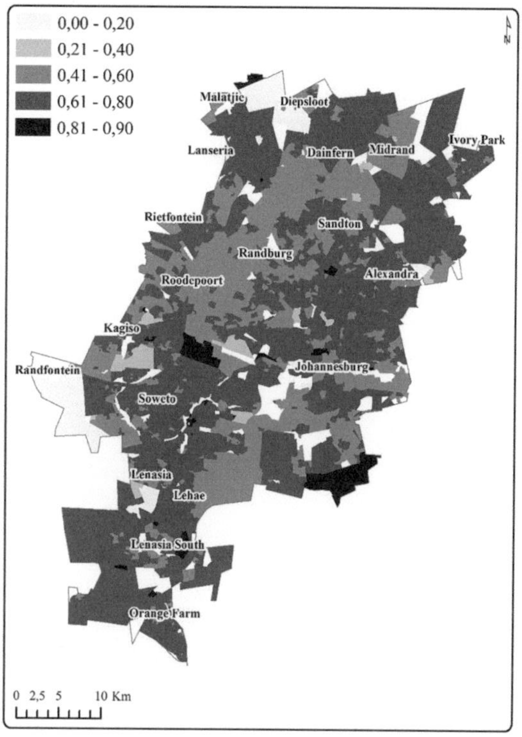

Fig. 7.3 Local exposure to income inequality (500 m radius around neighbourhoods). *Source* Authors 2021

around the polygons, the degree of economic inequality was captured, not just within neighbourhoods but also in nearby neighbourhoods and hence the local exposure.

Third, the Gini-like coefficients for each of the 5800 SAL polygons were calculated. The 2011 census contains information about household income but only as income categories. For each of the SAL polygons it is known how many households have an annual income between ZAR 0 and 4800 (USD 0 and 250), between ZAR 4801 and 9600 (USD 250 and 500), between ZAR 9601 and 19,600 (USD 500 and 1000), and so on. To calculate the coefficients, it was assumed that all of the households in each income category had an income equal to the midpoint of that category. This is obviously suboptimal. However, without information about the distribution within each category, it seems the most defensible solution.

Fourth, the coefficients were plotted, using heat maps. They are provided in Figs. 7.3, 7.4 and 7.5. The maps capture the local exposure to economic or specifically income inequality in each neighbourhood in the City of Johannesburg. Light grey means low exposure; dark grey means high exposure. Note that the exposure bands differ for each of the maps.

7.7 Empirical Findings

The three maps provide a fascinating sequence which is analysed below. Taken either individually or together, the maps suggest that high local exposure to economic inequality is most common in poor areas. Where the household income map suggested homogeneity in poorer areas and heterogeneity in better-off and wealthy areas, the choropleth maps of inequality show how the middle classes are cocooned in neighbourhoods with low population density and high homogeneity. Their exposure to different (lower) income groups is very low; neighbourhoods are dispersed because of low densities; and when traversing neighbourhoods, middle-class people are mainly being exposed to other middle-class people. The opposite is true of poorer areas, which

Fig. 7.4 Local exposure to income inequality (1000 m radius around neighbourhoods). *Source* Authors 2021

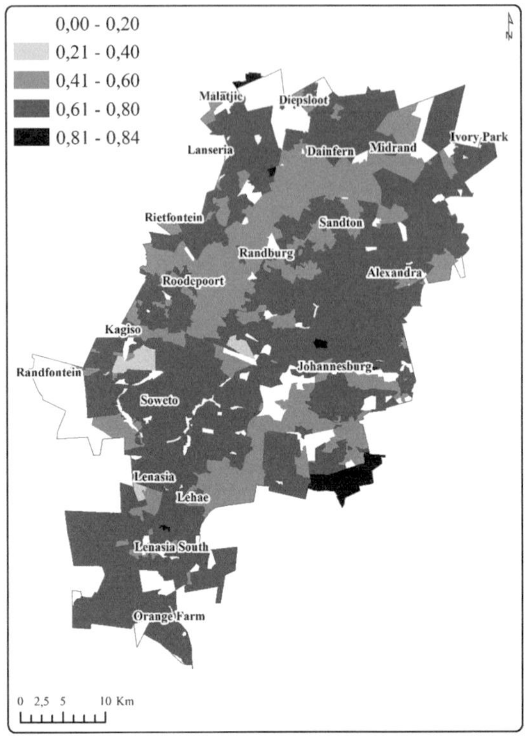

Figure. 7.5 Local exposure to income inequality (2000 m radius around neighbourhoods). *Source* Authors 2021

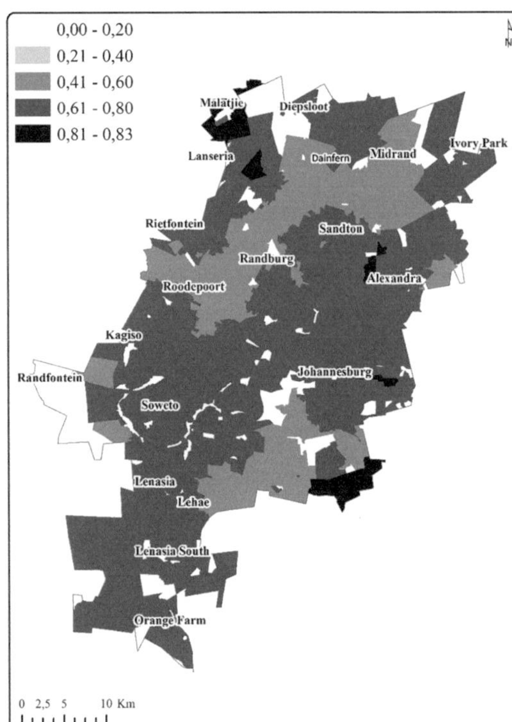

are made up of smaller plots with higher densities, so there is a proximate need to traverse many smaller neighbourhoods, encountering more or less poor people in nearby neighbourhoods.

Turning first to the final (2 km radius) map, which tends to emphasise city-wide patterns because of the 2 km distance, the pattern is clear: people living in the affluent northern suburbs of Johannesburg are least likely in the city to come into contact with people of a different class status where they live. Their exposure score is very low, and they live in and near consistently well-off areas. It should be noted that these suburbs tend to have houses on very large plots, with long distances to drive before encountering areas like Ivory Park or Diepsloot, the closest townships. The suburbs are cocooning residents in spaces of affluence, with little exposure to inequality. The pattern is less clear but still holds in the 500 m and 1 km radius maps, which show that some exposure does occur, although it is very slight, moving between the lowest (0–0.52) and second lowest (0.53–0.59) levels.

So, the first key finding is that well-off residents of Johannesburg's northern suburbs are least likely to be troubled by exposure to inequality. When this does occur, it is on the low side when measured for Johannesburg as a whole. There are other small pockets of low exposure dotted around the city, but they do not form a coherent pattern. The swathe of suburbia that stretches from Roodepoort to Midrand is home (as seen earlier) to affluent residents and who, as noted, are likely untouched by exposure to inequality.

Even as suburbia deracialises, it does so within clear class lines. The poor are not moving to the former white northern suburbs, but affluent Africans, coloureds and Indians are doing so. As a result, while exposure to people of different races may have increased at local level—in the form of neighbours—exposure to different classes barely occurs beyond domestic staff. The people who most benefited from apartheid are most shielded from the day-to-day challenges of democracy.

In all the maps, certain areas remain stubbornly dark (high exposure to inequality), namely townships and informal settlements. These include Diepsloot and Ivory Park (as above), Soweto, Orange Farm, Weilers Farm, Lenasia (an area designated for Indians under apartheid and now the site of significant informality and associated poverty), Alexandra, and virtually all townships: Tokoza, Evaton, Sebokeng, Boipatong, Katlehong, Vosloorus, Sharpeville, and more (not all are labelled in the map). Before analysing this in any detail, we should be clear that the poorer (and poorest) parts of the city are home to the greatest exposure to inequality where people live, be it measured at 500 m, 100 m or 2000 m.

This is a remarkable finding. Rather than being sites of undifferentiated and unchanging poverty, the townships and some informal settlements are home to change and vibrancy. Exposure to economic inequality is very high: and it is exposure to inequality, not sameness. This is not universally true, however: informal settlements, especially large ones such as Orange Farm and Diepsloot (which include formal and informal areas) and others which can be found on the fringes of virtually all formal areas (barring suburbia), show an undifferentiated but high exposure. That is, there are many encounters occurring, but like is encountering like, mirroring the experience of those in suburbia at the opposite end of the scale. Once inside the townships (and other informal settlements, as well as the city centre and elsewhere), the encounters have a difference, suggesting that very different energies are finding expression in these spaces.

A recent study (Everatt et al. 2023) of the city reached a similar conclusion—using standard survey methodology—finding that behind the well-known facades of the built environment exists the matchbox township house or the shack in an informal settlement.

> ... the experiences of neighbourhood types despite differences in race, cut across racial lines, such that the neighbourhoods (places) themselves become characteristic of class divisions. These appear to reflect the way large groups of people are treated by the state, or the way that large geographical places are met with the infrastructure of the state, to produce very similar outcomes in neighbourhoods all across the city. (Everatt et al. 2023: 41)

The study noted the need for granular data to unpack what the survey hinted at, and which this method (despite using 2011 census data) has now provided.

Look at the 500 m radius map. It shows, for example, the extent to which many residents of Lenasia—surrounded by Soweto's growth, and informal settlements such as Thembelihle—are also cocooned in their former Indians-only (and relatively affluent) township. Immediately around Lenasia, in Soweto, which has been the recipient of large investments by post-apartheid governments, the shade changes, and exposure to inequality rises steeply. Within Lenasia, the situation resembles suburbia.

But looking at Soweto in the 500 m map, it is seen that it is a nuanced picture. Soweto, with a population of some 1.8 million, boasts a robust black middle class, a powerful internal residential property market, and there appears to be a degree of homogeneity present. The fringes of Soweto—which is where informal settlements have taken root—show the darkest grey, but Soweto itself shows mid-range hues of grey. This suggests that there is a high degree of class homogeneity, and that within 500 m, exposure is more likely to be of someone of a similar class than elsewhere. This is clearly suburb-specific within Soweto (Fig. 7.6).

Equally fascinating is Orange Farm. This is one of the largest informal settlements in the province, let alone the city of Johannesburg, but is also the site of much service delivery, upgrading, and associated activity. As noted above, if Orange Farm is approached with expectations based on its informal status—the way it looks, in short—it should present a picture of solid dark: an unremitting sequence of high exposure (because of densities and small SALs) to sameness. But when smoothed out in the 2 km radius map, Orange Farm emerges in mid-range hues of grey, representing the mid-category (of the five available). But this apparent homogeneity of exposure is a function of the map: in the 500 m and 1000 m maps, a far more variegated picture emerges. In this massive, sprawling informal settlement which is going through the process of formalising there is no sea of unchanging poverty, and differentiation is more apparent than homogeneity. Encounters with economic inequality are unpredictable, and the 500 m radius map shows Orange Farm ranging from virtually no encounter with inequality to very high encounters—all within the same settlement. Again, behind the label—site and service, informal settlement, and so on—is a level of activity reflected in the uneven exposure to inequality.

This is not to ignore the negative results. The dark colour represents the highest local exposure to income inequality in newer informal areas such as Diepsloot, and in the informal fringes of virtually every township (most clearly shown by Alexandra township, and the Johannesburg inner-city centre, in all the maps). These spaces are commonly presented in the media as areas of unchanging poverty as well as crime, drugs use, gangs and criminality. The maps all suggest that this is not necessarily incorrect: there are high levels of encounter with unchanging sameness. There may be a high rate of encounter, but these are *not* encounters between neighbourhoods with different Gini-like coefficients.

The method has thrown up some fascinating findings. Most obviously, local context matters: one informal settlement may show high exposure to sameness, another might reflect high exposure to neighbourhoods with different exposure coefficients. In spaces such as Diepsloot, for example, there is a high exposure to undifferentiated neighbourhoods, as there is in some other informal areas. But other informal areas, such as Orange Farm, show high exposure to economic inequality across a range of Gini scores. These areas may include new migrant arrivals as well as spill-over from existing households; regardless of this, the high exposure to economic inequality suggests that these areas are home to people with a wide-ranging mix of economic statuses. Townships, on the whole, represent nuanced sites and the need to navigate a range of socio-economic statuses. Soweto is settling into a staid, class-based suburb, with mid-range exposure to inequality.

The engines of change and growth seem to lie in areas with high exposure to neighbourhoods with different Gini coefficients, which house residents across a wide range of socio-economic statuses, where people are both 'making it' and failing to do so or surviving on social grants, and live cheek by jowl. Over time, the more economically successful may move to townships such as Soweto—only the most successful can migrate to the pricey northern

Fig. 7.6 Local exposure to income inequality in Soweto. *Source* Authors 2021

Panel A:
500m radius around neighbourhoods

Panel B:
1,000m radius around neighbourhoods

Panel C:
2,000m radius around neighbourhoods

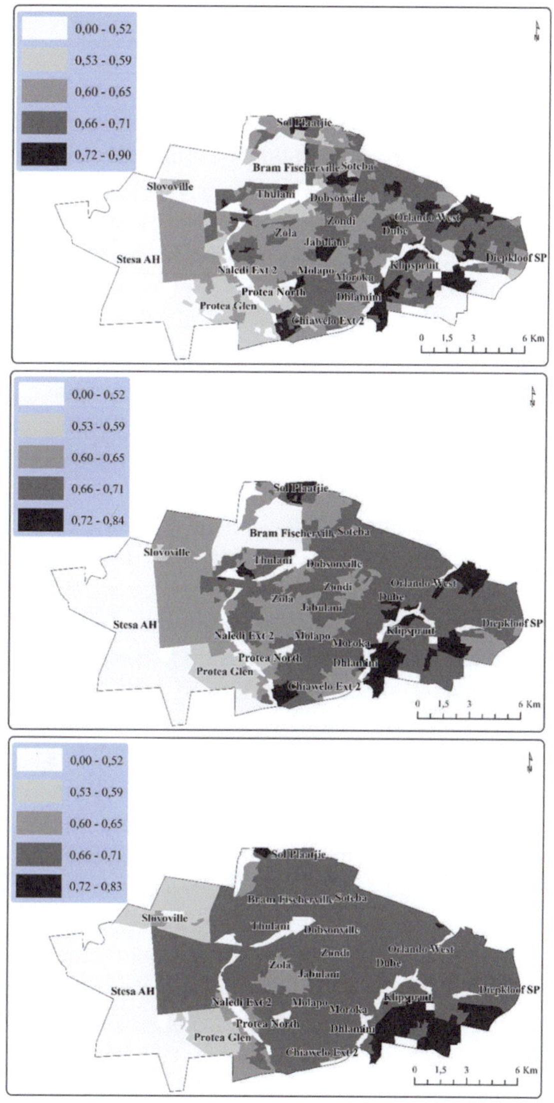

suburbs—but for now, they (and their energies) are locked in the spaces they can afford, or choose to afford to live in.

The methodology seems to have worked and provides the nuance and granularity that other (mainly survey-based) studies have called for. The methodology needs to be re-run on Census 2022 data, to bring everything up-to-date, and to check if the patterns still hold.

This is a clear if an unexpected outcome of this approach: socio-economic change is happening where exposure to inequality is highest. The literature suggests (as cited above) that economic inequality can lead to all sorts of social ills. Here what is seen is that high exposure to economic inequality is not happening in wealthy suburbs or settled townships. Rather, it is to be found in more recently established, mainly informal areas, where hustling, energetic residents are striving to attain the status enjoyed by those in suburbs and townships. They are driving change—as their built environment has brought change to their surrounds—and the socio-economic energies of Johannesburg are likely to be found in these spaces.

7.8 Conclusion

A first attempt at measuring local exposure to income inequality has been presented in this study. The measure partially addresses the checkerboard problem and the MAUP by capturing the income inequality within neighbourhoods as well as between nearby neighbourhoods. However, the measure has three shortcomings:

> First, it assumes all households in each neighbourhood are located at the centroid. A more refined measure would capture the distribution of households ideally, by overlaying census data with satellite imagery.
> Second, the measure does not fully capture the distance between neighbourhoods. Presumably, exposure to inequality within or against nearby neighbourhoods has a greater effect on human behaviour than exposure to inequality against neighbourhoods further away. This study captures this by drawing concentric circles around each neighbourhood with a radius of 500 m, 1000 m, and 2000 m. However, these radiuses are arbitrary and do not capture the differences, for example, between 100 and 500 m, between 501 m and 1000 m, or between 1001 m and 2000 m. Again, a more refined measure would weigh these distances and capture the declining exposure effect.
> Third, the measure does not take barriers into consideration. As explained, rivers, mine dumps, highways, railways, industrial zones, etc., are an important feature of South Africa and shape movement, especially in and out of the former townships. The barriers could be factored, again, by overlaying census data with satellite imagery and plotting 'hard' boundaries that restrict exposure.

Finally, the measure captures income inequality rather than economic inequality, which would also have to take into consideration asset ownership. The South African census captures asset ownership, to some extent, by asking whether households own refrigerators, washing machines, computers, motorcars, televisions, cell phones, etc. A more refined measure would incorporate asset ownership, although the inclusion in our Gini-like calculations would be complicated. One solution could be to convert each asset into a monetary value and use the sum of values as a proxy of household wealth. Future research will explore this solution further.

References

Abbott A (1997) Of time and space: the contemporary relevance of the Chicago school. Soc Forces 75(4):1149–1182. https://www.jstor.org/stable/2580667

Alesina A, Rodrik D (1994) Distributive politics and economic growth. Q J Econ 109(2):465–490. https://doi.org/10.2307/2118470

Banerjee AV, Duflo E (2003) Inequality and growth: what can the data say? J Econ Growth 8:267–299

Barro RJ (2000) Inequality and growth in a panel of countries. J Econ Growth 5(1):5–32

Bénabou R, Tirole J (2006) Belief in a just world and redistributive politics. Q J Econ 121(2):699–746. https://doi.org/10.1162/QJEC.2006.121.2.699

Berg AG, Ostry JD (2011) Inequality and unsustainable growth: two sides of the same coin? SDN/11/08; IMF Staff Discussion Note. https://www.imf.org/external/pubs/ft/sdn/2011/sdn1108.pdf

Burns JK, Tomita A, Kapadia AS (2014) Income inequality and schizophrenia: increased schizophrenia incidence in countries with high levels of income inequality. Int J Soc Psychiatry 60(2):185–196. https://doi.org/10.1177/0020764013481426

Chatterjee A, Czajka L, Gethin A (2020) Estimating the distribution of household wealth in South Africa. 2020/45; WIDER Working Paper. https://doi.org/10.35188/UNU-WIDER/2020/802-3

Côté S, House J, Willer R (2015) High economic inequality leads higher-income individuals to be less generous. Proc Natl Acad Sci USA 112(52):15838–15843. https://doi.org/10.1073/pnas.1511536112

Davidai S (2018) Why do Americans believe in economic mobility? Economic inequality, external attributions of wealth and poverty, and the belief in economic mobility. J Exp Soc Psychol 79:138–148. https://doi.org/10.1016/j.jesp.2018.07.012

Delhey J, Dragolov G (2014) Why inequality makes Europeans less happy: the role of distrust, status anxiety, and perceived conflict. Eur Sociol Rev 30(2):151–165. https://about.jstor.org/terms

Delmelle EC (2015) Five decades of neighborhood classifications and their transitions: a comparison of four US cities, 1970–2010. Appl Geogr 57:1–11. https://doi.org/10.1016/j.apgeog.2014.12.002

Delmelle EC (2017) Differentiating pathways of neighborhood change in 50 U.S. metropolitan areas. Environ Plan A 49(10):2402–2424. https://doi.org/10.1177/0308518X17722564

Dikeç M (2001) Justice and the spatial imagination. Environ Plan A: Econ Space 33(10):1785–1805. https://doi.org/10.1068/a3467

Easterly W (2007) Inequality does cause underdevelopment: insights from a new instrument. J Dev Econ 84(2):755–776. https://doi.org/10.1016/j.jdeveco.2006.11.002

Ebrahim Z, Everatt D (2023) Politics, planning and the future of the Gauteng city-region. Urbanisation. https://doi.org/10.1177/24557471231165431

Everatt D, Abrahams C, Lynge H (2020) Johannesburg: the growing importance of class in shaping neighbourhoods. www.centreforsustainablecities.ac.uk

Everatt D, Lynge H, Abrahams C (2023) Neighbourhood characteristics and inequality in the city of Johannesburg: a preliminary analysis

Galster GC (2001) On the nature of neighbourhood. Urban Studies 38(12):2111–2124. https://doi.org/10.1080/00420980120087072

Galster GC (2012). The mechanism(s) of neighbourhood effects: theory, evidence, and policy implications. In: Neighbourhood effects research: new perspectives. Springer, Dordrecht, pp 23–56. https://doi.org/10.1007/978-94-007-2309-2_2

Grannis R (2002) Discussion: segregation indices and their functional inputs. Sociol Methodol 32(1):69–84

Hincks S (2015) Neighbourhood change and deprivation in the greater Manchester city-region. Environ Plan A 47(2):430–449. https://doi.org/10.1068/a130013p

Hincks S (2017) Deprived neighbourhoods in transition: divergent pathways of change in the Greater Manchester city-region. Urban Stud 54(4):1038–1061. https://doi.org/10.1177/0042098015619142

Hsieh C-C, Pugh MD (1993) Poverty, income inequality, and violent crime: a meta-analysis of recent aggregate data studies. Crim Justice Rev 18(2):182–202. https://doi.org/10.1177/073401689301800203

Israel E, Frenkel A (2018) Social justice and spatial inequality: toward a conceptual framework. Prog Hum Geogr 42(5):647–665. https://doi.org/10.1177/0309132517702969

Kane JV, Newman BJ (2019) Organized labor as the new undeserving rich?: Mass media, class-based anti-union rhetoric and public support for unions in the United States. Br J Polit Sci 49(3):997–1026. https://doi.org/10.1017/S000712341700014X

Kondo N, Sembajwe G, Kawachi I, van Dam RM, Subramanian SV, Yamagata Z (2009) Income inequality, mortality, and self rated health: meta-analysis of multilevel studies. BMJ 339(nov10 2):b4471–b4471. https://doi.org/10.1136/bmj.b4471

Kuziemko I, Norton MI, Saez E, Stantcheva S (2015) How elastic are preferences for redistribution? Evidence from randomized survey experiments. Am Econ Rev 105(4):1478–1508. https://doi.org/10.1257/aer.20130360

Lynge H, Visagie J, Scheba A, Turok I, Everatt D, Abrahams C (2022) Developing neighbourhood typologies and understanding urban inequality: a data-driven approach. Reg Stud Reg Sci 9(1):618–640. https://doi.org/10.1080/21681376.2022.2132180

Massey DS, Denton NA (1988) The dimensions of residential segregation. Soc Forces 67(2):281–315. https://academic.oup.com/sf/article/67/2/281/2231999

McCall L, Burk D, Laperrière M, Richeson JA (2017) Exposure to rising inequality shapes Americans' opportunity beliefs and policy support. Proc Natl Acad Sci USA 114(36):9593–9598. https://doi.org/10.1073/pnas.1706253114

McLennan D, Noble M, Wright G (2016) Developing a spatial measure of exposure to socio-economic inequality in South Africa. S Afr Geogr J 98(2):254–274. https://doi.org/10.1080/03736245.2015.1028980

Messias E, Eaton WW, Grooms AN (2011) Economic grand rounds: income inequality and depression prevalence across the United States: an ecological study. Psychiatr Serv 62(7):710–712. https://doi.org/10.1176/ps.62.7.pss6207_0710

Mishra S, Son Hing LS, Lalumière ML (2015) Inequality and risk-taking. Evol Psychol 13(3):1–11. https://doi.org/10.1177/1474704915596295

Morrill RL (1991) On the measure of geographic segregation. Geogr Res Forum 11(1):25–36

Nishi A, Shirado H, Rand DG, Christakis NA (2015) Inequality and visibility of wealth in experimental social networks. Nature 526(7573):426–429. https://doi.org/10.1038/nature15392

Oishi S, Kesebir S, Diener E (2011) Income inequality and happiness. Psychol Sci 22(9):1095–1100. https://doi.org/10.1177/0956797611417262

Payne BK, Brown-Iannuzzi JL, Hannay JW (2017) Economic inequality increases risk taking. Proc Natl Acad Sci USA 114(18):4643–4648. https://doi.org/10.1073/pnas.1616453114

Persson T, Tabellini G (1994) Is inequality harmful for growth? Am Econ Rev 84(3):600–621

Philippopoulos-Mihalopoulos A (2010) Spatial justice: law and the geography of withdrawal. Int J Law Context 6(3):201–216. https://doi.org/10.1017/S174455231000011X

Pickett KE (2005) Wider income gaps, wider waistbands? An ecological study of obesity and income inequality. J Epidemiol Community Health 59(8):670–674. https://doi.org/10.1136/jech.2004.028795

Pickett KE, Wilkinson RG (2015) Income inequality and health: a causal review. Soc Sci Med 128:316–326. https://doi.org/10.1016/j.socscimed.2014.12.031

Piketty T, Saez E (2014) Inequality in the long run. Science 344(6186):838–843. https://doi.org/10.1126/science.1251936

Pirie GH (1983) On spatial justice. Environ Plan A: Econ Space 15(4):465–473. https://doi.org/10.1068/a150465

Posel DR, Casale DM (2011) Relative standing and subjective well-being in South Africa: the role of perceptions, expectations and income mobility. Soc Indic

Res 104(2):195–223. https://doi.org/10.1007/s11205-010-9740-2
Reardon SF, O'Sullivan D (2004) Measures of spatial segregation. Sociol Methodol 34:121–162. https://doi.org/10.1111/j.0081-1750.2004.00150.x
Sands ML, de Kadt D (2020) Local exposure to inequality raises support of people of low wealth for taxing the wealthy. Nature 586(7828):257–261. https://doi.org/10.1038/s41586-020-2763-1
Schneider SM (2016) Income inequality and subjective wellbeing: trends, challenges, and research directions. J Happiness Stud 17(4):1719–1739. https://doi.org/10.1007/s10902-015-9655-3
Sen A (1995) Inequality reexamined. Oxford University Press. https://doi.org/10.1093/0198289286.001.0001
Sen A (2008) The idea of justice. J Hum Dev 9(3):331–342. https://doi.org/10.1080/14649880802236540
Soja E (2009) The city and spatial justice. Justice Spatiale/Spat Justice 1(1):1–5
Solt F (2008) Economic inequality and democratic political engagement. Am J Polit Sci 52(1):48–60
Solt F (2010) Does economic inequality depress electoral participation? Testing the Schattschneider hypothesis. Polit Behav 32(2):285–301. https://doi.org/10.1007/s11109-010-9106-0
Soudien C, Reddy V, Woolard I (2019) South Africa 2018: the state of the discussion on poverty and inequality. In: Soudien C, Reddy V, Woolard I (Eds.), The state of the nation: poverty & inequality: diagnosis, prognosis and responses (pp. 1–28). HSRC Press. https://repository.hsrc.ac.za/handle/20.500.11910/13577
Subramanian SV, Kawachi I (2006) Whose health is affected by income inequality? A multilevel interaction analysis of contemporaneous and lagged effects of state income inequality on individual self-rated health in the United States. Health Place 12(2):141–156. https://doi.org/10.1016/j.healthplace.2004.11.001
Sulla V, Zikhali P, Cuevas PF (2022) Inequality in Southern Africa. http://documents.worldbank.org/curated/en/099125303072236903/P1649270c02a1f06b0a3ae02e57eadd7a82
Trump K-S (2018) Income inequality influences perceptions of legitimate income differences. Br J Polit Sci 48(4):929–952. https://doi.org/10.1017/S0007123416000326
Trump K-S (2021) Public opinion and reactions to increasing income inequality. In: Rosenbluth FM, Weir M (eds) Who gets what? The new politics of insecurity. Cambridge University Press, pp 29–79
Uslaner EM, Brown M (2005) Inequality, trust, and civic engagement. Am Politics Res 33(6):868–894. https://doi.org/10.1177/1532673X04271903
White MJ (1983) The measurement of spatial segregation. Am J Sociol 88(5):1008–1018
Wilkinson RG (1992) Income distribution and life expectancy. BMJ 304(6820):165–168. https://doi.org/10.1136/bmj.304.6820.165
Wilkinson RG, Pickett KE (2009) Income inequality and social dysfunction. Ann Rev Sociol 35(1):493–511. https://doi.org/10.1146/annurev-soc-070308-115926
Wong DWS (1993) Spatial indices of segregation. Urban Studies 30(3):559–572. https://doi.org/10.1080/00420989320080551
Wong DWS (2002) Spatial measures of segregation and GIS. Urban Geogr 23(1):85–92. https://doi.org/10.2747/0272-3638.23.1.85
World Bank (2023) Overview. The World Bank in South Africa. https://www.worldbank.org/en/country/southafrica/overview

8 Socio-Spatial Inequality in Urban Neighbourhoods in Johannesburg: Place, Belonging, and Agency

Caryn Abrahams

Abstract

This chapter paper discusses socio-spatial inequalities in neighbourhoods from a qualitative perspective specifically examining three concepts which define place-making, namely, sense of place, belonging, and agency. The chapter argues that these concepts give a perspective of how people experience life in their neighbourhoods and how inequalities in neighbourhood experiences then relate to different qualities of life in the different types of neighbourhoods in Johannesburg, South Africa. The data for this chapter comes from rich, in-depth focus group discussions conducted by SHLC researchers, and the presentation of data is categorised by neighbourhoods made up of different income quintiles. The paper concludes that sense of place, belonging, and agency allow us to see inequality through a different lens, adding a nuance to neighbourhood scale research that balances the quantitative and morphological approaches that often characterise it.

C. Abrahams (✉)
Wits School of Governance, University of the Witwatersrand, Johannesburg, South Africa
e-mail: Caryn.Abrahams@wits.ac.za

Keywords

South Africa · Johannesburg · Social inequality · Spatial inequality

8.1 Introduction

This chapter considers socio-spatial inequality as it plays out in different neighbourhoods in Johannesburg, South Africa. According to Abrahams and Everatt (2019) Johannesburg is the most unequal city in the world and measures of inequality here as elsewhere are typically identified spatially across different urban areas in the larger Johannesburg area and the inner city. These measures include transport access, property prices, distance from the city centre and other economic hubs in the city and region, infrastructure provision, and in-service delivery disparities (Von Fintel 2018). Although there are several studies of spatial inequality in Johannesburg, and between Johannesburg and other metropolitan municipalities in the Gauteng city region (see Todes 2014; Everatt 2014; von Fintel 2018; Cole et al. 2018), there are few that go beyond specific indicators of inequality at neighbourhood level as defined through infrastructure and service provision. 'Spatial inequalities and policies in South Africa: Place-based or people-centred?' by Todes and Turok (2018) is an exception.

Y. P. Wang et al. (eds.), *New Drivers of Division*, Sustainable Development Goals Series,
https://doi.org/10.1007/978-981-96-5629-5_8

The chapter considers how spatial inequality plays out at the neighbourhood level, by zooming in on how local residents experience life in different neighbourhoods and public space. It does so by considering focus group discussions from five neighbourhood types and highlighting elements of neighbourhood inequality that plays out spatially. These provide data that would otherwise be subsumed within quantitative data approaches that compute inequality across different variables and through different statistical analyses. The approach in this paper does not seek to replace, but rather to sit alongside analyses drawn from the data that has a quantitative approach to understanding, reading, and analysing inequality. The approach in this paper also sits alongside other qualitative approaches to spatial inequality in neighbourhoods that seek to understand it through the built environment and morphological elements, such as green spaces, housing, access to public services, and this would fail to capture the way that residents engage with urban public life more broadly. For this reason, the focus in this chapter is not essentially on public amenities like parks, hospitals, schools, or libraries, but also on the experience of public life in a particular neighbourhood. While some commentary about the public amenities, services, and space is necessary and present here, it is used to enter into a conversation about what it means for belonging and agency in those neighbourhoods.

8.2 Spatial Inequality

Space, according to Han (2022) 'is constructed socio-economically', meaning that the inequalities that are built into everyday life are inscribed upon the places which people inhabit. Spatial inequality has typically been framed as the inequalities between places and regions (Abdul 2022), based on the quality of the built infrastructure (Rigolon et al. 2018), the level of services like health care (Bravo 2018) or water (Cole et al. 2018), access to green space (Chen et al. 2022), how space interacts with ethnicisation (Kanbur and Venables 2005), with gendered and queer identities (Chant 2013), and with the socio-economic status of residents in different areas (Everatt et al. 2020).

Crucially, measures of inequality relate not only to built environment, infrastructure, or economic access; they also relate to place-making: which is the qualitative sense of relational and associative elements which relate to belonging of place and agency in creating that space according to rights of self-determination (Kaplan 2023). Here, measures of place-making are about how different communities experience life in the city through their neighbourhoods, and whether they feel a sense of inclusion/belonging or agency in what happens in these neighbourhoods, and how these neighbourhoods develop (Gokce and Chen 2018; Switalski et al. 2023; Alkire 2008; Kaplan 2023; Wafer and Rameetse 2020).

This view has its roots in the 'right to the city' discourse (Harvey 2003). Place-making has re-emerged as an important conversation in urban studies in response to social anomie, ethnic polarisation, and increasing economic exclusion and marginalisation of the majority of city-dwellers (Kaplan 2023). This chapter therefore locates itself in a body of work that argues that the exercise of highlighting everyday experiences of urban residents opens up new pathways for understanding urban dynamics (Miraftab et al. 2015).

8.3 Methodological Considerations: Seeing from the Neighbourhood

In the South African case, the racist apartheid regime was built on separate developments for racial groups and the spatial fractures mirrored the socio-economic fissures, which were cyclically responsible for decades of dispossession, disenfranchisement, and economic suppression (Von Fintel 2018).

Neighbourhood level analyses in South Africa tend to focus on discussions of (disparate) access to services or participatory governance in local administrative areas as a way of understanding distributive justice (Everatt et al. 2020). There are few studies in South Africa that use the neighbourhood as a scale of analysis for understanding urban inequality (Lynge et al. 2022). This is problematic because local policy and infrastructure provisioning happens through ward level participatory processes, which is also the scale at which political governance happens at a local level. Analysis at the ward level does not take into consideration the disparities in neighbourhoods even in one ward, as wards are not made up of homogeneous neighbourhoods. As Everatt (2014: 73) has argued '... [wards] are not an ideal unit of analysis, given that they are political/voting constructs and not developmental units, thus may include very different community types in a single ward'.

The neighbourhood level, as Lynge et al. (2022: 619) argue:

> has altered our conception of the city, depicting it as an agglomeration of different neighbourhoods that are constantly being changed by multi-scalar social, economic, political, and environmental processes. [Structural] inequalities in class, race, and gender manifest themselves across neighbourhoods and [these] influence the sustainability of cities.

Various approaches internationally have been used to define the neighbourhood as a socio-spatial unit in urban research, which may focus on the territorial boundaries of places which share infrastructure, public space, or shared sets of identities (Alkan and Maksudyan 2020; Baffoe 2019; Lynge et al. 2022). As social constructs, perceptions and subjective factors shape the meaning of neighbourhood and these depend on individuals' historical background and positionality. They may encompass fluid, multiple, and dynamic boundaries, including both physical and emotional elements or neighbourhood intimacies (Alkan and Maksudyan 2020; Kearns and Parkinson 2001). This paper, sitting alongside the chapter on inequalities ('Measuring Local Exposure to Economic Inequality in Johannesburg' by Lynge et al. 2024), uses the neighbourhood approach to explore dynamics on inequality in a different way, but seen from the neighbourhood scale. These chapters consider how urban analyses might be deepened by 'seeing from' the neighbourhood scale, particularly if life and experiences in these neighbourhoods are the focus. Understanding urban dynamics at the neighbourhood level will in turn serve to rethink urban governance and the sets of priorities that make urban life meaningful and sustainable.

8.3.1 Research Methods

The study on which this chapter is based used focus group discussions in Johannesburg drawn from a selection of neighbourhoods, as a result of a detailed a cluster analysis of urban data and development of neighbour typologies (see detailed methodology in Lynge et al. 2022, and the chapter on Measuring Local Exposure to Economic Inequality in the City of Johannesburg by Lynge, Everatt, and Katumba in this book). There were ten focus groups studied in South Africa, five each in Johannesburg and Cape Town. This paper uses material only from the Johannesburg focus groups. The methodology included identifying neighbourhood typologies by income quintiles, housing infrastructure patterns, and geographical location. The result was a series of neighbourhood typologies that could be reproduced across the city, which showed a more granular picture of spatial differences. Neighbourhood typologies in turn have been categorised according to neighbourhood descriptors in Table 8.1.

Focus groups comprised eight to ten respondents in each of the five study areas, made up of residents who had lived in the area for more than two years, and thus were familiar with the infrastructure experiences, public amenities and services, as well as the characteristics of the neighbourhood. Each focus group ensured a gender mix and a mix of those who had or did not have children of school-going age. The general themes covered in the focus groups related to health, education, and governance experiences in those neighbourhoods, and the elements that made living there desirable or undesirable. The focus group discussions in Johannesburg took place in January 2022.

Table 8.1 Neighbourhood types, neighbourhoods, and descriptions

Descriptors	Neighbourhoods	Neighbourhood type
Upper-income neighbourhood (UIN or Upper IN)	Weltervreden Park Windsor Glen Blairgowrie	Freestanding homes, often fully serviced, in gated communities (community funded boomed-off areas)
Upper middle-income neighbourhood (UMIN upper middle IN)	Paulshof Jackal Creek Golf Estate Paulshof	Cluster dwelling, in formal gated complexes
Middle-income neighbourhood (MIN middle IN)	Klipsruit West Noordgesig Eldorado Park	Freestanding, single dwelling housing units, in established townships[1]
Lower middle-income neighbourhood (LMIN lower middle IN)	Orlando West, Soweto	Post-1994 freestanding state-provisioned homes for mass distribution. matchbox style (less than 30 sq. m.) housing units
Low-income neighbourhood (LIN low IN)	Rabie Ridge Ext 4 Kaalfontein Ext 1 Mayibuye SP	Freestanding houses; informal dwellings

Source Author 2021

8.3.2 Conceptual Framing

The following themes around place-making frame the presentation of data and discussion:

1. Sense of Place—this includes what the attitudes and perceptions are about a particular place; whether residents identify with that place; are dependent on that place; and have a sense of attachment to that place (Gokce and Chen 2018; Wafer and Rameetse 2020).
2. Belonging—this includes relational elements including whether residents as members of the neighbourhood have a sense of belonging and inclusion: whether they have a sense of familiarity with neighbours; whether there is social bonding or associational strength; whether there is a sense of trust of neighbourhood members; and whether residents experience recognition (they feel seen, their lives valued, and are deemed to matter by external policy-makers (see Switalski et al. 2023; Gokce and Chen 2018; Alkire 2008).
3. Agency—this includes whether people have the freedom to engage in civic activity that allows for mutual thriving and neighbourliness; whether residents can advance the goals of a good life; whether they can exercise effective power to make choices that affect their well-being; and whether they exercise their responsibility to control procedures or processes that affect them (Alkire 2008).

8.4 Discussion

8.4.1 Sense of Place

Common among all neighbourhood types with reference to *Sense of Place*—those comprising households of the lowest IN, lower middle IN, middle IN, upper middle IN, and high IN—was nostalgia about the past, and the features of the neighbourhood. The neighbourhoods comprising households with the lowest income levels tend to be newer township areas and informal settlements and there was little nostalgia about the place, but about experiences of the past: 'It is not like the time when we went to primary school' (Participant 5, LIN).

In the established township areas, in Soweto particularly, in the first two quotes, there are vibrant memories of place and a fond nostalgia related to those historical places, and obvious comparisons with how it is now:

> For me, because I came to Soweto when I was four years old, so I grew up there. It was very vibrant and people were well focused in the areas of education [...] the world-famous magician, (Tita), who was from Orlando West. That was a very good society, but as the years went on from 1970 to 1980, it went into another direction. [...] We were talking about that, the community. I remember, once every three

[1] Townships are former group areas that were demarcated for distinct demographic groups away from the city centre.

> months, the Johannesburg Orchestra would come there. And the conductor, Edgar Cree, would come there to the Community Hall, issue tickets and that is how I got introduced to classical music and jazz. (Participant 8, LMIN)
> I would say that a long time ago, [...], it used to be a place where you would find so many activists who would be working towards employing the communities. But now, because we are so congested, things are not going the way they used to. (Participant 4, LMIN)

Those from the upper INs (UIN) related their deteriorating sense of place to quality of life and crime, not the characteristics or amenities of the neighbourhoods:

> I would like to say on that point even though the area where I am living is established and a fairly safe area, but at nighttime, I think very few areas in this country are safe. [...]And because of that the quality of life seems to have deteriorated a little because you tend not to go out at night anymore. You do not want to accept invitations to visit friends at night because of the fear of driving home at night [...] Yes, the area has gone down unfortunately. So yes, I definitely would not allow my kids out at night. (Participant 4, UIN)

Other participants from the upper IN agree:

Participant 1: I remember when I was a child we used to go there [Botanical Gardens] and play.
Participant 6: Or sokkies (dances).
Participant 1: Yes, not anymore, nothing. Nobody does nothing.
Moderator: When people have events like weddings then what happens?
Participant 3: They go to hotels.

The sense of place in these privileged spaces is not only about preservation of quality of life, but also about the preservation of historical spatial privilege.

Participant 3, UIN: I just feel like they keep on adding more and more developments. And there is low-cost development that is coming in as well, which also brings a new crowd into the area where you thought you were living in a better area that is safer. And then you see the low class coming in, and that creates a lot of problems too.

In lower-middle IN and middle IN, similar to the low IN, residents also report sense of place as a memory of the past replete with vibrancy, activism, and conviviality. Here, while they mention deteriorating neighbourhoods and services, there is yet a fondness of place and everyday life experienced in those places. In the most deprived neighbourhoods there is what can be termed *lament of resignation* which characterises residents' sense of place, yet some describe the actual neighbourhood as having such deplorable living and service conditions as pertaining to housing, environment, health care and infrastructure, that they choose not even to complain.

Participant 1, LIN: You must not write to complain. Now what is the solution? You have a lot of problems, but you can't moan [...] That is what is happening. The bridges have fallen, the streets have potholes [...] It is just from my sense of life; I just hate to complain. I just explained my situation. [...] Crime is everywhere. [...] if you are at the low point. You will go nowhere. You are just going to make a lot of noise unnecessarily [...] that this is just how it is.)

In other responses in the lower IN (LIN), residents lament the total lack of services in their neighbourhoods.

Participant 4: But on our side, Kaalfontein, we do not have any parks.
Participant 5: We do not have a playground. We have nothing.

Participant 4: We are still fighting for those things. We don't have those things. There are no big grounds.
Participant 5: Not even a library.
Participant 4: Nothing.
Participant 5: In our area, we don't have a library at all.
Participant 8: We also don't have it.
Participant 5: We only get to go to the library at Ivory 2 which is out–you would have to take a taxi.
Participant 6: It is difficult to walk to. The last time that I was there ... we didn't even have books in there. There is nothing there.

Similarly, this conversation in the lower middle IN (LMIN) describes the sense of place by what is absent:

Moderator: Tell me what facilities do we have around us?
Participant 6: The facilities that we have?
Moderator: Yes.
Participant 6: Or what we don't have?! [all participants laugh]

Even in middle-income township neighbourhoods there are deleterious characteristics of place. When asked to describe the characteristics of the neighbourhoods they live in, residents said:

> Bad, very bad, nightmare, more service, drug infested, high crime rate, no governance, poverty, unemployment, living in fear, lack of activities. (all participants, MIN)

Participant 7, MIN: This is what I'm saying: is my happy life in the township? During the day, like the lady said, you have to be a guard to watch your washing line. [...] because the social health in Eldorado Park is hell, the social problem is hell.

Participant 5, MIN: Even the kids like the place where they were supposed to be safe; it's somewhere they can play but they are not even safe there, so that means you as a parent are afraid to look for a job because you have to leave your home to go watch your child play in a public park and when you come back your house is broken into, so it's safe nowhere.

The picture is rather different in upper middle and upper INs. When asked to describe their sense of place, the characteristics of the place, and life in the neighbourhood, residents in the UMIN presented an idyllic, safe life, where the only complaints were the inconvenient times of garbage removal or grass-cutting. The situation is so idyllic and all-contained, it seems, that there is no reason to venture out of the secure estate neighbourhood.

Participant 8: Quiet, safe, clean, not a lot of noise.
Participant 7: But otherwise, we like a little family. Everyone knows everyone. So, if you cook mince and curry the whole of that side is going to be known. So, yes, for me, I definitely feel secure leaving my kids with the caretaker because I know everything is there, it is closed, no one is coming in and out; they have to sign if they want to come in and out. And when I leave for work, I feel happy to leave, I do not have to stress about them at home, so that is most important.
Participant 8: It is always nice to stay in a complex. I think if I stay outside a complex or in a house, I do not think my kids will feel like they can play with the other kids, so it is nice for them as well.
Moderator: Okay, any other dislikes guys, anything at all?

Participant 1: The garbage removal.
Moderator: Tell me about the garbage removal, what is going on?
Participant 1: It is either too early in the morning and we miss it or it is too late in the afternoon and you still miss it.
Moderator: Anything else?
Participant 4: The awkward times they cut the grass (laughing).
Participant 7: There is not much that we dislike about the area because at the pool, it is quite big.
Participant 4: Tennis courts, action cricket, and obviously an open field where you can play soccer, cricket, rugby. Like an open park.
Moderator: And are there places where they have those gyms that are built in the parks?
Participant 7: Yes, in our community there is. I think there are like two, I saw but outside, it is not too far out.
Moderator: And the swimming pools outside?
Participant 7: We do not go outside.

In the upper IN, there are several other ways the neighbourhood is described, lending to the sense of place in these areas despite reservations about night time driving. Interestingly, this also profoundly shapes the expectations or norms in these areas. Better-off neighbourhoods relate these elements as though they are normal and unexceptional as characteristics of neighbourhood. When prompted by the moderator, these words and phrases were mentioned by residents:

> Safe; Friends; Established; Old, like forever; Secure; And nature; Groups; Walking with puppies. Gym, Parks, Golf course; Local parks; Well, just your normal. There is nothing extra. (Participants, UIM)

Yet there are areas in the neighbourhood, particularly around low-cost housing in neighbouring areas—described by these residents as unsafe, where they cite lack of police presence. Unlike upper IN which can reportedly opt out of neighbourhood safety services provided by the state and higher private security, those in lower, lower middle and middle INs cannot.

What is seen in the concept of place-making is that there are fundamental inequalities when it comes to different neighbourhoods' sense of place, the characteristics of those spaces, and the norms associated with those neighbourhoods. While nostalgia about the past was similar across all neighbourhood types, the sense of ease of everyday life was less so. In upper and upper-middle INs, there is a sense of safety, norms of access to quality amenities, ability to use the neighbourhood's amenities freely, and a preservation of place that comes with its use. These are not seen as exceptional in any way, but very much the norm of how it should be. In middle, low-middle and low INs, there is a lamentful and resigned feeling which characterises the neighbourhoods because of the deterioration or non-existence of services, increasing crime, and decreasing safety in these areas. What this section presented in terms of this aspect of place-making is a sobering reflection of the differences in how residents related their sense of place.

8.4.2 Belonging

As stated earlier, the element of *Belonging*, as seen in the place-making literature comprises a sense of affinity with others in that neighbourhood, as well as a sense of external recognition by others. In the Johannesburg case there are peculiarities related to its history. Belonging, as it pertains to residents' sense of place, has an apparent 'us' versus 'them' undercurrent, and a severe feeling of being unrecognised by (mainly) the state. Belonging, after all, relates to how people feel in relation to others, and a perceived sense of how 'they' are recognised and 'their' needs met, but external actors—in this case external recognition—would mean the state, or local policy actors.

The 'us' versus 'them' undercurrent is perhaps unsurprising given the spatial divisions along racial lines in South Africa's history. In better-off areas there is a social sense of belonging to the neighbourhood, but this is not relative to one's person, nor recognition, nor to needs being met; instead it is mobilised as a set of norms about who belongs and who does not belong. Here we see preservation of the neighbourhood on the basis of who is seen to belong and who is not. In less well-off areas, while there is a greater sense of social solidarity as the section on *Agency* will depict, there is little sense of exclusion, despite even recognising unsavoury behaviours in the neighbourhood. Take these contrasting views, for instance, between an upper IN respondent and a middle IN:

Participant 4, UIN: You see a car that looks suspicious, then you just say, listen, there is a car parked, does anybody know that guy? If nobody knows him then one of the responders will go check him out.)

Participant 7, MIN: At night they do not sleep. You can open your window at 2 o'clock and you will see people walking around like it's broad daylight, normal' and there is no police visibility. So how do you sleep at night when people are walking up and down in the streets chatting like it's day break [...], but tonight I want to say I blame the government for neglecting coloured people and that's why we are in a social mess.

In middle-income neighbourhoods this sense of being unrecognised is not just about one's own needs, but also stated in comparison to the groups, and the 'us' versus 'them' sentiment that underlies social experience in the country—that is being unrecognised in comparison to another group:

Participant 4, MIN: Let me tell you what happened, most of our coloured areas are surrounded by black areas, so, for instance Noordgesig was under ward 68. So now they built a [new] area, which is moving toward 68 which was supposed to benefit our people. [...] And to our surprise they brought people from somewhere else and gave them the [state provisioned] houses, so strategically you are surrounded by black areas. It's not a race thing, so now you block them and move from ward 68 to another block, now you are in another ward which falls under Soweto, you understand. So now these few coloureds here do not matter, you can vote [for minority opposition] until you turn blue, so where does Noordgesig stand a chance?

Similarly, another respondent from the same neighbourhood type notes:

Participant 8, MIN: You know, what are the choices that we face? I don't want to sound like we are racial, I know our community leaders are doing so much [...] out of their own pockets. Ask us what our country can do for you, not what we can do for our country. They do it but they do not even recognise us.

The sense of belonging is also curiously formed as a shared sense of not *mattering*, not having one's needs met, and not feeling recognised by local state actors. While feeling unseen is endemic in less well-off areas, it is not even apparent in better-off areas. In the low and low-middle IN, this sense of not mattering or being recognised by the state is articulated as not being 'cared about'. Care is a fundamental element of belonging, and where residents perceived a systemic lack of care was seen through lack of meeting the basic needs of people. In the lower-middle IN focus group, there was agreement of perceived systemic exclusion by focus group participants from similar neighbourhoods, naming the lack of care as abuse by the government.

Participant 4, LMIN: Our government is not actually focusing on the grassroots. They are only focusing on what they want and not focusing on what people are actually crying for. There are so many grievances but no one cares about that.

This time, the responses are not framed as 'us' versus 'them', but even in lower-middle IN, the sentiment about not been seen or recognised by the state appears frighteningly common. It is a shared sense that the government simply does not care about 'people like themselves' or 'our people', despite (even) representation through the franchise.

Participant 8, LMIN: I am a member of the ANC [African National Congress]. I support the ANC, but I am very disappointed in my party. My party is very arrogant, that I will say. They don't consult. They consult amongst the inner court, their cabals. And then they decide on our people and hear their arrogance when they say "our people" so our people, they do not have brains as far as they are concerned. They will decide for our people on what is good for you and that is it. For a lot of things, if they just had the decency of coming to the community and listening to us with honesty, right.

In better-off areas 'belonging' is not ascribed to recognition or care given by the state, or others, instead it is a purely surface social phenomenon, where 'neighbours greet and wave to each other' (UIN participants), and the sense of belonging to this neighbourhood (or gated community) is not related to issues of identity or recognition of needs—which, it may be argued, is a feature of privilege.

In sum, spatial inequality plays out differently in different neighbourhoods in a very fundamental sense of belonging, but not in the positive and convivial notions invoked by place-making scholars. What gives this group coherence is a profound sense that their needs do not matter, and they themselves do not matter, especially as seen in comparison with that of other groups. Yet those self-same 'other groups' also feel the same profound lack of care and recognition of needs.

On the one hand the sense of belonging relates to entitlements of inhabiting space, which includes determining who is included and who is excluded from well-off neighbourhoods (manifesting in safety concerns and preservation of entitled space). On the other hand, in less privileged neighbourhoods the sense of belonging includes reflections of racial divides, experiences, and the effect of not being recognised, of being forgotten or not being cared about by the state.

8.4.3 Agency

The final element that qualitatively examines place-making and gives insights into how spatial inequality plays out in different neighbourhoods is *Agency*. There are different ways that agency is exercised in different places. Barolsky (2016) terms activities that exercise social agency in a community as *collective efficacy* irrespective of whether actions are deemed to be acceptable or not. The agencies and social efficacy expressed in different neighbourhoods relate closely to the lament and resignation around sense of place and the recognition and care as it challenges the notion of belonging—and all these manifest themselves spatially. Moving away from a neighbourhood remains an option, but this is dependent upon economic and social mobility and privilege. If self-determination cannot be achieved because of systemic socio-economic issues, then resignation, lament, and despondency may set in. As one resident in the worst-off neighbourhood type who lamented the lack of choice notes:

Participant 1, LIN: They are not privileged enough to go and challenge other things. Like now, if it is in your area [...] That is why we say, if you want something better, go out of your area. Now, you can't do it financially, to go out. And that is why we are back in and have to stay in the area.

This, more stark response, is sobering, and relates to the inability to either escape one's circumstances, or one's inability to even exercise self-determination, becoming infantilised in one's ability to even make choices:

Participant 4, LMIN: If you look at the social ills that have been created by the government, it has put the parents in a tight corner. By becoming a child in your own house. [...] Some of our parents are now depressed, there is so much depression. Even thinking about killing themselves. People are killing themselves and women are getting mental illnesses. Too many sisters can't look after the children. It has put so much pressure on us as parents. And at the end of the day, we look like we can't take care of our children. We are so down and out.

Despite these sobering responses by some, agency and social efficacy is evident; in other cases, it is exercised specifically and optimistically in relation to the state's failure to provide social adequate amenities and protection, or the perceived failure of the state to take care of the needs of all residents. This plays out in uneven ways in better and worse-off neighbourhoods. In worse-off neighbourhoods, agency is exercised in ways that are fundamental to social existence and well-being, in better-off neighbourhoods, in fairly cosmetic ways—relating to comparatively fewer 'life or death' situations. The reasons that agency emerges are multiple, but they all tend to relate to the perceived failure of the state. Take these sets of disparate responses where respondents from different neighbourhood types discuss how they confront the social and other challenges that beset their neighbourhoods:

Participant 5, LIN: Those people who come from those backgrounds [say] "I am just going to do it for my family and if this one is going through what I went through, then so be it.

In this quote, street justice, as a component of social efficacy, is evident:

Participant 8, LMIN: Orlando Police Station is the worst corrupt police station. [...] They are very, very corrupt. So we didn't report [a specific crime]. We dealt with it, my way, because my way is very effective. We found out who it was, and we dealt with it [...] when you report a case in Orlando, it is just a waste of time. So, deal with it your own way.

Again, here, in response to an incident of crime:

Participant 5, MIN: I'm going to take the law into my own hands. It's like the police want us to take those measures to help our community because they don't do anything.

More generally, the sense of agency is where residents have to make life work in neighbourhoods every day, apart from the state. This quote reflects guidance from church leadership from the pulpit, also responding to calls from civil society to join community protest action:

Participant 8, MIN: And what the bishop said when the government is not doing it—we got to do it. So, what we also need to do is empower ourselves but then again, it's not easy.

The previous quotes relayed sentiment and inferred examples of exercising agency. The following gives a practical example of what 'exercising agency' looks like, which had to do with storm water drainage blockages and sewerage impact on the streets of a particular neighbourhood. The problem was with one resident's infrastructure, and there was a need to collectively determine where the issue was:

Participant 2, MIN: You know what happened? It was on the news last week about the water. Then I got up and I went door to door, and I said, look at your drain blockage. […] we got such a beautiful park just behind my neighbour but it's a dumping spot, now I do not know where the sewage comes from. There is a sewage hole there that blocks the whole street's drains and then they open the pipe and the children are playing there. It might not be my child but all the children are my children.

In the upper IN (UIN), there is less of the direct exercise of agency, rather, a 'remote-control' and more passive form of engagement with the state—through raising complaints on social media:

Participant 8: In our neighbourhood there is normally just a lot of moaning.
Moderator: But with the moaning does it help to fix anything?
Participant 6: No, but it helps to sort our problem out.
Moderator: To sort the problem out, okay. I got you.
Participant 6: So, we got the grass cut, we got everything.

Agency is exercised through different means, one is direct—individual efficacy, as above—and other forms of social organising, as below: To keep the community safe during school time, one resident started a community initiative to chaperone kids and parents walking to school:

Participant 4, MIN: The change I want to make is the change I want to take; what we did is we volunteered. That's how I started. I volunteered and the people started following […] it's mostly ladies in the area that go to the schools in the morning, say, from 7 o'clock until 10 o'clock, then the other shift starts from 10 o'clock until the schools come out. So that's how our community stays in a better space where you know when crime happens, we can nip them.

Another participant, demonstrating similar tenacity, organised community action for a specific purpose:

Participant 5, MIN: We have the school and next to a *veldt* [Afrikaans for field with grass that is high]. Each parent of the team put their heads together. We went and we took space from it and we don't care if it is the government's, we are not caring anymore. We took a space and they [paid labourers] cut the grass for us. We paid them and our kids play soccer there every day. Those were the measures we had to take.

Protest remains an important mechanism of expressing agency through making it impossible for the state to ignore demands and entitlements of residents.

Participant 4, MIN: Sometimes you get to the boiling point knowing you don't have a choice; you just go to the street. That is how the government will understand and listen to you. That is the only way they listen to you, then they will come and send someone they were supposed to send three months ago.

Participant 5, MIN: Then it is more than enough to say 'more than enough' and they want to teargas the people and rubber bullet the people, then three days later they want to help the community.

In this next case agency is linked to a xenophobic form of social efficacy: Operation Dudula was a wide-reaching community mobilisation strategy, initiated by community organisers, based on 'taking back the neighbourhood' from foreign Africans who were accused of undue economic control of neighbourhoods:

Participant 4, LMIN: It is actually coming closer, they are going from township to township to organise, there will be the biggest meeting to shake our government. [...] It is on a Sunday and please don't tell us that you are going to church because this problem is for us all. So, everyone is supposed to be there. Because we are trying to fix up the agenda [...] The things that we do for ourselves, it is not the government.

The next few excerpts show that in better-off neighbourhoods, agency is not expressed directly, but is mediated through corporate bodies, resident associations, and landlords, or by remote action of online complaints. While there is activity to let one's voice be heard, seldom would the (even direct) exercise of agency here be considered dangerous or risky to do so, unlike in less well-off neighbourhoods where the stakes of direct mobilisation are high.

In well-off gated estates, participants said the following in response to challenges concerning the delivery of public services such as water or storm-water drainage:

Participant 1, UMIN: Well, the manager logs it on behalf of the resident, so the more we complain the more they call.

Also, in community-funded gated/boomed-off areas, agency is exercised by UIN participants mainly through the provided customer-engagement systems:

Participant 6: Well, we normally just report it to the City, Joburg Water […] and then we keep on.

Participant 8: Yes, we must just keep on reporting.

Participant 6: Now you can register on Twitter also. You can register on your account, the City of Joburg [app], everything. And you can register there. And if you do not have the meter, you can just fill it in all the time to follow up. That is what we do.[2]

In the upper IN (UMIN), residents noted that for safety concerns community agency is exercised because of incapacities in public safety sectors. Unlike the gated estates, this agency does not result in mediated action through a corporate body, but in a privatised community action body a layer of market services that are paid for.

Participant 2: What is happening is they are not interested. Because of their lack in policing and response, (addressing participant 6) you spoke about a community thing, we got a community thing too.

Participant 7: Yes, it is under CPI [Community Policing Initiative].

Participant 2: We got an attorney that is heading this thing up and he is very proactive. And he gets all these security companies, you know like Legal Watch and all these guys, all to be part of this. So, if there is a burglary or a break-in any of the properties, if you are on this community watch you can put a report on there and all the security companies get it and they will react.

Participant 8: Yes, there is more reaction there than the actual police.

Participant 3: They are very active.

Participant 8: They are visible. They are there.

Participant 6: If you request an extra guard in your street, they will send him.

The stakes of exercising agency or pointing to gaps in service delivery are not high or risky for the active resident in privileged neighbourhoods. Yet, this is not so for those in disadvantaged neighbourhoods, where agency expressed in opposition to the state is met with intimidation as the response below argues. Community members who repeatedly raise complaints to the city, escalating these and pointing to the ineptitude of local councillors, find themselves at risk of retribution (even death threats) by local political representatives who can lose their jobs for inefficiencies in their constituencies.

Participant 7, MIN: With the drains the water that is running, the community leader is in great danger. I was shocked that they wanted to take me out because at the end of the day I was not allowed to talk out loudly against these things. You become a problem because how do you do this things? […] They do not want to work with us they think we are a threat rather than a helper, because when I'm on the ground dealing with the community, like the leaders here, we give you feedback [...] and tell you exactly what is happening on the ground, but in Eldorado Park you become a threat and they do not like to hear the truth of Eldorado Park […]

Although there are indeed individualist tendencies, agency is seen to be exercised for the good of the neighbourhood. There may also be some extra-legal ways that agency is exercised in these cases such as 'taking the law into one's own hands'

[2] This mode of citizen engagement has become more difficult since the City of Johannesburg closed its civic offices mid-2023, and the call centre service also rarely runs.

or 'doing it for ourselves'. In well-off neighbourhoods, agency is exercised for the preservation of a particular quality of life, and it is often mediated through market provisioning or technology 'remote' engagement with the state. In all cases, agency is exercised in response to the state—either because of the (perceived) ineptitude of the state, or because of materially declining public service provision.

8.5 Conclusion

Using in-depth responses from residents across five neighbourhood types, this chapter has looked at three concepts which help to identify spatial inequalities at the neighbourhood level: sense of place, belonging, and agency. There are several clear juxtapositions where neighbourhood inequality is evident—lament and resignation at the poor quality of life in deprived neighbourhoods, and relative safety and conviviality in those well-off; a sense of belonging in a secure living environment and a sense of being forgotten or uncared for in those places which are less secure; and direct, mediated or 'marketised' action or exercised agency for those in deprived, middle-class, and wealthy areas, respectively. These translate into unequal ways that residents live in the city, can directly influence their own well-being or have a say in the way they live. The findings also demonstrate that urban life is characterised by the experience of life and interaction with others. Public infrastructure, services, and amenities is only the stage upon which these interactions and associations play out. Life in the city is about how life is lived, how agency is exercised, and whether people have the freedoms to be able to exercise neighbourliness or live a good life. More importantly, it is about whether residents in neighbourhoods have the ability to stave off the harsh effects of the market, like displacement, and the slide into precarity and anomie. Indeed, the findings indicate the need for continued salience of public infrastructure, amenities, and service provision as a marker of the material conditions that profoundly affect urban life. Yet, the social fractures that are inscribed onto neighbourhoods, historically—such as the legacy of racialisation, the divisions between communities and demographic groups, and the distance between people and those who govern—remain. Many of the responses here relay the disillusionment and alienation whole neighbourhoods experience because of the deplorable infrastructure, amenities, and services they face every day. It is also seen how disillusionment is life and death for some, and not others. The differences between neighbourhoods illustrate spatial inequality in ways that are visceral and nuanced. So, what does it mean to have this visceral and nuanced account of neighbourhood inequality? There are two arguments here. One, it contributes to academic discourse on both spatial inequality and the scale of urban analysis. Two, it conveys a message for urban governance.

1. *Academic discourse*: Spatial inequality in the literature is typically considered from a quantifiable perspective of the built environment, infrastructure, or morphology of places in comparison to other places. While this is a crucial perspective, it presents an incomplete picture. Indeed, it assumes a lived reality, but by rarely including socio-spatial aspects of inequality, it also equates the fact of provisioning as a good quality of life, and it makes the policy or governance challenge transactional—the state will simply deliver more for inequality to lessen. Seen as this set of transactional relationships between the state and groups of beneficiaries, the effect of material inequality on life and neighbourhoods simply becomes an inconvenient by-product or 'unintended consequence' of the slowness of reach or the distributive priorities of the state. The governance challenge then becomes to speed up equitable distribution of resources. Or, building on Lemanski's (2022) argument that infrastructure itself becomes a means of controlling and disciplining citizens and conferring citizenship, mainly focuses on infrastructure and relegates residents' experiences to servility.

Yet the way people experience life in neighbourhoods is a product of the processes of governance acting on those spaces (Charlton 2018) and place-making, as argued here. Indeed this is what Wafer and Rameetse (2020) argue distinguishes a neighbourhood from just a place. Place-making then, represents the opposite of the centralist impulse to create servitude in the way Lemanski (2022) has argued. It is the self-determination of neighbourhood members themselves, in addition to infrastructural elements, where the latter is then understood not as the end to the problem of inequality, but as a means to reduce inequality. How people inhabit, use and experience space is an important counterpoint to the tendency of subsuming lived realities to the activities of a benefactor.

There are also insights that relate to the methodological consideration of understanding urban processes and form from the neighbourhood scale. Residents do not experience place as a voting constituency, nor as those with flush toilets or those without, and the sense of place is not an individual or every household activity. The neighbourhood as a scale of insight and analysis, allows us to understand the city as the interface between infrastructure provision, governance, and place-making, not determined by provision, but perhaps only constrained by it. There are sets of socio-spatial behaviours and practices that are mediated by class, constrained by infrastructure, but overall determined by the residents themselves, their sense of place, belonging, associational strength, and agency.

2. *Urban governance*: An approach of socio-spatial inequality at the neighbourhood level balances a state provisioning model based on voter districts. In the first instance it raises questions about the centrality of the state as the benefactor or patron of public good, and relatedly cautions the patronage and intimidation associated with such a politicised state-society engagement. The failures of the state in this frame, are not only hyper visible, but also unconscionable. If the state is indeed developmental, with a strong social policy that underpins its politics, then the material provisioning of infrastructure must be in service of the public good. Material provisioning or access must allow for the capacities, associations, and agency of the residents to emerge as the determining factor of their own quality of life. And if it does not, and remains as an 'end' to the question of urban development rather than its means, then it produces mass dissatisfaction and discontent, and reproduces social inequality.

References

Abrahams C, Everatt D (2019) City profile: Johannesburg, South Africa. Environ Urban ASIA 10(2):255–270

Abdul S (2022) Spatial development and inequalities in the Global South. Reg Sci Policy Pract 14:211

Alkan H, Maksudyan N (eds) (2020) Urban neighbourhood formations: boundaries, narrations and intimacies. Routledge

Alkire S (2008) Concepts and measures of agency—GSDRC. https://gsdrc.org/document-library/concepts-and-measures-of-agency/, https://gsdrcwebsite-pjp5ov869n.live-website.com/document-library/concepts-and-measures-of-agency/

Baffoe G (2019) Understanding the neighbourhood concept and its evolution: a review. Environ Urban ASIA 10(2):393–402. https://doi.org/10.1177/0975425319859115

Barolsky V (2016) 'Loving thy neighbour' in times of violence: social cohesion and collective efficacy in South Africa. Psychol Soc 51:1–27. https://doi.org/10.17159/2309-8708/2016/n51a1

Bravo L (2018) We the public space. Strategies to deal with inequalities in order to achieve inclusive and sustainable urban environments. J Public Space 3:163–164. https://doi.org/10.5204/jps.v3i1.326

Chant S (2013) Cities through a "gender lens": a golden "urban age" for women in the Global South? Environ Urban 25(1):Article 1. https://doi.org/10.1177/0956247813477809

Charlton S (2018) Spanning the spectrum: infrastructural experiences in South Africa's state housing programme. Int Dev Plan Rev 40(2):97–120. https://doi.org/10.3828/idpr.2018.6

Chen B, Wu S, Song Y, Webster C, Xu B, Gong P (2022) Contrasting inequality in human exposure to greenspace between cities of Global North and Global South. Nat Commun 13(1):Article 1. https://doi.org/10.1038/s41467-022-32258-4

Cole MJ, Bailey RM, Cullis JDS, New MG (2018) Spatial inequality in water access and water use in South

Africa. Water Policy 20(1):37–52. https://doi.org/10.2166/wp.2017.111
Everatt D (2014) Poverty and inequality in the Gauteng city-region. In: Todes A, Wray C, Götz G, Harrison P (eds) Changing space, changing city: Johannesburg after apartheid. Wits University Press, pp 63–82
Everatt D, Abrahams C, Lynge H (2020) Johannesburg: the growing importance of class in shaping neighbourhoods. SHLC City Reports, pp 1–44 [City Project Report]. University of Glasgow Centre for Sustainable, Healthy, Learning Neighbourhoods, Glasgow. http://www.centreforsustainablecities.ac.uk/research/johannesburg-the-growing-importance-of-class-in-shapingneighbourhoods/
Gokce D, Chen F (2018) Sense of place in the changing process of house form: case studies from Ankara, Turkey. Environ Plan B: Urban Anal City Sci 45(4):772–796. https://doi.org/10.1177/0265813516686970
Han S (2022) Spatial stratification and socio-spatial inequalities: the case of Seoul and Busan in South Korea. Hum Soc Sci Commun 9(1):Article 1. https://doi.org/10.1057/s41599-022-01035-5
Harvey D (2003) The right to the city. Int J Urban Reg Res 27(4):939–941. https://doi.org/10.1111/j.0309-1317.2003.00492.x
Kanbur R, Venables A (2005) Rising spatial disparities and development. World Institute for Development Economic Research (UNU-WIDER), Working Papers
Kaplan S (2023) Fragile neighborhoods: repairing American society one zip code at a time. Little Brown
Kearns A, Parkinson M (2001) The significance of neighbourhood. Urban Stud 38(12):2103–2110. https://doi.org/10.1080/00420980120087063
Lemanski C (2022) Infrastructural citizenship: conceiving, producing and disciplining people and place via public housing, from Cape Town to Stoke-on-Trent. Hous Stud 37(6):932–954. https://doi.org/10.1080/02673037.2021.1966390
Lynge H, Visagie J, Scheba A, Turok I, Everatt D, Abrahams C (2022) Developing neighbourhood typologies and understanding urban inequality: a data-driven approach. Reg Stud Reg Sci 9(1):618–640. https://doi.org/10.1080/21681376.2022.2132180
Miraftab F, Wilson D, Salo K (2015) Cities and inequalities in a global and neoliberal world. Routledge
Rigolon A, Browning MHEM, Lee K, Shin S (2018) Access to urban green space in cities of the Global South: a systematic literature review. Urban Sci 2(3):Article 3. https://doi.org/10.3390/urbansci2030067
Switalski M, Torres MG, Grêt-Regamey A (2023) The 3P's of place-making: measuring place-making through the latent components of person, procedures and place. Landsc Urban Plan 238:104817. https://doi.org/10.1016/j.landurbplan.2023.104817
Todes A (2014) The impact of policy and strategic spatial planning. In: Todes A, Wray C, Götz G, Harrison P (eds) Changing space, changing city: Johannesburg after apartheid. Wits University Press, pp 83–100
Todes A, Turok I (2018) Spatial inequalities and policies in South Africa: place-based or people-centred? Prog Plan 123:1–31. https://doi.org/10.1016/j.progress.2017.03.001
von Fintel D (2018) Long-run spatial inequality in South Africa: early settlement patterns and separate development. Working Paper No. 16/2018; Issue 16/2018. Department of Economics, Stellenbosch University. https://econpapers.repec.org/paper/szawpaper/wpapers309.htm
Wafer A, Rameetse K (2020) What makes a township a neighbourhood? The case of Eldorado Park, Johannesburg. In: Alkan H, Maksudyan N (eds) Urban neighbourhood formations: boundaries, narrations and intimacies. Routledge, pp 17–35

9 Urban Expansion and Neighbourhood Inequality in Kigali

Gilbert Nduwayezu, Vincent Manirakiza, Josephine Malonza, Leon Mugabe, Aimable Nsabimana, Pierre Claver Rutayisire, Manasse Nzayirambaho, and Kato Jonas Njunwa

Abstract

Since the 1990s, Kigali has experienced rapid urbanisation, leading to uncontrolled sprawl and challenges in providing basic amenities to all residents, exacerbating urban inequality. This chapter presents empirical evidence of rapid growth over a 30-year period (1987–2018), with an average annual spatial growth rate of 10%, converting most agricultural land into built-up areas. It also describes Kigali's 130 neighbourhoods, including 9 mixed-use quarters and 121 residential neighbourhoods categorised as planned, unplanned, and mixed or semi-planned, and explores inequalities in housing, income, employment, and household amenities. Findings show that unplanned areas face overcrowding, low incomes, and limited services, although disparities in electricity access and waste collection are less pronounced across neighbourhoods. Data is drawn from the literature and the SHLC Household Survey.

Keywords

Kigali · Urban expansion · Neighbourhood · Inequality · Livelihood

G. Nduwayezu (✉)
College of Science and Technology, School of Engineering, University of Rwanda, Kigali, Rwanda
e-mail: g.nduwayezu@ur.ac.rw

V. Manirakiza · L. Mugabe
College of Education, University of Rwanda, Kigali, Rwanda
e-mail: v.manirakiza@ur.ac.rw

L. Mugabe
e-mail: l.mugabe@ur.ac.rw

J. Malonza
College of Science and Technology, University of Rwanda, Kigali, Rwanda
e-mail: jmalonza@ur.ac.rw

A. Nsabimana
Department of Agricultural Economics, University of Rwanda, Kigali, Rwanda

P. C. Rutayisire
College of Business and Economics, University of Rwanda, Kigali, Rwanda

M. Nzayirambaho
School of Public Health, University of Rwanda, Kigali, Rwanda

K. J. Njunwa
University of Rwanda, Kigali, Rwanda

9.1 Introduction

Developing countries are recording rapid urbanisation of which close to 90% is expected to take place in Asia and Africa by 2050 (United Nations 2018). People are migrating to cities at a faster rate, with consequent uncoordinated spatial

Y. P. Wang et al. (eds.), *New Drivers of Division*, Sustainable Development Goals Series,
https://doi.org/10.1007/978-981-96-5629-5_9

expansion of urban areas (Burak et al. 2017). Such urban sprawl gives rise to a host of challenges, including landlessness, socio-economic conflicts, and the degradation of livelihoods, affecting approximately 60% of the urban poor (World Bank 2022). Moreover, this rapid urbanisation significantly impacts urban planning, resulting in the emergence of both planned and unplanned settlements, which exhibit stark differences in living conditions and contribute to growing urban inequality.

The situation in Rwanda, particularly in the city of Kigali mirrors the urbanisation trend observed in cities across developing countries. Rwanda has experienced a notable shift in urbanisation patterns over the years. Initially, urbanisation progressed at a slow rate in the early decades leading to independence in 1962, with a rate of 2.6% at that time. It accelerated thereafter coinciding with the onset of industrialisation and significant rural-to-urban migration, particularly towards Kigali. The period between 1990 and 2000 witnessed another surge in urbanisation rates, increasing from 5.4% to 14.9% (Macrotrends 2024). This spike was largely attributed to individuals seeking refuge in urban areas during the liberation war and following the tragic events of the 1994 genocide against the Tutsi in Rwanda. This trend has persisted, with urbanisation rates reaching 27.9% in 2022. Factors driving this continued urbanisation include economic growth, particularly in urban centres, which attracts rural residents, as well as natural population increase (National Institute of Statistics of Rwanda—NISR 2022).

Established in 1907, Kigali became the capital city of Rwanda in 1962. Initially a small town with approximately 6000 residents occupying 0.08 km^2 in 1962, Kigali has since transformed into the most rapidly urbanising city in Rwanda (Manirakiza et al. 2019). Over the past three decades, the city has undergone significant demographic and spatial changes. Its population increased from 235,664 in 1991 to 1.3 million in 2014, further reaching 1.6 million in 2017, and finally 1,745,555 in 2022 (NISR 2018, 2022). The average annual growth rate of the built-up area has exceeded 10% in the last 20 years (Manirakiza et al. 2020). Referring to the national administrative framework as defined in the Republic of Rwanda's Organic Law No. 29/2005, Kigali City is divided into three districts: Gasabo, Kicukiro, and Nyarugenge. These districts are further divided into 35 sectors, which are then subdivided into 161 cells and a total of 1061 villages. Presently, its administrative boundaries span across 730 km^2.

The rapid urbanisation has brought challenges such as land scarcity and emergence of unplanned settlements despite the strict building regulations. Consequently, uncontrolled urban sprawl has disrupted the equitable distribution of basic amenities, leading to increased spatial inequality, evident in the mosaic of neighbourhoods.

It is worth understanding the concept of a neighbourhood before delving into the context of analysing the neighbourhoods of Kigali, despite the ambiguity surrounding its definition. While there is no universally accepted definition of a neighbourhood, many definitions share common elements, emphasising the presence of people and services within a defined physical area, or simply an ultra-local community within a broader setting. According to Park and Rogers (2015), a neighbourhood can be defined as a group of individuals who are typically identified by shared services and social cohesion within a geographically bounded space. Baffoe (2018) further elaborates that a neighbourhood is an area where individuals reside and spend a significant portion of their time. These definitions underline the importance of space, people, interactions, and activities in shaping the fabric of a neighbourhood.

A neighbourhood can be studied using a combination of a variety of physical and social characteristics. From a geographic perspective, a neighbourhood can be understood from a top-down administrative structure or from a social perspective as mental maps and subjective identifications (Young Foundation 2010). Depending on the scope of the study, neighbourhoods can also be understood by their shapes and sizes, the housing status and quality, different levels of socio-economic inequality of their residents, income and cultural preferences, streets and buildings to explore residential segregation (Crooks

2010). Antoni (2009) proposes a method for investigating a neighbourhood that is both practical and effective. According to him, a neighbourhood can be studied in terms of content: the socio-economic standing of its residents (rich or poor) or the specialisation of activities. It can also be investigated in terms of its geographical position as well as its physical structure or characteristics (urban fabric, date of construction, typology of buildings and constructions). A central, peri-central, or peripheral geographic location might help thus in identifying an old or new, a suburban, or a spontaneous neighbourhood.

Considering the process of urbanisation of Kigali, which has been mainly spontaneous, making different categories of neighbourhoods with specific characteristics of their residents, both perspectives—content, and location and physical structure—can apply to the study of the neighbourhoods in Kigali city. The approach can be applied to study the characteristics of households and also help to establish a neighbourhood's typology.

This chapter describes the expansion of the city of Kigali over the past three decades, which has led to the development of various types of neighbourhoods reflecting urban inequality. The expansion is analysed through changes in land cover and the extension of built-up areas. Urban inequality is depicted from the categories of the neighbourhoods and the socio-economic characteristics of their residents. The methodology used to assess these aspects involved two main steps. Firstly, urban growth over time was quantified, and neighbourhoods were mapped using remote sensing imagery. Secondly, a survey was conducted to collect data on the residents' characteristics in different neighbourhoods, allowing for a description of the inequalities present within the urban fabric.

The study provides empirical evidence of the increase of built-up area from 26 km^2 in 1987 to 115 km^2 in 2018 and the city's 130 neighbourhoods which include 121 residential and 9 mixed-use quarters. The residential neighbourhoods comprise three categories that are planned, unplanned, and mixed. Inequality among the neighbourhoods is evident from their physical characteristics but also the services offered to their residents. For example, most residents from the planned neighbourhoods (high and middle income) live in their own flats and detached houses, whereas those in unplanned neighbourhoods (lower income) dwell in cluster houses in a complex, with a high population density. As a result, these unplanned neighbourhoods often experience overcrowded living conditions and have limited access to basic amenities. However, the inequalities are not significant when it comes to some aspects, such as the supply of electricity and door-to-door garbage collection.

This chapter is structured into three main sections, in addition to the introduction, methodology, and conclusion. The first section focuses on the expansion of Kigali. The second explores the different types of neighbourhoods and their spatial distribution. The third section provides a detailed description of the characteristics of residents, emphasising the aspect of inequality among the neighbourhoods.

9.2 Methods and Data

Two types of data were collected for the study: primary and secondary. Secondary data consists of published papers and scientific reports. The primary data consists of spatial facts collected using remote sensing imagery to study the nature and pattern of urban sprawl over a 30-year period (1987–2018) and the distribution of the spatial neighbourhoods and quantitative data obtained from the survey.

Neighbourhood analysis was conducted using a mixed-method approach, which involved two main components: (1) a participatory approach, consisting of field observations of the neighbourhoods' fabric and interviews with residents to understand the context of their creation; (2) a remote sensing-based approach to digitise neighbourhood boundaries using various Google images (Fallatah et al. 2022; Kohli et al. 2012).

The entire process involved three steps: The first consisted of field observations of the main characteristics of the neighbourhood and interviews with the key informants (local leaders, local and other residents). Both observation and interviews helped to get data about the site occupancy, whether planned (presence of a physical plan) or unplanned, the predominant types of buildings, the site topography, the road network, residential density, and access to basic services and infrastructure.

The second step consisted of placing the neighbourhoods in four main categories, including planned, unplanned, and mixed for residential neighbourhoods, and mixed-use quarters (mostly for businesses). Planned neighbourhoods are developed following a physical plan. Unplanned neighbourhoods are those settled illegally without abiding to urban construction regulations, such as having a construction permit. Mixed neighbourhoods comprise a mixture of both planned and unplanned settlements on one site.

The third step, which was the remote sensing-based approach, consisted of digitalising neighbourhood boundaries using high-resolution printed images. The technique used was to first refer to the administrative village boundaries where applicable, then digitalise the neighbourhood boundaries in case they do not correlate with the existing village administrative boundaries, and finally conduct a field visit for validation.

Quantitative data was collected from a sample of households selected from 15 neighbourhoods, which were classified into five categories of income bands (Table 9.1). The sample size was determined using the formula of Yamane (1967: 888) based on a target population of 46,634 households, resulting in a sample of 1092 participants, assuming a precision level of 3%. Purposive sampling was employed to select the 15 neighbourhoods, ensuring representation from all income bands. The sample size per neighbourhood was determined according to the number of households of the selected neighbourhoods. Household selection involved on-site systematic sampling of every tenth household from each direction to be included in the survey.

Table 9.1 Sampled neighbourhoods and sample size

Main types	Income bands	Sampled neighbourhoods	Sample size
Planned	1. High income	Estate 2020, Rebero, Kibagaba	122
	2. Upper middle-income	Kagunga Estate, Niboye, Gasave	120
Mixed	3. Middle income	Kabeza, Nyamirambo, Masaka	251
Unplanned	4. Low income	Rwarutabura, Nyagatovu, Karembure	299
	5. Lowest income	Gatenga, Gitega, Kanserege	300
Total Kigali	**5**	**15**	**1092**

Source Authors 2021

9.3 Urban Expansion

9.3.1 Identifying Land Cover, Urban Sprawl

During the years 1987–1999 and 1999–2009, Kigali doubled in size due to increased urban activities, whereas the years 2009–2018 saw a steadier change (see Fig. 9.1). Between 1987 and 2018, the average annual growth rate was about 10%. The land-cover changes documented in this study illustrate that Kigali's natural environment has been impacted by urban expansion. By 2018, most of the agricultural land shown in 1987 spatial maps had been converted into built-up areas. Similarly, Kigali's forested lands were cut in half. The construction of many administrative institutions, schools, and industries in the last 30 years has fuelled land demand considerably. Furthermore, residents have built bungalows instead of high-rise apartments, and new city sub-centres have been created.

Our analysis shows that by 2018, Kigali had a total of new urban area of 90 km^2. In 1987, the densely built-up areas were around the city centre, but settlements began to grow in the hinterland

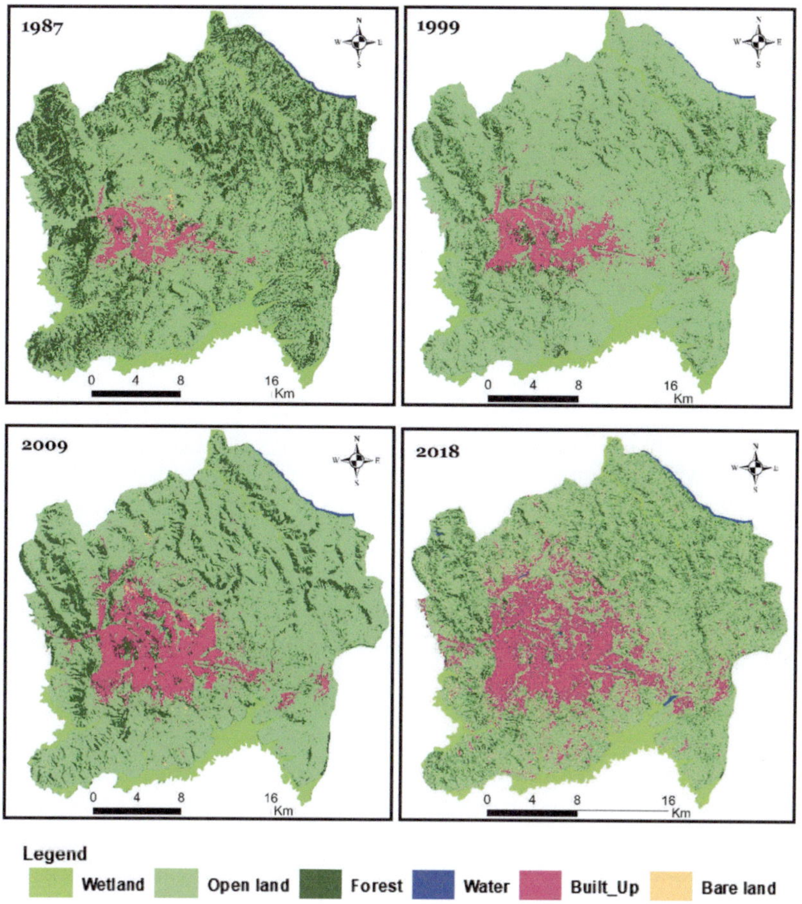

Fig. 9.1 Land cover trends in Kigali city (1987–2018)

from one hill to the next in the following years, leaving wetlands in between. Gasabo district, with 40.6 km^2 of built-up area in 2018, was denser than Nyarugenge (12.52 km^2) and Kicukiro (20.36 km^2) districts. This shows some degree of dispersion (Table 9.2).

9.3.2 Driving Factors of Urban Growth

Kigali's rapid spatial expansion is a consequence of demographic growth in the city since Rwanda's independence in 1962, driven by internal migration as the population searched for jobs. The city also experienced unexpected demographic growth during 1990–2000 caused by spontaneous waves of immigrants related to the liberation war and genocide perpetrated against the Tutsi. Since then, the urban population has kept on growing, propelled by rural–urban migration coupled with natural growth (Uwimbabazi and Lawrence 2011; Nduwayezu et al. 2017).

Given their major impact on the city's renewal and renovation during the post-genocide period, urban plans and policies have contributed to its growth. Zoning restrictions are used to facilitate the segregation of land uses in the plan. They contain rules for predicting growth and development and serve as a road map for the long-term goals of the city. However, informal settlements often emerge beyond the designated zones outlined for planned development, contributing to urban sprawl. In fact, throughout that dynamic,

Table 9.2 Land cover trends in Kigali city (Km^2)

land cover type	1987 (Km^2)	1999 (Km^2)	2009 (Km^2)	2018(Km^2)
Built-up	25.959	39.67	76.051	115.653
Bare land	1.286	0.00	4.4514	0.526
Open land	433.972	567.827	485.157	464.949
Forest	228.147	82.132	123.3801	104.689
Wetland	36.621	36.484	36.621	36.552
Water	1.6362	1.287	1.9638	3.117
Total	**727.62**	**727.40**	**727.62**	**725.49**

Source Authors 2021

Kigali's rapid expansion forced the design of a city plan in 1982, which was again reviewed in 2001. The Kigali Master Plan elaborated in 2007, amended in 2013, and again in 2020 for the 2050 prediction, was primarily intended to halt the city's informal urban development. However, the construction of small sites for planned settlements has been a persistent challenge in all these schemes.

Overall, achieving effective land use to control urban sprawl faces challenges due to the historical housing culture in Rwanda. Residents tend to construct their own homes in cities, somehow replicating rural housing patterns (Manirakiza 2014). This involves building bungalows within large compounds rather than opting for shared living in high-rise apartments. Furthermore, the ongoing changes in city development have resulted in an increased demand for land to accommodate various infrastructure needs, such as institutions, schools, and industries.

9.4 Characteristics and Spatial Distribution of the Neighbourhoods

9.4.1 Characteristics of the Neighbourhoods

The main indicator to identify the neighbourhoods was the manner in which they emerged or were created. The description and mapping process allowed us to identify 130 neighbourhoods in Kigali, including 121 residential and 9 mixed-use quarters (mostly for businesses). The residential neighbourhoods comprise three main categories: 44 planned, 42 unplanned, and 35 mixed or semi-planned neighbourhoods (Fig. 9.2). Each of these three main categories has its characteristics and includes different sub-categories that are spatially identifiable based on their appearance (the modernity of the architecture), predominant types of buildings (storey/s or single floor) residential density, the period of their creation (pre/post genocide), and their location (Fig. 9.3).

Planned neighbourhoods are defined by separated and marked plots, robust structures, and relatively low residential and population density, typically around 2000 inhabitants per km^2. These neighbourhoods often feature single-family homes, occasionally constructed in a uniform manner following specific standards. Planned neighbourhoods comprise three main tiers: high, medium, and low-standing, each further sub-categorised. High standing encompasses individual houses and estates, often developed by private entities, presenting a uniform aesthetic with detached or semi-detached villas and upscale apartments equipped with modern amenities. It comprises two types: *luxurious multi-storey houses* (e.g. Nyarutarama, Rebero), and *private estates* (Vision City, Estate 2020 Gacuriro). These are residences of certain people of high social and economic status, including expatriates, businessmen, and senior government officials.

The medium-standing comprises *neighbourhoods with modern villas* (Niboye, Kagarama), and *old fashioned planned neighbourhoods*

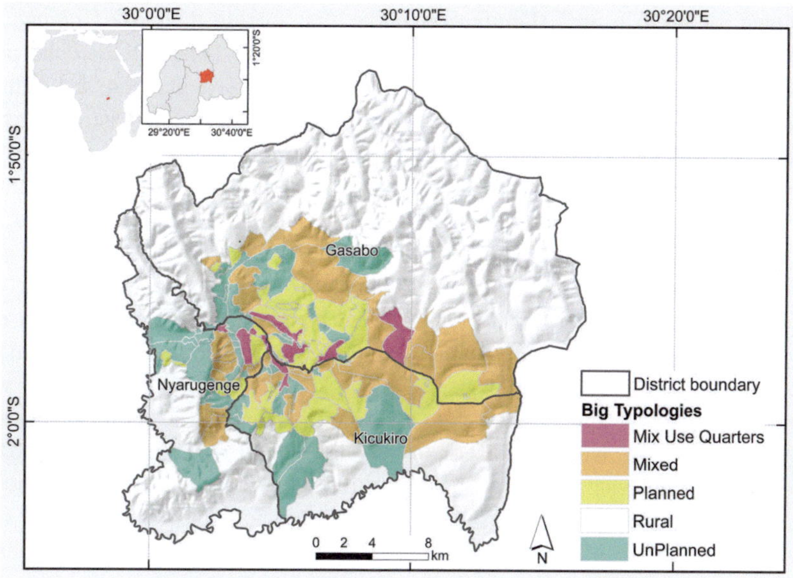

Fig. 9.2 Spatial distribution of the main categories of neighbourhoods. *Source* Authors 2021

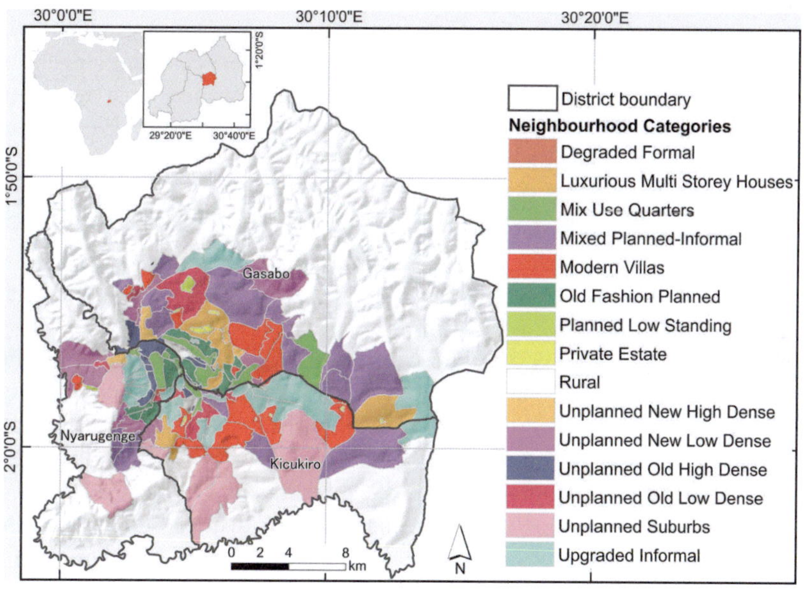

Fig. 9.3 Spatial distribution of the sub-categories of the neighbourhoods. *Source* Authors 2021

(Kiyovu, Kimihurura, Remera). They are built with durable materials but with or without a low proportion of multiple floors.

The low-standing refers to *planned low-standing community dwellings* constructed on developed sites with access to basic infrastructure for economically vulnerable families and those from disaster-prone areas or high-risk zones. They are commonly known as Model Villages built in the framework of the Integrated Development Programme (IDP) being implemented under Rwanda's National Human Settlement Policy of 2009. Some comprise multiple buildings, one being occupied by four families and therefore commonly known as 'Four in one' (Urumuri in Kicukiro). Others are high-rise condominiums for multiple families (Karama in Nyarugenge and Busanza in Kicukiro).

Unplanned neighbourhoods are characterised by severe land scarcity, with houses densely packed on plots often below the minimum legal size, resulting in a high population density reaching up to 10,000 inhabitants km^2. Construction materials and techniques are typically non-durable, and the neighbourhoods lack adequate infrastructure and services. Occupants are predominantly from the low-income class, including mostly daily wage workers, informal traders, and new rural migrants. While house owners may hold legal land tenure, the houses themselves were constructed without legal authorisation, leading to the designation of these neighbourhoods as unplanned. Notably, landlords may own multiple houses, choosing to reside in one while renting out the others to tenants. City authorities make efforts to regulate construction and demolish new (unauthorised) buildings in these unplanned areas, albeit facing challenges in enforcement. Some of these neighbourhoods have emerged in peripheral zones beyond the official city boundaries, driven by land affordability and attempts to circumvent stringent urban construction regulations. Despite this, they have become integrated into the administrative boundaries of the city over time.

Regarding the sub-categories, unplanned neighbourhoods include *old high dense* (Cyahafi), *new high dense* (Nyabisindu), *old low dense* (Kimisagara), and *new low dense* (Nyagatovu, Rwarutabura), and *suburbs or semi-urban neighbourhoods* (Karembure, Gasanze). The suburbs are considered as neighbourhoods-in-transition from rural settings and from informal to formal settings. Original landowners sell plots to individuals, who then privately build new houses or replace existing ones in the suburbs of Kigali. However, only a few such houses adhere to the city's construction code, resulting in unplanned neighbourhoods.

Mixed neighbourhoods comprise three sub-categories depending on how they were formed. The first one is the *mixed planned-unplanned* neighbourhood (Masizi) that has progressively merged around a planned site. The second is the *upgraded unplanned* neighbourhood. These have been upgraded with paved roads, newly built or renovated houses, and the provision of water and electricity facilities. Upgrading is done because of the initiative of the residents themselves (Kabeza) or by the government as a community-based urban regeneration approach. An example of government intervention is Agatare in Nyarugenge district, where wider roads are opened, basic drainage is improved, and electricity and water supply are enhanced. Residents are now authorised to build new houses or upgrade their old units. The third sub-category is the *degraded planned* neighbourhood. This encompasses former planned neighbourhoods that have become degraded due to increasing density. Despite this shift, they retain planned features such as road networks, albeit with multiple small, old houses occupying single plots. Biryogo, originally designated as a residential area for local public servants in 1960, exemplifies this phenomenon.

9.4.2 Spatial Distribution of the Neighbourhoods

In terms of spatial distribution, all neighbourhood types are scattered in the city with certain areas being predominantly characterised by particular types. For instance, planned and mixed neighbourhoods are most prevalent in Kicukiro and Gasabo districts—Kagarama, Niboye, and Kigarama sectors in Kicukiro, and Kimironko, Remera, and Kimihurura sectors in Gasabo. The favourable topography of these two districts attracts real estate investors who construct residential houses and estates aimed at high-income urban residents. Many unplanned neighbourhoods are found around the CBD in Nyarugenge and at the edge of the city. Unplanned neighbourhoods such as Gitega, Muhima, Rwezamenyo, Cyahafi, and Kimisagara in Nyarugenge attract low-income individuals employed in the informal sector in the CBD and Nyabugogo commercial hub. Due to financial constraints, these residents prefer to reside close to their workplaces to minimise transportation costs relative to their salaries.

Another notable observation is the existence of small unplanned neighbourhoods surrounding planned neighbourhoods within the inner city, and large unplanned neighbourhoods in the suburban areas. This is because when land is expropriated for a planned residential project, not all expropriated families leave the area. Instead, those who rent and those who receive low compensation use their limited resources to create new unplanned neighbourhoods in the surrounding zone (e.g. Nyagatovu and Nyabisindu around Kibagabaga and Kimironko, respectively). The large unplanned neighbourhoods in the city suburbs and peripheries indicate informal expansion of the city. In fact, the construction of a house without authorisation in the inner city is currently difficult due to the strict urban planning regulations and control and implementation of the Kigali City Master Plan, therefore, unplanned neighbourhoods have emerged in the peri-urban areas.

The identified categories mainly from spatial occupancy and housing types reflect the socio-economic characteristics of their residents. Research suggests that individual well-being can be significantly influenced by the physical, social, and economic environment in which one lives (Loschiavo 2019; Baffoe et al. 2020). Therefore, the identified characteristics bear evidence that living conditions in Kigali city neighbourhoods are not distanced from the socio-economic structure of each one of them. Those categories replicate five wealth groups, ranging from the poorest to the richest: lowest, low, middle, upper-middle, and high-income households (Manirakiza et al. 2020). Generally, the lowest and low-income groups reside in the unplanned neighbourhoods. Middle-income households mainly occupy the mixed and a few planned neighbourhoods, while most upper-middle and high-income groups occupy planned neighbourhoods. The distribution of these neighbourhoods according to income bands reflects their respective main categories of planned, unplanned, and mixed (Fig. 9.4).

9.5 Neighbourhood Inequality

Inequality among the five neighbourhood categories is analysed across four groups of indicators: (1) income and employment; (2) housing conditions as reflecting safety and satisfaction; (3) household amenities; and (4) access to education and health services.

9.5.1 Income and Employment

The household monthly income data reveals significant wealth disparity across neighbourhoods. Over 70% of residents in the lowest and low-income bands earn less than 100,000 Rwf[1] per month, which is insufficient to sustain their livelihoods, due to their employment in casual jobs. The percentages of 11.4%, 22.5%, and 33.4% among residents in the high-income, upper middle-income, and middle-income brackets, respectively, who earn less than 100,000 Rwf can be attributed to household dependents who lack income-generating activities and rely on the household head's earnings (Table 9.3).

The income reflects the employment status (Table 9.4). Many residents in high and upper-middle classes are respectively self-employed with employees (43% and 25%) compared to 6.7% and 11.3% for low and lowest income, respectively. For the first two groups, many are in the formal sector (45% and 63.3%) compared to 33.9% and 34.6% for the lowest two groups which even have a considerable proportion of employees in the informal sector—31.5% and 28.6%, respectively. Formal employment gives more security than informal employment.

Many residents in high (85.9%), upper middle (86.2%), and middle-income (75.8%) groups have permanent jobs compared to those of low (45.1%) and lowest income (51.5%). Even those doing

[1] 1 USD equals 1,276.507203 Rwanda francs (Rwf) https://www.bnr.rw/currency/exchange-rate/ (12.02.2024).

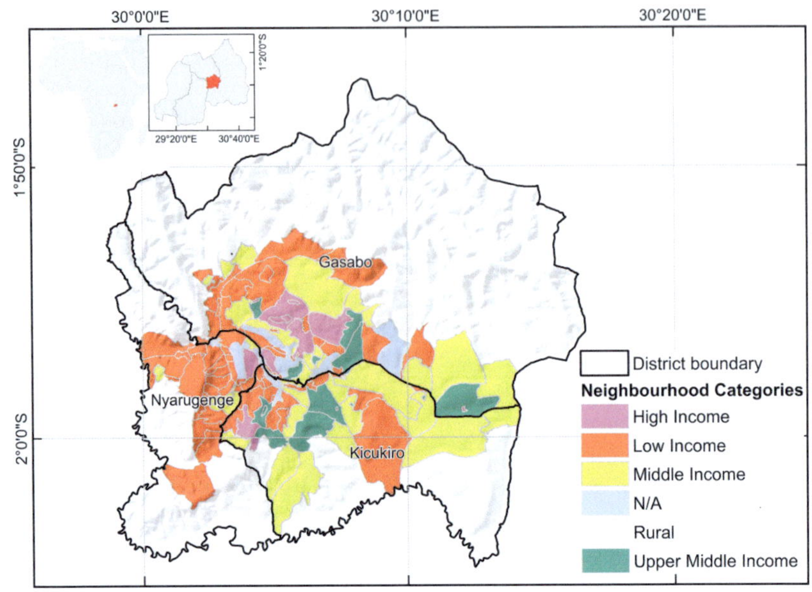

Fig. 9.4 Spatial distribution of the neighbourhoods per socio-economic bands. *Source* Authors 2021

Table 9.3 Monthly income (in 1000 Rwf) across neighbourhood categories (%)

Income bands	0–100	101–200	201–300	301–400	401–600	601–800	801–1M*	>1 M*
High-income	11.4	2.5	3.3	5.7	11.5	13.1	13.9	23.8
Upper middle-income	22.5	9.2	11.7	8.3	15	7.5	13.3	10.8
Middle-income	33.4	18.7	19.5	12	6.8	4.8	4	0.4
Low-income	78.3	13	3.3	3.7	1.3	0	0	0.3
Lowest	70.7	15.3	8	1.7	2.3	1.7	0.3	0
City level	52.3	13.4	9.2	5.8	5.5	3.8	4	4

Note *M = million
Source Primary data, 2021

Table 9.4 Employment status

Income bands	Self-employed, no employees	Self-employed, with employees	Contributing family worker	Formal employee	Informal employee	Others
High-income	6	43	3	45	3	0
Upper middle-income	8.3	25	0	63.3	1.7	3.4
Middle-income	17.4	15.5	1.9	58.1	7.1	0
Low-income	22.4	6.7	3	33.9	31.5	2.4
Lowest	21.8	11.3	0	34.6	28.6	3.8
City level	17	17.6	1.8	44.9	17.1	1.6

Source Primary data, 2021

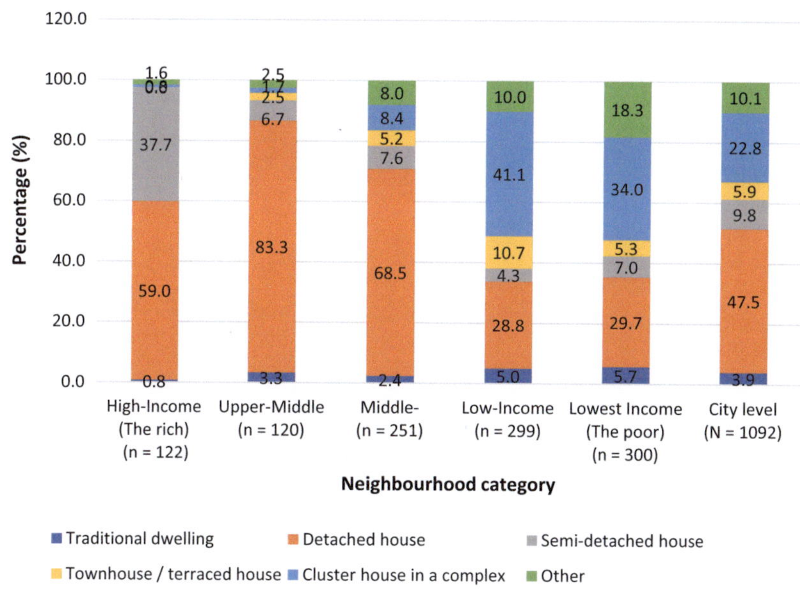

Fig. 9.5 Type of household dwelling. *Source* Primary survey, 2021

permanent jobs do not get a good wage because, as mentioned above, more than 70% earn less than 100,000 Rwf, and this is very little to support living in Kigali. The proportion of casual workers is relatively high for the two last groups (29.9% and 28.8%, respectively) compared to 1% and 1.7% for the high and upper middle-income groups.

9.5.2 Housing Type, Ownership, Safety and Satisfaction

The type of dwelling and ownership status are crucial factors for households in the city, significantly influencing residents' comfort levels both within their homes and in their neighbourhoods.

9.5.2.1 Type of Dwelling and House Ownership

The types of dwellings are different across neighbourhoods. The majority (up to 59% and 83.3%) of households who live in a detached house were found in the high and middle-income wealth category neighbourhoods, respectively, while most (up to 41.1%) of those who lived in a cluster house in a complex were in the low-income neighbourhood categories, (Fig. 9.5). Typically, affordable houses are not easily available in Kigali due to rapid and unplanned urbanisation whereby residents try to build their houses without respecting the building code (i.e. informally), which tends to create overcrowding, especially in the low- and lowest-income categories.

Regarding house ownership, 53% own their houses in full, while most (43%) of the remainder houses are rented from private landlords (Fig. 9.6). However, there is a clear contrast with 60–70% of middle- to high-income neighbourhoods owning homes compared to 41% and 46% for the low- and lowest-income categories, respectively. Quite striking is the fact that across all the neighbourhood categories the payment for ownership of houses by mortgage is negligible. This is mainly due to high building costs, and low availability of housing loans because of the growing demand particularly from the priority high-income earners and salaried individuals. Hence, most of those owning houses use family savings to build or buy them. As a result, unlike the middle- and high-income residents (i.e. 26–36%), over half (up to 56%) of the low-income categories are tenants of private landlords or institutions.

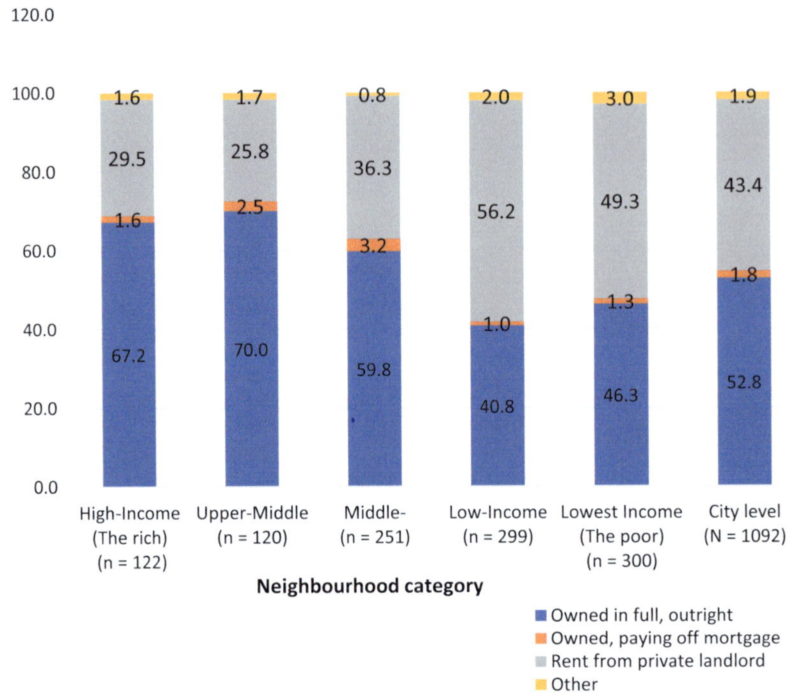

Fig. 9.6 Ownership of dwellings by householders. *Source* Primary survey, 2021

9.5.2.2 Safety and Satisfaction

Residents generally feel safe and satisfied in their neighbourhoods, although there may be slight variations across locations. Safety of residents in their own dwellings was reported to be good ranging between 90% in the lowest category up to 100% in the upper middle-income category, and 96.4% at city level. However, as expected, while out alone during daytime, reports of safety declined from 95% in the high-income category to 62.7% in the lowest category; and to 73.5% at city level. Even so, the individuals that reported to have been victims of crime in the previous two years were few, at 3.0% in the high-income community, and between 6.4% and 10.3% for the rest of the categories, without any specific trend. At city level, victims of crime were reported at 8.3%. These figures indicate that Kigali city is generally safe, even across the neighbourhood categories, but mostly so in the high-income communities.

With respect to satisfaction, it has been observed that the majority of residents are well satisfied in their neighbourhoods. However, high-income, upper middle-income and middle-income categories reported a higher satisfaction rate (respectively at 93.5%, 86.7%, and 81.6%, respectively) compared to the low (59%) and lowest (71.5%) income groups. This discrepancy in satisfaction among the income categories reflects some corresponding variability in conditions in their respective neighbourhoods, such as housing conditions which may disfavour the low-income categories in some way.

9.5.3 Household Amenities

The primary sources of drinking water for households, the type of toilet facilities used and their ownership, methods of garbage disposal, and the type of cooking energy are key indicators that reflect the livelihood level of a household.

9.5.3.1 Main Source of Drinking Water for the Households

There is a sharp contrast in available water sources whereby 82.2% of the high-income and 79% of the upper-middle categories source potable drinking

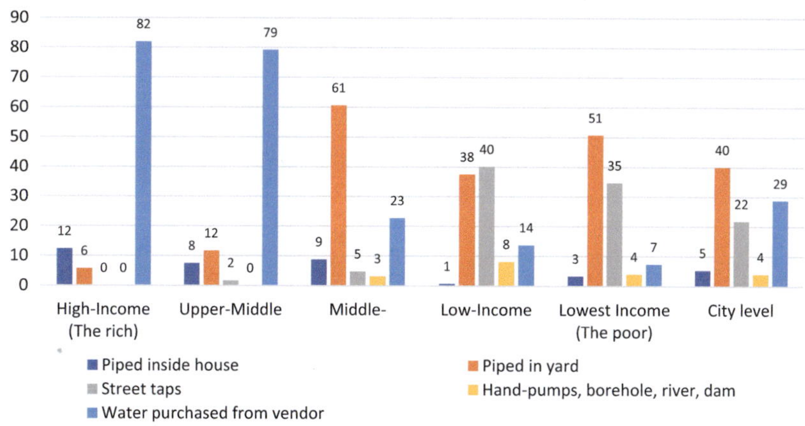

Fig. 9.7 Main source of drinking water used by the households. *Source* Primary survey, 2021

water from the vendor (mainly the city water supply company of distilled water), compared to just up 23% in the lower categories (Fig. 9.7). Many residents in other groups drink piped water either from the garden yard or street as do the majority (62%) of all the respondents. Although this water may not be completely safe like distilled water, it reflects a commendable achievement by the city in terms of ensuring access to drinking water.

9.5.3.2 Toilet Facilities Used and Their Ownership

In the city of Kigali, 70% of households own a toilet. Flush toilets are used by 30% and pit latrine by 70%. A comparison of neighbourhoods shows the big difference with flush toilets used—up to 72% by the high-income category and 68% for the upper middle-income groups, and only 33% using pit latrines; on the other hand, 79% and 90% in the lowest and low-income groups use pit latrines, with much fewer (less than 21%) using flush toilets. Most (100%, 98%, and 88%, respectively) of the high-, upper-middle, and middle-income categories own toilets. However, those in the low- and lowest-income categories owning toilets comprised as low as 41% and 58%, respectively. Just a little above the city-level rate of 30%, the remainder (59% and 41%, respectively) of the low-income categories shared toilets (Fig. 9.8). Sharing toilet facilities among families can be a big challenge exposing users to various health risks due to unhygienic conditions. It requires considerable effort to ensure that the toilet facilities are not likely to spread infections, a service which can be a little above the reach for the majority of users in the lower income categories.

9.5.3.3 Disposal of Garbage

With respect to waste disposal, door-to-door is the main garbage collection method which at city level is 86.1% (Fig. 9.9). In the neighbourhoods, apart from the low-income category, which had a rate of 68.6%, the rest had door-to-door garbage collection ranging from 82 to 100%. What is surprising is how the lowest category reached 96% door-to-door collection while the low-income and middle-income categories have just 68.6% and 82.5%, respectively. The possible explanation for these two last categories is that many of them are in the suburbs where solid waste can be used for family farming. The city of Kigali has made remarkable strides in improving hygiene and sanitation, earning recognition as one of the cleanest cities in Africa. This achievement is attributed in part to the widespread participation in monthly community clean-up activities known as ‘umuganda’. This concerted effort has positively impacted residents’ attitudes towards garbage and solid waste management.

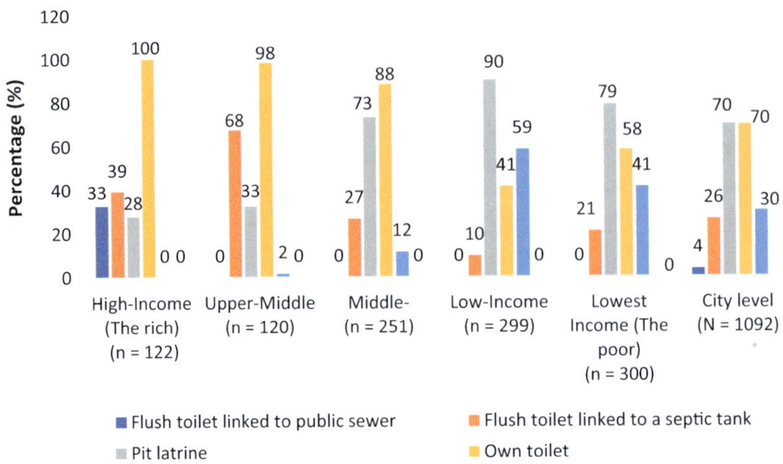

Fig. 9.8 Main toilet facility used by the household and its ownership. *Source* Primary survey, 2021

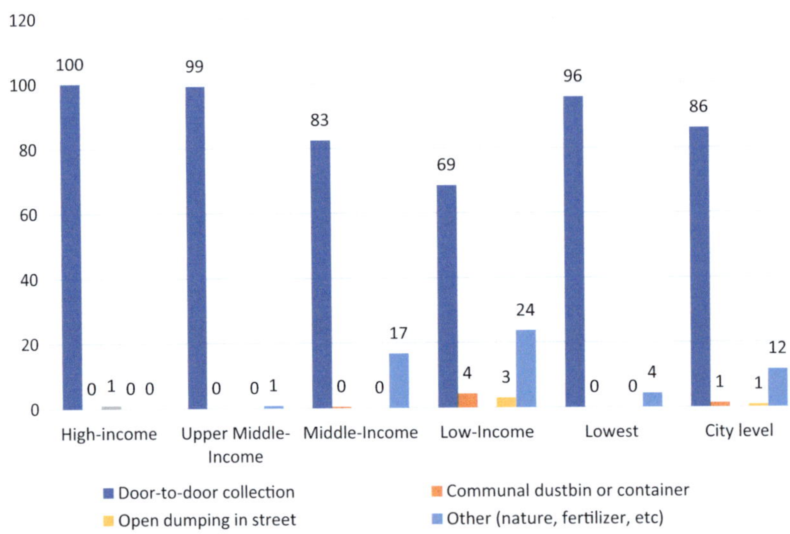

Fig. 9.9 Household garbage disposal method (%). *Source* Primary survey, 2021

9.5.3.4 Electricity Supply and Main Type of Cooking Energy

Although there is a slight variation among neighbourhood categories, with those in the low-income bracket having the highest proportion (4.7%) of households without any electricity, overall it can be concluded that electricity supply is reaching most households (97.5%) across different neighbourhood categories.

Concerning the source of energy used for cooking, disparities are evident across neighbourhoods, with many high-, upper-middle, and middle-income categories primarily using gas (86%, 76.6%, and 54.6%, respectively). In contrast, a significant portion of the low and lowest income groups predominantly rely on charcoal (68.8% and 61.7%, respectively) for cooking, which implies clean energy.

9.5.4 Education and Health Services

The education aspect was examined based on the type of schools attended by children, the mode of transportation used, and the time taken for

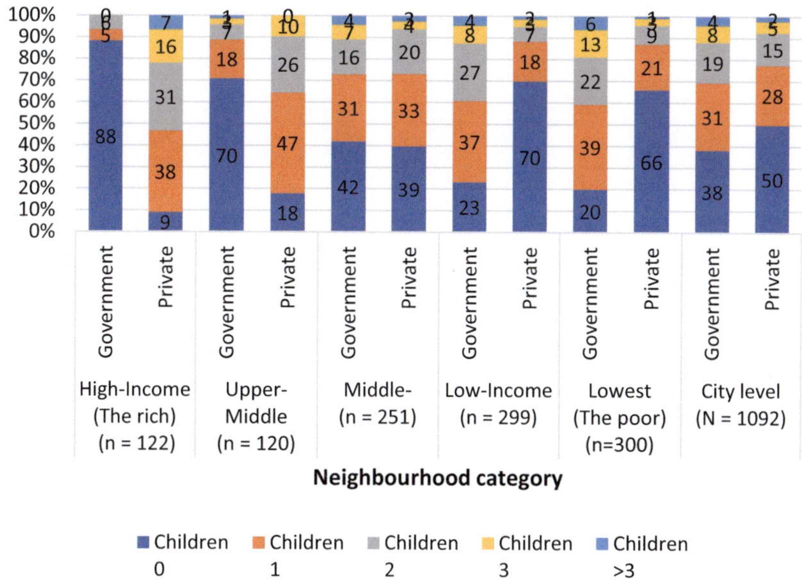

Fig. 9.10 Number of children in government or private school. *Source* Primary survey, 2021

students to reach school. For health, the assessment of service appreciation across neighbourhoods consisted of analysing residents' overall well-being, the accessibility of health facilities, and the methods of payment for medical treatment.

9.5.4.1 Education Service

The type of school attended by children varies significantly across different income categories of neighbourhoods (Fig. 9.10). Notably, only 11% of children from high-income categories attend government schools, with the majority enrolled in private schools.

Private schools are typically more expensive than government-owned ones, which is why they are predominantly attended by high-income earners. Moreover, private schools often offer better quality education, particularly at the primary level.

Conversely, 80% of children from the lowest-income category attend government schools. This trend persists across other income categories, with the middle-income category having a nearly equal number of children attending both government (42%) and private schools (39%).

Regarding the mode of transport to school for children, there are notable transport variations among income categories, with 53.2% of high-income children using motor vehicles compared to only 10.0% in the low-income category. Interestingly, the majority of children in the low- and lowest-income categories walk to school, with many spending up to 30 min commuting. This suggests that schools are easily accessible for children in these categories, reducing disparities among income groups.

9.5.4.2 Health Service

Overall city level results show that 64.7% of respondents are in very good or good health, (Fig. 9.11). At neighbourhood level, however, the residents with good or very good health in the middle-, upper-middle, and high-income categories range from 76 to 87%, while the corresponding rates for the lower categories are much lower, at 51% and 57% in the low and lowest-income categories, respectively. These results show a clear association between health status and income level, with the lowest health status among the poor. It is therefore not surprising that those who feel cheerful all the time or most of the

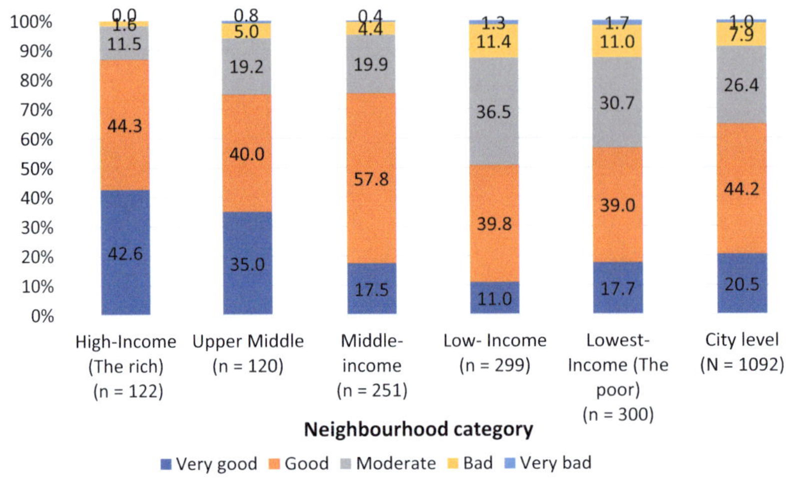

Fig. 9.11 Appreciation of personal health. *Source* Primary survey, 2021

time in the high-income category comprise 77%, compared to just 47% in the lowest category and 50% at the city level.

As for the health facilities, in normal situations during common illnesses such as colds and flu, 23–28% across all the categories visited the pharmacy for medication, although 38.5% of the high-income category went to the local doctor as opposed to 14% of the lowest category. On the other hand, when affected by a serious disease, 55–60% of the high- and upper-income categories went to a private hospital, while only 9–17% could do so from the low- and lowest-income categories. The majority of low-income categories attended either the local health centre or public health facility, which at the city level comprised 48.4% and 16.3%, respectively.

Regarding paying for treatment, the use of household resources for treatment does not differ significantly across neighbourhoods, ranging between 32.3% in the lowest category and 43% in the upper middle-income level. On the other hand, significantly more (47.5%) in the high-income categories received government aid or free health care than those in the low-income bracket (19%). However, it is worth noting that the majority, up to 79% of the low-income categories, use the community-based health insurance commonly known as 'Mutuelle de sante' supported by the government. Community insurance was initiated to especially help people in low- and middle-income categories to have access to affordable health care. It is encouraging to find that health care was generally ranked well by all the categories, good or very good, ranging from 55% (in the lowest category) to 75% (in the high-income category); it was 62% at the city level, indicating that even the poor are receiving good health care.

9.6 Conclusion

Like many other cities in developing countries, Kigali is experiencing substantial demographic and spatial growth, particularly over the last 30 years. This is primarily driven by internal migration and natural population increase. Rapid urbanisation is causing urban sprawl and informal settlements overloading the capacity of the city to ensure adequate housing and equal distribution of basic amenities to all neighbourhoods despite notable developments such as income diversification activities, new business infrastructure, and improved housing facilities.

This study helps to explore the process of urban urbanisation in Kigali and understand the living conditions of the city's residents from the bottom to the top of the economic spectrum. It permitted not only the analysis of progress made by the city in improving livelihoods of the people, but also depicting the inequalities among various strata of the urban residents. Income inequality is a common evil of urbanisation in the world, from which Kigali is not exempt. Despite the disparity in terms of dwelling categories, the high rate of housing ownership and predominantly detached houses can be an indicator of relative liveability and quality of life. However, this is a threat to sustainable urban planning as it is the cause of rapid spatial expansion.

As discussed, Kigali residents enjoy access to various amenities with a high rate of access to water and electricity and improved waste disposal. This may be the basis for most of the residents to be satisfied with the living conditions in their neighbourhoods. The quality of education is quite good although very expensive among private schools, which makes them affordable mainly to students from high-income neighbourhoods. Health care conditions are also provided to the satisfaction of most residents, most of who have some form of health insurance.

In conclusion, the study highlighted a degree of inequality across neighbourhoods, especially in housing, income, and employment. However, this inequality is not as pronounced in other aspects, as many residents still have considerable access to urban living facilities.

References

Antoni J-P (2009) Lexique de la ville. Ellipses

Baffoe G (2018) Understanding the neighbourhood concept and its evolution. Environ Urban Asia 10(2):393–402

Baffoe G, Ahmad S, Bhandari R (2020) The road to sustainable Kigali: a contextualized analysis of the challenges. Cities 105:102838. https://doi.org/10.1016/J.CITIES.2020.10283

Burak G et al (2017) Urbanisation in Africa: challenges and opportunities for conservation. Environ Res Lett 13:015002

Crooks AT (2010) Constructing and implementing an agent-based model of residential segregation through vector GIS. Int J Geogr Inf Sci 24(5):661–675

Fallatah A, Jones S, Wallace L, Mitchell D (2022) Combining object-based machine learning with long-term time-series analysis for informal settlement identification. Remote Sens 14:1226

Kohli D, Sliuzas R, Kerle N, Stein A (2012) An ontology of slums for image-based classification. Comput Environ Urban Syst 36:154–163

Loschiavo D (2019) Big-city life (dis)satisfaction? The effect of living in urban areas on subjective well-being. Temi di discussione (Economic working papers), 1221. Bank of Italy, Economic Research and International Relations Area

Macrotrends (2024) Rwanda urban population 1960–2024. Retrieved from https://www.macrotrends.net/countries/RWA/rwanda/urban-population

Manirakiza V (2014) Promoting inclusive approaches to address urbanisation challenges in Kigali. Afr Rev Econ Financ 6(1):161–180

Manirakiza V, Mugabe L, Nsabimana A (2019) City profile: Kigali, Rwanda. Retrieved from https://doi.org/10.1177/0975425319867485

Manirakiza V, Nduwayezu G, Malonza J, Mugabe L, Rutayisire PC, Nzayirambaho M, Njunwa JK (2020) Kigali: the rapid urban growth and neighbourhood dynamics of a prosperous city. Retrieved from https://11nq.com/eQyeV

Nduwayezu G, Sliuzas R, Kuffer M (2017) Modeling urban growth in Kigali city Rwanda. Rwanda J 1(2). https://doi.org/10.4314/rj.v1i1S.7D

NISR (2018) Rwandan Integrated Household Living Conditions Survey (EICV5) 2007/2018. Main Indicators Report. National Institute of Statistics of Rwanda. Retrieved from https://acesse.dev/vUwWu

NISR (2022) Fifth population and housing census: 2012. National Institute of Statistics of Rwanda, Kigali. Retrieved from https://www.statistics.gov.rw/publication/main_indicators

Park Y, Rogers GO (2015) Neighborhood planning theory, guidelines, and research: can area, population, and boundary guide conceptual framing? J Plan Lit 30(1):18–36

UN (2018) Revision of World Urbanisation Prospects. United Nations, Department of Economic and Social Affairs. Retrieved from https://www.un.org/en/desa/2018-revision-world-urbanisation-prospects

Uwimbabazi P, Lawrence R (2011) Compelling factors of urbanisation and rural-urban migration in Rwanda. Rwanda J Ser B 22, Social Sciences

World Bank (2022) Urban development. Retrieved from https://www.worldbank.org/en/topic/urbandevelopment/overview

Yamane T (1967) Statistics: an introductory analysis, 2nd edn. New York: Harper and Row

Young Foundation (2010) How can neighbourhoods be understood and defined? Retrieved from https://encr.pw/qiYnT

Neighbourhood Inequalities and Access to Housing, Amenities, and Services: A Comparative Analysis of Delhi and Madurai

10

Debolina Kundu, Tania Debnath, and Pragya Sharma

Abstract

Drawing on findings from the SHLC Household Survey conducted across 21 neighbourhoods in Delhi and Madurai, the study investigates the extent and nature of inequalities in housing, basic amenities, and social infrastructure such as education and healthcare, both within and among neighbourhoods. The comparative analysis reveals that Madurai exhibits higher levels of inequality than Delhi, particularly in housing, amenities, and services like transport, healthcare, and education. This underscores the need for smaller regional cities to enhance infrastructure and service coverage. The study also highlights significant disparities in networked services between the core and periphery areas in both cities, emphasising the role of location in shaping inequality. Additionally, intra-neighbourhood inequalities were observed, especially between different income groups, with unequal access to services, facilities, and infrastructure.

Keywords

Urban inequality · Housing · Basic amenities · Social infrastructure · Neighbourhood · Delhi · Madurai

10.1 Introduction

India, the most populated country in 2023, surpassing China, is also the epicentre of southward shifting of the mean latitude of urbanisation (UN DESA 2018, 2022). As per the UN DESA (2018) projection, 50% of the Indian population will be urban by 2046, accounting for 825.9 million people. This demographic weight of projected urban population is expected to put enormous pressure on urban infrastructure, if not planned. Moreover, the Indian pattern of urbanisation always had a big-city bias, with concentration of population, as well as resources and infrastructure in the big cities, and a stark inequality in terms of provision of infrastructure across states and size-class of the cities (Kundu et al. 1999). In fact, inequality and informality are the inherent characteristics of urbanisation in the Global South and India is no exception (Roy 2009). The colonial legacy, lack of adequate and efficient planning, and lack of economic dynamism are often

D. Kundu (✉) · P. Sharma
National Institute of Urban Affairs, New Delhi, India
e-mail: dkundu@niua.org

P. Sharma
e-mail: psharma@niua.org

T. Debnath
Department of Geography, University of Allahabad, Allahabad, India

Y. P. Wang et al. (eds.), *New Drivers of Division*, Sustainable Development Goals Series,
https://doi.org/10.1007/978-981-96-5629-5_10

cited as the main causes of such stark inequalities in these cities (Simone 2020).

If urbanisation can effectively tackle sustainability challenges, such as inequality, then it can serve as a means to achieve global sustainable development (Seto et al. 2017). Sometimes, the increasing inequalities among urban communities are closely linked to the disparity in spatial accessibility (Nicoletti et al. 2023). Spatial inequalities in the form of residential segregation, and inequalities in income, living standards, and access to infrastructure along the lines of caste, class, and religious groups are quite visible in the Indian cities (Sidhwani 2015; Kumar 2015; Bhan and Jana 2015; Haque 2016; Haque et al. 2021). Moreover, there are higher levels of intra-city inequality in bigger cities like Delhi compared to secondary cities or cities of smaller size, and often these intra-city inequalities have spatial expressions of a well-developed core and under-serviced peripheries (Bhan and Jana 2015). However, there are stark differences in health and well-being, education, employment, and life chances and exposure to crime among residents in different neighbourhoods (Wang and Kintrea 2019), which is difficult to capture at aggregate level. Most of the existing empirical research conceptualises 'wards' as the units for studying intra-city spatial inequalities as these are the smallest officially recognised administrative units in a city, unlike in Western countries, where neighbourhoods are the lowest units of study for similar socio-economic analysis. Unfortunately, paucity of official data beyond ward level limits neighbourhood level analysis in studying intra-city inequality. As a result, micro-level inequality is under-researched in most of the Indian cities.

To fill this research gap, the present study intends to explore neighbourhood level inequality in Indian cities that shapes everyday lived experiences of the residents. Based on the findings of a large-scale household survey covering 21 neighbourhoods in Delhi and Madurai, the present study aims to explore the nature and extent of inequalities in various areas that exist in different types of neighbourhoods and within similar types of neighbourhoods. This chapter focuses primarily on studying existing spatial inequalities in housing, basic amenities, and provision of public amenities such as education, health, and other social infrastructure at the neighbourhood level.

Following the introductory section, the second section of the chapter discusses the study area and provides a comprehensive picture of Delhi and Madurai and different types of neighbourhoods in these two cities. The next section discusses the data and methods used for analysing inequality. The fourth section discusses the condition of housing and different types of amenities and services across different types of neighbourhoods. The fifth provides an empirical measurement of existing inequalities in access to housing and amenities across as well as within different types of neighbourhoods. The final section discusses the findings and includes the conclusion.

10.2 Study Area: Delhi and Madurai in a Comparative Lens

Akin to the other chapters of this book this study has been carried out under a international consortium titled 'Sustainable, Healthy and Learning Cities and Neighbourhoods', anchored by the University of Glasgow and funded by United Kingdom Research and Innovation's Global Challenges Research Fund (GCRF). Two metropolitan cities, Delhi and Madurai, have been selected as case study cities, the former being the national capital of India, while the latter represents a regional city of comparatively smaller size. These two cities differ in population size, area, historicity, levels of socio-economic development, and economic growth prospects. These diversities helped to draw attention towards the role of city size and prosperity in affecting intra-city inequality in access to housing and services.

Delhi is the main administrative and political centre of India. Moreover, the Delhi urban agglomeration is the second largest in the world, following Tokyo, and one of the most rapidly growing urban agglomerations in the country. On the other hand, Madurai is a regional city in the southern state of Tamil Nadu, which shows characteristics of a 'shrinking city' with depopulation in the core along with economic stagnation and out-migration (Kundu et al. 2019).

Although Delhi is the national capital city with global standard infrastructure and economic prosperity, it is also an unequal city (Chakravorty 2021; Sircar 2021). It is often referred to as 'a city of multiple cities' (McFarlane et al. 2017: 12). The city is characterised by wide heterogeneity of built forms and infrastructure ranging from elite bungalow zones and gated housing societies to unauthorised and unplanned *jhuggi-jhopris* (JJ) or slums. Only a quarter of its population live in planned colonies as per the zoning of master plans. The remaining city is often considered as unplanned and residential pockets are manifested as unauthorised regularised colonies, unauthorised colonies, rural villages, urban villages, resettlement colonies, and slums (Bhan 2013). While the planned colonies and Lutyens bungalow zone accommodate the wealthiest sections of the population, the slums, including JJ clusters and resettlement colonies, provide cheap housing for the poor. The unauthorised colonies and urban villages often house the less well-to-do people.

Madurai, on the other hand, is among the oldest cities in India and has immense historical and regional importance in the southern state of Tamil Nadu. Once at the peak of religious and economic prosperity, the city started experiencing de-industrialisation at the beginning of the twenty-first century, which rigorously modified the city's socio-economic and spatial structure. In addition, the de-densification of the city core, followed by urban sprawl and the emergence of wealthier suburbs, led to a substantial spatial restructuring of the city (Kundu et al. 2019). Also, the city experienced resettlement drives of slums along the river Vaigai to the newly constructed resettlement colonies in the peripheries. Like Delhi, residential areas of Madurai can be categorised into groups as per their socio-economic status ranging from high-class bungalow zones to slum clusters and peripheral villages.

In terms of spatial growth, both cities exhibit a planetary pattern of expansion towards the peripheries. Like any other global city, Delhi experienced rapid growth in terms of built-up area in the last few decades. However, the sprawl in its periphery was not associated with corresponding provision of basic infrastructure in the newly built areas, resulting in a stark difference in services between the core and the peripheries (Bhan and Jana 2015). Similarly, Madurai went through boundary expansion in 2010 to accommodate sprawling townships and peripheral villages inside the jurisdiction of the Madurai Municipal Corporation (Kundu et al. 2019).

10.3 Data and Methods

10.3.1 Data

The chapter is based on a primary survey of 2378 households in Delhi and Madurai conducted during August–December 2021. Based on a previous exercise of neighbourhood mapping and auditing, different types of neighbourhoods were identified for the case study of these two cities. Two neighbourhoods from each of the categories were selected for the survey. In total, 1335 households from 11 neighbourhoods in Delhi and 1043 households from 10 neighbourhoods of Madurai were interviewed. The sampling ensured good representation of all main types of neighbourhoods as far as possible; and within each type of neighbourhood there was a good representation of the people. It is observed that the residential areas of Madurai can be categorised into several types according to their socio-economic status ranging from high-class bungalow zones to slum clusters, and peripheral villages and slums for the urban poor. Therefore, all the neighbourhoods selected for the survey were further categorised into five classes based on the average income of the residents of the neighbourhoods: (1) high-income, (2) upper middle-income, (3) middle-income, (4) low-income, and (5) poor neighbourhoods.

10.3.1.1 Indicators

The chapter focuses on selected indicators of housing and basic amenities. To address the issue of housing conditions in the study, the focus was primarily on the adequacy of living space and indicators such as floor area and availability of a separate kitchen facility. This study divided all amenities into two categories, namely, basic or lower order, and higher order. Basic or lower order amenities include access to piped water, toilet facilities, and to any form of garbage collection. On the other hand, amenities of a higher order include facilities like access to tap water within premises, toilet facilities within premises, access to flush toilets connected to public sewer, and door-to-door garbage collection. All the indicators, except for the total floor area, were categorical and were converted into binary variables.

Apart from household amenities, the chapter considers neighbourhood level amenities such as availability and access to open space, proximity to different services such as medical facilities, market places, sports facilities, and different institutions. All of these variables are binary in characteristic. On the other hand, proximity to schools, which is a categorical variable, was also used in the study.[1] Households were asked the distance (in kilometres) of the school for the eldest child, and households were categorised according to the distance travelled to schools. This indicator roughly indicated accessibility and choice in terms of distance to school. However, measuring access to good quality schools is beyond the scope of the paper. Apart from satisfaction levels of the households about the schools their children were attending as well as neighbourhood schools, no question was asked regarding the quality or nature of schools. Therefore, this study limits itself to accessibility. This categorical variable has been converted to a binary variable by combining the categories into two groups, namely, 'up to 2 km' (indicating a close proximity) and 'more than 2 km' (otherwise).

10.3.2 Methods to Assess Inequality

This study used the Gini coefficient, one of the most popular measures to estimate overall inequality (Malakar et al. 2018). The value of the Gini coefficient lies between 0 and 1. When the value lies closer to 0, it indicates a fair distribution of income or resources; when it is close to 1, it indicates a concentration of income or resources, or inequality. The present study utilised this index to calculate city-level and neighbourhood level inequalities in terms of housing and access of amenities and social infrastructure using household-level data collected through the survey.

10.4 Access to Quality Housing, Amenities, and Services Across Different Neighbourhoods

Access to quality housing and adequate basic amenities and infrastructure such as safe drinking water, sanitation and hygiene determines quality of life. Good health and quality education are also two important pillars of sustainability. The United Nations Sustainable Development Goals (SDG 3 on good health, SDG 4 on quality education, SDG 6 on water and sanitation, and SDG 11 on sustainable cities and habitats) have provided special focus on access to these facilities for all. Moreover, the New Urban Agenda (NUA) put forward by HABITAT III (2016) emphasised equitable access to 'quality housing as a component of the right to an adequate standard of living' and 'universal access to safe and affordable drinking water and sanitation', particularly in the rapidly urbanising Global South. Reducing inequalities and ensuring no one is left behind are integral to achieving the Sustainable Development Goals. However, unequal availability and accessibility of facilities within and among countries is a persistent cause for concern in the Global South, including India.

[1] IN05: On an average day, how much distance in km does this child have to travel from home to school? This categorical variable gives the following options: 'Less than 1 km', '1–2 km', '2–5 km', '6–10 km', '11–20 km', '21–30 km', '31–40 km', '40 km, and more'. What is INO5? It is not in the text.

Table 10.1 Housing characteristics of surveyed households across different types of neighbourhoods

Types of amenities	High income	Upper middle income	Middle income	Lower middle income	Poor	All samples
	Delhi					
Mean floor area sq. m.	123.0	102.7	112.4	33.9	26.4	81.7
Congestion factor	0.0	0.4	10.8	16.4	32.4	10.4
Access to sole use of Kitchen	100.0	100.0	93.6	72.9	45.7	85.0
	Madurai					
Mean floor area sq. m.	199.2	164.2	85.4	39.3	33.9	110.6
Congestion factor	0.0	9.0	10.0	15.6	63.7	18.2
Access to sole use of Kitchen	100.0	100.0	98.2	87.7	82.0	94.8

Note Congestion Factor = Distribution of HHs with three or more members in one room or less
Source Primary survey in Delhi and Madurai, 2021

Although the condition of housing and basic service provision in urban India has improved substantially over the years, there is still wide disparity across caste, religious, and economic groups. Often the weaker socio-economic groups, that is, Scheduled Castes (SC) and Scheduled Tribes (ST) lack access to quality housing and basic amenities (Kumar 2015). Existing studies on Delhi highlight a stark inequality between core and peripheral areas in terms of housing quality and infrastructure (Bhan and Jana 2015) and a similar pattern can be noted in Madurai (Kundu et al. 2019). The present section discusses the condition of housing and basic amenities across neighbourhoods and within neighbourhoods of different economic status.

10.4.1 Housing

In terms of the provision of quality housing, the study focuses on the availability of adequate living space, which has been analysed on the basis of total floor area, congestion factor,[2] and access to a separate kitchen (Table 10.1). A city-level comparison indicates that the average floor area of the surveyed households in Delhi was 81.7 sq. m. compared to 110.6 sq. m. in Madurai. However, in spite of a higher mean floor area in Madurai, 18.2% of the surveyed households lived in congested dwellings[3] compared to 10.4% households in Delhi. Notably, 60% of households in Madurai live in rented accommodation and congestion was much higher among the tenants (Kundu et al. 2019).

A further analysis of availability of living space across neighbourhoods of different income categories indicates a wide disparity in both the cities. In the case of Delhi, mean floor area varied from 123 sq. m. in high-income neighbourhoods to 26.4 sq. m. in poor neighbourhoods. On the other hand, mean floor area had a wider range, from 199.2 sq. m. in the high-income neighbourhoods to 33.9 sq. m. in poor income households in Madurai. Although average floor area is higher in Madurai, 63.7% of households in poor income neighbourhoods live in congested houses. On the other hand, 16.4% and 32.4% of the households in low-income and poor neighbourhoods, respectively, in Delhi live in congested dwellings

[2] Congestion factor has been defined as share of households having more than 2 members living in no exclusive room or just 1 room.

[3] Dwellings with one or less room for a family of three members and more.

Table 10.2 Access to basic amenities across different neighbourhoods in Delhi and Madurai

Types of amenities	High income	Upper middle income	Middle income	Lower middle income	Poor	All samples
	Delhi					
Piped drinking water	98.6	82.2	86.6	86.5	53.5	82.4
Access to individual toilet	100.0	100.0	100.0	96.3	88.9	97.5
Access to garbage collection	100.0	99.8	97.6	96.4	63.7	93.0
	Madurai					
Piped drinking water	60.4	87.9	44.2	53.3	70.8	66.7
Access to individual toilet	100.0	100.0	100.0	80.7	98.8	97.4
Access to garbage collection	99.2	98.5	94.6	98.5	84.3	95.3

Source Primary survey in Delhi and Madurai, 2021

(Table 10.1). This is mainly due to larger average household size.

It is seen that, 94.8% of households in Madurai had access to a separate kitchen compared to 85% households in Delhi. Even 82% of households in poor neighbourhoods in Madurai had a separate kitchen while the corresponding share was only 45.7% for Delhi (Table 10.1). It is a common sight in a majority of the slums of Delhi that a significant proportion of the households, used a single room to perform all day-to-day activities, which include cooking and bathing.

The above findings indicate that a larger metropolitan cities like Delhi find it difficult to provide adequate living space to their poor residents compared to a regional city like Madurai, which is more sustainable and inclusive in terms of housing.

10.4.2 Access to Basic Amenities

Piped water supply: This section discusses the condition of basic amenities like piped water supply, basic toilet facilities, and garbage collection versus higher order amenities such as piped water within houses, exclusive access to toilet facilities, and door-to-door collection of garbage in different neighbourhoods.

It is observed that 82.4% of surveyed households used piped water for drinking in Delhi compared to only 66.7% in Madurai. Poor quality and irregular supply are considered as reasons behind reliance on other sources of drinking water. In fact, only 60.4% of the surveyed households in the high-income group of Madurai used piped drinking water. On the contrary, a larger share (70.8%) of residents living in poor neighbourhoods, such as the slums along the river Vaigai, relied on piped drinking water. In Delhi, almost 100% households in high-income neighbourhoods, and more than 80% of the households from upper middle, middle, and low-income neighbourhoods used piped water supply. However, only 53.5% of slum households in Delhi used piped water for drinking (Table 10.2).

The situation is even more unequal if the coverage of piped water supply within dwelling units is considered. Only 36.1% in Delhi and 20.5% households in Madurai from poor neighbourhoods had access to piped drinking water within the dwelling units (Table 10.3). Half of the households in Madurai from the poor neighbourhoods were dependent on public taps for drinking water. Similarly, in the absence of adequate quantity and quality of drinking water supply on a regular basis, 45.3% of surveyed households in the poor neighbourhoods of Delhi depended on purchased water supplied through public tankers.

Toilet facilities: The findings indicate that households in both Madurai and Delhi had almost universal access to toilet facilities, barring a small share of households living in low-income and poor neighbourhoods in both the cities. However, access to flush toilets connected to public sewerage networks seems to be equally

Table 10.3 Access to higher order basic amenities across different neighbourhoods of Delhi and Madurai

Types of amenities	High income	Upper middle income	Middle income	Lower income	Poor	All sample
	Delhi					
Piped tap water inside house	98.6	82.2	78.4	79.6	36.1	76.9
Flush toilet linked to public sewer	91.6	49.8	46.5	52.0	25.6	54.2
Own toilet/exclusive use	100.0	98.4	93.1	91.4	65.3	91.2
Door-to-door collection	99.5	90.8	86.9	64.7	12.6	74.3
	Madurai					
Piped tap water inside house	54.4	85.3	38.0	53.3	20.5	54.0
Flush toilet linked to public sewer	72.9	72.0	29.6	0.0	70.9	55.9
Own toilet/exclusive use	100.0	99.1	100.0	80.7	67.0	91.1
Door-to-door collection	73.2	74.8	76.8	98.5	64.4	75.7

Source Primary survey in Delhi and Madurai, 2021

unsatisfactory in both the cities. The situation is highly unequal across different types of neighbourhoods. In both the cities, where the access to flush toilets with sewage network is very wide in high-income neighbourhoods (more than 90% in Delhi and around 72% in Madurai), it is deplorable in poor neighbourhoods in Delhi (25.6% households) and in low-income neighbourhoods of Madurai there is a complete absence. Compared to universal access to exclusive toilet facilities in high, upper middle and middle-income neighbourhoods in both the cities, only two-thirds of the households living in poor income neighbourhoods had this facility (Table 10.3). It is noted that 19.8% and 24.7% households of poor neighbourhoods in Delhi and Madurai were dependent on community toilets, which often lack cleanliness and running water facilities.

Solid waste collection: There are also wide variations in solid waste collection mechanisms in these two cities. Although more than 90% of the total households in Delhi had some arrangement for garbage disposal, it was noted that only 63.7% households living in poor neighbourhoods had access to it (Table 10.2). However, only around three-quarters of the surveyed households of these two cities had access to door-to-door collection of garbage. In Delhi, this service is noted to be mostly limited to high and middle-income neighbourhoods and only 12.6% households from poor neighbourhoods had similar access. However, this facility is more inclusive in Madurai with 64.4% of the households from the poor neighbourhoods reporting access to the service (Table 10.3).

10.4.3 Other Physical and Social Infrastructure Services

Like housing and basic amenities, access to other services such as transport, shopping, and medical facilities within a convenient distance varied widely across different types of neighbourhoods in these two cities, barring universal proximity to bus stops, grocery shops, and pharmacists in all types of neighbourhoods (Table 10.4). In Delhi, the high-income neighbourhoods enjoyed universal access to most of the amenities and social infrastructure. On the contrary, the situation was less than satisfactory for neighbourhoods

Table 10.4 Access to physical and social infrastructure and services across neighbourhoods of Delhi and Madurai

Access to Different Facilities		Delhi						Madurai					
		High Income	Upper Middle Income	Middle Income	Lower Income	Poor Income	All Sample	High Income	Upper Middle Income	Middle Income	Lower Income	Poor	All Sample
Transport	Bus Stop	100.0	99.6	98.4	100.0	79.8	96.4	92.3	97.3	90.4	100.0	96.7	95.2
	Train/Metro/Subway Stations	96.7	68.8	41.5	31.3	27.3	55.9	66.1	79.6	32.3	0.7	55.1	54.1
Health Care Facilities	Doctor's surgery/clinic	100.0	84.7	22.4	31.5	40.0	59.7	94.3	58.2	84.6	85.2	68.7	76.2
	Pharmacy	100.0	100.0	98.6	93.4	98.6	98.3	98.6	99.1	90.0	100.0	95.7	96.9
Different Institutions	Bank Branch	100.0	99.4	98.2	81.7	71.6	91.5	98.7	92.8	86.9	95.5	76.0	90.3
	Post Office	100.0	97.6	97.7	66.4	46.2	83.9	42.2	86.1	73.0	3.5	72.2	61.0
	Police Station	100.0	93.7	40.0	81.4	80.7	81.2	78.4	96.6	34.6	37.1	66.5	68.9
Market	Grocery/Small Food Shop	100.0	100.0	100.0	97.6	96.5	99.0	98.6	99.8	91.2	96.7	85.7	95.0
	Supermarket	100.0	86.4	27.4	25.5	5.6	54.3	91.7	69.7	39.3	85.0	47.9	67.3
Recreation	Sports Facilities	84.8	16.0	75.1	18.5	10.2	39.7	67.4	30.0	29.4	2.8	19.6	33.1
Open Space/Park	Availability	99.7	28.5	82.7	68.8	23.3	59.1	78.1	46.2	46.4	39.2	48.0	53.0
	Accessibility	82.8	17.0	59.8	37.5	7.3	40.1	58.1	27.6	17.7	8.2	19.8	29.0

Note Colour schemes

Colour	Range
(green)	*80–100%*
(yellow)	*50–80%*
(orange)	*<50%*

Source Primary survey in Delhi and Madurai, 2021

of similar background in Madurai as only 42.2% and 67.4% households, respectively, had access to a post office and sports facilities.

Health care: Notably, the rich-poor divide in access to some amenities and services was noted to be higher in Delhi. To elucidate, access to basic health care facilities within neighbourhoods is observed to be very limited in low-income and poor neighbourhoods of Delhi as only 31.5% and 40%—households living in these neighbourhoods had access to a doctor's clinic. In Madurai, on the other hand, access to basic health care is noted to be much better in the poor neighbourhoods, indicating more equal distribution of services. Also, the slums and resettlement colonies in Madurai are often provided with basic health care facilities owing to interventions of government and non-state actors. However, the low-income neighbourhoods particularly located at the peripheries of Madurai city, were found to have limited access to services among all the categories of neighbourhoods (Table 10.4). Most of the non-slum low-income neighbourhoods in Madurai lack all the basic facilities, mostly owing to their disadvantageous locations.

Open spaces and parks: The situation is less than satisfactory in the case of availability of open space in both the cities. Only 53% of surveyed households in Madurai and 59.1% in Delhi had the availability of open space/parks. Moreover, the situation is highly unequal across different types of neighbourhoods. While availability of open space is noted to be universal in high-income neighbourhoods in Delhi, the figure was as low as 23.3% in poor neighbourhoods. However, disparity in access to open space is noted to be marginally lower in Madurai as the share varied from 78.1% in high-income neighbourhoods to 39.2% in low-income neighbourhoods. In fact, the situation is noted to be worse when accessibility to these open spaces/parks is taken into consideration as only 40% and 29% surveyed households in Delhi and Madurai confirmed their accessibility to the same. Notably, the accessibility to open space is dismal in poor neighbourhoods in both the cities as only 7.3% residents of poor income neighbourhoods of Delhi and 8.2% low-income neighbourhoods in Madurai had access to these open spaces (Table 10.4).

10.4.3.1 Physical Proximity to Schools

Access to schools is an important social infrastructure and has been measured through the maximum distance travelled to school by a student. The city-level distribution reveals that the children from

Table 10.5 Distance travelled to schools in different types of neighbourhoods in Delhi and Madurai

Distance travelled	High income	Upper middle income	Middle income	Lower income	Poor	All sample
	Delhi					
Less than 1 km	10.6	27.4	50.3	46.3	45.4	37.6
1–2 km	6.3	22.3	21.6	27.0	48.3	26.1
2–5 km	43.9	36.1	11.1	10.4	5.4	20.3
6–10 km	25.3	14.2	10.7	12.5	0.9	12.0
>10 km	13.9	0.0	6.2	3.9	0.0	4.0
	Madurai					
Less than 1 km	10.0	31.6	15.7	35.1	23.6	22.1
1–2 km	10.2	24.5	27.2	13.3	33.5	21.4
2–5 km	39.4	14.0	27.2	15.1	32.7	26.6
6–10 km	26.3	25.9	26.2	24.6	7.7	22.6
>10 km	14.2	4.0	3.7	12.0	2.5	7.4

Source Primary survey in Delhi and Madurai, 2021

63.7% households in Delhi travelled up to 2 km to school, while only 4% commuted more than 10 km for the same purpose. The respective figures for Madurai were 43.5% and 7.4% (Table 10.5).

A detailed analysis shows that more than 45% of the households from middle, low income and poor neighbourhoods in Delhi sent their children to schools located in close proximity to their homes (within a kilometre). In total, more than 70% children from these three types of neighbourhoods travel less than 2 km to school (Table 10.5). A further analysis indicates that about 80% of children from low-income and more than 90% of children from poor neighbourhoods in Delhi were attending government schools with free education, mostly located in close proximity to these neighbourhoods. On the other hand, only 16.9% children from high-income neighbourhoods of Delhi attended schools within their neighbourhoods (within 2 km) while around 40% children travelled beyond 5 km. A majority of the children from these neighbourhoods were enrolled in a private school facility and used school buses or personal motorised vehicles to commute.

Similarly, in Madurai, it is seen that only 20.2% households from high-income neighbourhoods sent their children to schools within close proximity (2 km), whereas about 14.2% households reported that their children travelled more than 10 km to school. The travel pattern in other neighbourhoods is mixed. It is noted that the children from 57.1% households of poor neighbourhoods and 56.1% from upper middle-income neighbourhoods sent their children to schools within 2 km from their neighbourhoods. On the other hand, due to lack of access to schools in close proximity, 36.6% households in low-income neighbourhoods reported that their children commuted more than 5 km (Table 10.5). Unlike Delhi, the commuting pattern in Madurai was not linked to socio-economic status of the neighbourhoods as 12% children from low-income neighbourhoods commute more than 5 km. Also, a large proportion of school-going children, even those belonging to poor income neighbourhoods, attended private schools.

Overall, one can notice a visible class disparity in terms of access to different types of amenities, facilities, and services in both the Indian cities. It is observed that residents of low-income and poor neighbourhoods lack adequate basic amenities, particularly the networked infrastructure. In the absence of adequate toilet facilities and an efficient garbage collection mechanism, access to a hygienic environment is a major concern in these neighbourhoods. Similarly, access to other amenities and services such as market facilities, transport, medical services, recreation, and access to open and green space varied widely across the different neighbourhood categories. Importantly, the poor neighbourhoods lack adequate access to all of these in both the cities. However, the access to education facilities showed a haphazard pattern. In Delhi, the households from neighbourhoods of lower socio-economic status generally availed free education in government schools which were within close proximity to their place of residence. On the other hand, distance of school from the neighbourhood was not a significant factor in selection of the educational facility across the different neighbourhoods in Madurai.

Table 10.6 Gini coefficients in housing conditions and basic amenities in Delhi and Madurai

Housing conditions and types of amenities	Delhi	Madurai
Housing		
Floor area (sq. m.)	0.417	0.553
Availability of separate kitchen	0.150	0.052
Basic amenities		
Access to piped drinking water	0.176	0.333
Access to toilet facilities	0.025	0.026
Access to garbage collection facilities	0.070	0.047
Amenities of higher order		
Access to piped drinking water within dwelling	0.231	0.460
Access to flush toilet connected to public sewer	0.458	0.441
Exclusive access to toilet	0.088	0.089
Access to door-to-door collection of garbage	0.257	0.243

Source Primary survey in Delhi and Madurai, 2021

10.5 Inequality Across and Within Neighbourhoods

This study measures inequality in terms of adequate living space and access to different amenities at city-level and between neighbourhoods using the Gini coefficient (G). The value of the G lies between 0 and 1. When the value lies closer to 0, it indicates a fair distribution of income or resources; when it is close to 1, it indicates a concentration of income or resources, or inequality.

A comparison of Gini coefficient values measured at the city-level indicates that the Gini coefficient values are much lower in both the cities in terms of access to basic amenities like piped drinking water, toilet facilities, and garbage collection facilities. It is noted that coefficient values are even less than 0.1 in case of access to toilet and garbage collection facilities, indicating wide coverage of these services in both the cities and near absence of class divisions. On the contrary, access to piped drinking water is more unequal, particularly in Madurai (G = 0.333) (Table 10.6). Poor and irregular supply of piped drinking water was more responsible for this disparity than the lack of a piped water network, barring a few localities in the newly merged peripheral wards.

In contrast to the basic amenities, accessibility to higher order amenities have been found to be more unequal in both the cities. Most of these services are either self-owned or paid services, which explains their skewed distribution. For example, access to flush toilet facilities connected to a public sewer network was highly unequal in both the cities (G = 0.458 in Delhi and G = 0.441 in Madurai). In most cases, public sewer networks in Indian cities are limited to the city core, excluding the rapidly sprawling peripheries. In the peripheral areas of the cities, the existence of public sewerage networks can only be found in high-income gated housing societies, where such networked infrastructure is mostly privatised (Sawyer et al. 2021). Similarly, unequal access to paid services like door-to-door garbage collection was considerably high in these two cities (G = 0.257 for Delhi and G = 0.243 for Madurai) (Table 10.6). In contrast, exclusive access to toilet facilities was widely available

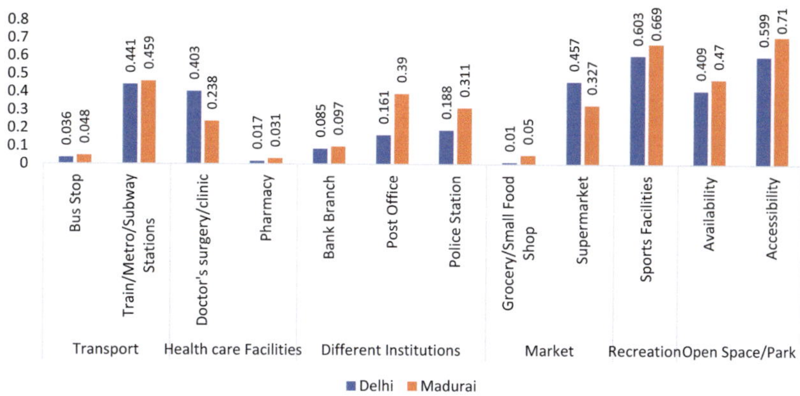

Fig. 10.1 Inequality (Gini coefficients) in access to different physical and social infrastructure and services in Delhi and Madurai. *Source* Primary survey in Delhi and Madurai, 2021i

in both the cities with G values less than 0.1. This might be an outcome of the investments under the central government flagship programme Swachh Bharat Mission—Urban, targeted to eliminate open defecation by constructing individual toilets and public toilets.

Like higher order amenities, access to adequate living space was noted to be highly unequal in Madurai (G = 0.553) and Delhi (G = 0.417). The G values indicate that inequality in terms of living space is higher in Madurai compared to Delhi. However, inequality in terms of access to separate kitchen facilities was low in both the cities (Table 10.6).

It is observed in both the cities that access to amenities and infrastructure such as bus stops, pharmacies, grocery shops, and banks was the same compared to doctors' clinics, supermarkets, train/metro stations, and sports facilities. To elaborate, proximity to bus stops (G = 0.036 in Delhi and 0.048 in Madurai), pharmacies (G = 0.017 in Delhi and 0.031 in Madurai), grocery shops (G = 0.010 in Delhi and 0.050 in Madurai), and banking facilities (G = 0.085 in Delhi and 0.097 in Madurai) was mostly available across all categories of neighbourhoods with lower coefficient values. On the contrary, amenities like proximity to train/metro stations (G = 0.441 in Delhi and 0.459 in Madurai), supermarkets (G = 0.457 in Delhi and 0.327 in Madurai), sports facilities (G = 0.603 in Delhi and 0.669 in Madurai) were observed to be highly unequal. Moreover, availability of doctors' clinics within the neighbourhood was noted to be highly unequal across neighbourhoods of Delhi (G = 0.403) as many poor ones lacked certified health care facilities in proximity. Further, availability as well as accessibility of open spaces/parks was highly unequal in both the cities (Fig. 10.1). Likewise, access to schools measured in terms of distance from the neighbourhood was unequal in both the cities and particularly high in Madurai (G = 0.566 compared to G = 0.363 in Delhi) (Table 10.9).

An analysis of the data for different types of neighbourhoods indicates that intra-neighbourhood inequalities increased with decreasing socio-economic status. Intra-neighbourhood inequalities were much higher in low-income and poor neighbourhoods in both the cities of Delhi and Madurai in case of access to adequate living space and basic as well as higher order amenities, with a few exceptions (Tables 10.7 and 10.8).

It is noted that inequalities are in terms of adequate housing and basic amenities higher where the share of tenants is higher. Inequality within same categories of neighbourhoods for floor area displayed a heterogenous pattern in both the cities. It is observed that inter-household inequality within similar categories of neighbourhoods was high among high- (G = 0.415) and upper middle-income (G = 0.602) neighbourhoods in Madurai. A similar pattern was

Table 10.7 Inequality (Gini coefficients) in housing and basic amenities for different types of neighbourhoods in Delhi and Madurai

Types of amenities		High income	Upper middle income	Middle income	Lower income	Poor
		Delhi				
Housing	Floor area (sq. m.)	0.109	0.291	0.429	0.303	0.328
	Availability of separate kitchen	0.000	0.000	0.064	0.271	0.543
Basic amenities	Access to piped drinking water	0.015	0.178	0.134	0.135	0.465
	Access to toilet facilities	0.000	0.000	0.000	0.037	0.111
	Access to garbage collection facilities	0.000	0.002	0.024	0.036	0.363
Amenities of high order	Access to piped drinking water within dwellings	0.015	0.178	0.216	0.204	0.639
	Access to flush toilet connected to public sewer	0.084	0.502	0.535	0.480	0.744
	Exclusive access to toilet	0.000	0.016	0.069	0.086	0.347
	Access to door-to-door collection of garbage	0.005	0.092	0.131	0.353	0.875
		Madurai				
Housing	Floor area (sq. m.)	0.415	0.602	0.340	0.295	0.390
	Availability of separate kitchen	0.000	0.000	0.018	0.123	0.180
Basic amenities	Access to piped drinking water	0.396	0.121	0.558	0.467	0.292
	Access to toilet facilities	0.000	0.000	0.000	0.193	0.012
	Access to garbage collection facilities	0.008	0.015	0.054	0.015	0.157
Amenities of high order	Access to piped drinking water within dwellings	0.456	0.147	0.620	0.467	0.795
	Access to flush toilet connected to public sewer	0.271	0.280	0.704	0.000	0.291

(continued)

Table 10.7 (continued)

Types of amenities		High income	Upper middle income	Middle income	Lower income	Poor
	Exclusive access to toilet	0.000	0.010	0.000	0.193	0.330
	Access to door-to-door collection of garbage	0.268	0.252	0.232	0.015	0.356

Source Primary survey in Delhi and Madurai, 2021

found in middle-income neighbourhoods in Delhi (Table 10.7).

Inter-city comparison indicates that intra-city inequality is highest in poor neighbourhoods in Delhi in case of access to adequate living space and basic as well as higher order amenities. In contrast, the coefficient values are much lower for the similar type of neighbourhood in Madurai (Table 10.7). The situation is similar in case of access to other amenities and social infrastructure as poor neighbourhoods in Delhi were found to be highly unequal, not only compared to other types in the city but also compared to the same type in Madurai (Table 10.8).

Unlike Delhi where inter-household inequality is found to be worst in poor neighbourhoods, in Madurai, inter-household inequality in access to amenities and services like train/metro station, post office, sports facilities, and accessibility to open space is noted to be worst in low-income neighbourhoods (Table 10.8).

Moreover, it is observed that a high level of inter-household inequality prevails in middle-income neighbourhoods in terms of overall access to networked infrastructure like piped drinking water (G = 0.558), particularly within the dwelling units (G = 0.620), and access to flush toilet system connected to a public sewerage network (G = 0.704). A similar situation was noted in the case of access to flush toilets connected to public sewers in upper middle (G = 0.502) and middle-income (G = 0.535) neighbourhoods in Delhi (Table 10.7). These upper middle and middle-income neighbourhoods of Delhi fall outside the purview of master plan zones and therefore, are yet to be connected with networked infrastructure such as public sewerage. Moreover, both of these types of neighbourhoods in both the cities accommodate a large number of tenants, which explains the high inequality in access to these amenities.

Compared to access to basic and other amenities and services, inequality in access to schools is often determined by choice, as found in both these cities. It is observed that inequality was higher in high-income neighbourhoods on the basis of distance to schools. In Delhi, inequality decreased with decreasing socio-economic status of neighbourhoods. This was because most households in poorer neighbourhoods send their children to close-by government schools. However, there is no particular pattern found in Madurai. The coefficient value was considerably high in low-income neighbourhoods (G = 0.516), while it was much lower in upper middle-income neighbourhoods (G = 0.439) (Table 10.9). The mixed pattern of inequality indicates two facts: less density of schools as a whole in the city, and a higher affordability of private schools, as even a large share of households in poor neighbourhoods sent their children to private schools.

10.6 Discussion

The results from this primary field based study highlight prevailing inequalities across different types of neighbourhoods in Delhi and Madurai in terms of adequate living space and access to various amenities and services. The inequality is low in terms of access to basic services across the neighbourhoods although both of these cities had high coverage of basic amenities like drinking water, toilet facilities, and garbage collection,

Table 10.8 Inequality (Gini coefficients) in access to different physical and social infrastructure across different types of neighbourhoods in Delhi and Madurai

Access to different facilities		High income	Upper middle income	Middle income	Lower income	Poor
		Delhi				
Transport	Bus stop	0.000	0.004	0.016	0.000	0.202
	Train/metro/subway stations	0.033	0.312	0.585	0.687	0.727
Health care Facilities	Doctor's surgery/ clinic	0.000	0.153	0.776	0.685	0.600
	Pharmacy	0.000	0.000	0.014	0.066	0.014
Different institutions	Bank branch	0.000	0.006	0.018	0.183	0.284
	Post office	0.000	0.024	0.023	0.336	0.538
	Police station	0.000	0.063	0.600	0.186	0.193
Market	Grocery/small food shop	0.000	0.000	0.000	0.024	0.035
	Supermarket	0.000	0.136	0.726	0.745	0.944
Recreation	Sports Facilities	0.152	0.840	0.249	0.815	0.898
Open SPACE/Park	Availability	0.003	0.715	0.173	0.312	0.767
	Accessibility	0.172	0.830	0.402	0.625	0.927
		Madurai				
Transport	Bus stop	0.077	0.027	0.096	0.000	0.033
	Train/metro/subway stations	0.339	0.204	0.677	0.993	0.449
Health care Facilities	Doctor's surgery/ clinic	0.058	0.418	0.154	0.148	0.313
	Pharmacy	0.014	0.009	0.100	0.000	0.043
Different institutions	Bank branch	0.013	0.072	0.131	0.045	0.240
	Post office	0.578	0.139	0.270	0.965	0.278
	Police station	0.216	0.035	0.654	0.629	0.336
Market	Grocery/small food shop	0.014	0.002	0.088	0.033	0.143
	Supermarket	0.083	0.303	0.607	0.150	0.521
Recreation	Sports facilities	0.326	0.700	0.706	0.972	0.804
Open space/park	Availability	0.219	0.538	0.536	0.609	0.520
	Accessibility	0.419	0.724	0.823	0.918	0.802

Source Primary survey in Delhi and Madurai, 2021

Table 10.9 Inequality (Gini coefficients) in distance to schools across different types of neighbourhoods in Delhi and Madurai

	High income	Upper middle income	Middle income	Lower income	Poor	Total
Delhi	0.831	0.503	0.280	0.267	0.063	0.363
Madurai	0.799	0.439	0.571	0.516	0.429	0.566

Source Primary survey in Delhi and Madurai, 2021

owing to successful implementation of national flagship programmes like Atal Mission for Rejuvenation and Urban Transformation (AMRUT) and Swachh Bharat Mission-Urban (SBM-U). Besides networked infrastructure, access to open space is also highly unsatisfactory and unequal in these two cities. The highly dense slums hardly had provision for open space. Even if open spaces are available, maintenance is poor.

There was a stark inequality across neighbourhoods of different socio-economic status in both the cities especially on the basis of access to housing, amenities, and services, particularly networked infrastructure such as public sewerage. Inter-city comparison reveals a higher level of inequality in Delhi compared to Madurai where, distribution of amenities and services, except for piped drinking water, are more inclusive along with piped drinking water. Madurai exhibits higher levels of inequality in terms of access to adequate living space. Apart from class-based inequalities, each type of neighbourhood exhibits different characteristics of inequality.

The high-income bungalow zones or gated communities enjoyed all major facilities in both the cities, particularly in Delhi. The residents enjoyed all the basic, higher-level and other services, and were willing to avail of distant infrastructure as they could afford long commutes. For example, the high-income households in the city preferred sending their children to private schools, even if they were situated far from the neighbourhood. The same pattern is observed in the high-income neighbourhoods of Madurai.

On the other hand, the low-income and poor neighbourhoods in both the cities lacked adequate living space and a majority of basic amenities including regular supply of drinking water, safe sanitation and solid waste management. Moreover, intra-household inequality was found higher in poor neighbourhoods, particularly in Delhi, indicating their differences in negotiating and purchasing power. It is observed in many cases that tenants were more dependent on shared resources such as public drinking water stands and community toilets. On the contrary, slums and resettlement colonies of Madurai are well served with all networked infrastructure like piped water, toilets, and garbage collection facilities. With the higher occurrence and easy availability of government schools in almost every poor neighbourhood of Delhi, accessibility to government schools was found to be higher in the city.

In Delhi, a high degree of inter-household inequality was noted within upper middle and middle-income neighbourhoods in case of access to higher order amenities such as public sewerage, exclusive toilet, and piped drinking water. A large share of upper middle and middle-income neighbourhoods in Delhi fall outside the master plan zones of the city and therefore lack networked infrastructure and planned growth. These neighbourhoods also provide a low cost rental housing, and therefore are marked by unequal access to amenities. The situation is similar in Madurai as upper middle and middle-income neighbourhoods had a high share of tenants and these neighbourhoods showed higher levels of inter-household level inequality.

Moreover, some of the inequality within the same type of neighbourhood can be explained in terms of locational differences of neighbourhoods of similar socio-economic status. For example, the upper middle-income neighbourhood of Uttaranchal Enclave in Delhi, which is in the northern periphery of the city, lacked access to a public sewerage network owing to its locational disadvantage. A similar pattern is observed for low-income and poor neighbourhoods in the case of piped drinking water supply. For example, a resettlement colony and a slum cluster located in the periphery of Delhi, lacked access to both piped water supply and public sewerage. This situation was prevalent in the poor socio-economic peripheral newly merged wards in Madurai. The latter lacked all networked infrastructure like piped drinking water supply and a piped sewerage network since these areas were rural in nature until recently, with no networked infrastructure and educational and healthcare facilities.

The present analysis reveals the presence of a stark class inequality in access to quality housing and services across different types of neighbourhoods. The chapter also sheds light on the inequalities in accessing services and facilities among residents living in neighbourhoods with a similar socio-economic background. However, a limitation of the study remains as the categorisation of the neighbourhoods is based on the status of household income and does not include the geographical location, whether in the core or periphery, which brings in the probability of 'aggregate error' in the estimation of the inequalities. Therefore, such categorisation undermines the differences and unequal situation arises from existing core–periphery differences in infrastructure. It leaves scope for conducting a separate study to measure inequality in access to services, facilities, and infrastructure at the neighbourhood level in order to unravel the intra-neighbourhood inequalities as well as inequalities between the neighbourhoods located in different parts of the cities. This will portray a more accurate measurement of the reality that exists. Also, it is important to highlight the locational aspects of neighbourhoods while determining the intra-neighbourhood inequalities within a city.

10.7 Conclusion

Inequality in accessing services, facilities, and infrastructure is very much the character and not the outcome of urbanisation in the Global South. This chapter attempted to highlight intra-city inequalities among different types of neighbourhoods in accessing quality housing, services, facilities, and infrastructure. It not only emphasised the existing inequalities across different types of neighbourhoods, but also discussed intra-neighbourhood inequalities in the two Indian cities of Delhi and Madurai. A comparative analysis revealed that higher levels of inequality existed in Madurai compared to Delhi in housing, basic amenities, and other services including transport, health care, and education. This illustrates the fact that smaller and regional cities need to develop a better coverage of infrastructure and amenities.

It was also observed that there is acute shortage of networked infrastructure in the peripheral neighbourhoods of Madurai, as these areas were very recently brought under the city administration. This also brought to light the responsibility of the city administration to provide major public services to the residents of the newly added areas of the city, which in many cases were overlooked. Consequently, such unequal patterns can ultimately shed some light on the advancing distributive equity, particularly the differences between the core and periphery of the city. Other than that, there was enough evidence of stark inequalities between different income classes, especially in accessibility of services, facilities, and infrastructure prevalent in urban India. The intra-neighbourhood categorical inequalities also highlighted the significance of location in shaping the inequality framework as well as individual wealth status in determining accessibility to services, facilities, and infrastructure. From the urban policy perspective, it is essential to understand the emergent inequalities that arise when urban infrastructure lags with respect to the increment in urban population. Other than the problems of addressing the distribution of infrastructure in an urban set-up, the disparity in accessing urban infrastructure depends considerably on the class-caste-communal divide. These inherent divisions pose fundamental constraints on the path of sustainable urban development. The research results strongly recommend an urgent need for inclusive policies towards reduction of disparities in accessing basic urban infrastructure and improvement of the available infrastructure in Indian cities.

References

Bhan G (2013) Planned illegalities: housing and the 'failure' of planning in Delhi: 1947–2010. Econ Pol Wkly 48:58–70

Bhan G, Jana A (2015) Reading spatial inequality in urban India. Econ Pol Wkly 50:49–54

Chakravorty S (2021) Introduction: what is Delhi? In: Chakravorty S, Sircar N (eds) Colossus: the anatomy of Delhi, vol 15. Cambridge University Press, pp 1–26

Haque I (2016) Infrastructure development and access to basic amenities in Class-I cities of West Bengal, India: insights from census data. J Infrastruct Dev 8(1):36–84

Haque I, Das DN, Patel PP et al (2021) Towards an enhanced understanding of caste-based residential

segregation in Indian cities: reflections from Kolkata and Bengaluru. Spat Demogr 9:187–211. https://doi.org/10.1007/s40980-021-00085-8
Kundu A, Bagchi S, Kundu D (1999) Regional distribution of infrastructure and basic amenities in urban India: issues concerning empowerment of local bodies. Econ Pol Wkly 34:1893–1906
Kundu D, Lahiri B, Pandey A, Sharma P (2019) City profile: Madurai, India. Environ Urban Asia 10(2), 219–234. https://doi.org/10.1177/0975425319867487
Kumar A (2015) Housing shortages in urban India and socio-economic facets. J Infrastruct Dev 7(1):19–34
Malakar K, Mishra T, Patwardhan A (2018) Inequality in water supply in India: an assessment using the Gini and Theil indices. Environ Dev Sustain 20:841–864. https://doi.org/10.1007/s10668-017-9913-0
McFarlane C, Silver J, Truelove Y (2017) Cities within cities: intra-urban comparison of infrastructure in Mumbai, Delhi and Cape Town. Urban Geogr 38(9):1393–1417
Nicoletti L, Sirenko M, Verma T (2023) Disadvantaged communities have lower access to urban infrastructure. Environment and Pl anning 8: Urban Analytics and City Science 50131:831–849
Roy A (2009) Why India cannot plan its cities: informality, insurgence and the idiom of urbanization. Plan Theory 8(1):76–87
Sawyer L, Schmid C, Streule M, Kallenberger P (2021) Bypass urbanism: re-ordering center-periphery relations in Kolkata, Lagos and Mexico City. Environ Plan A: Econ Space 53(4):675–703
Seto KC, Golden JS, Alberti M, Turner BL (2017) Sustainability in an urbanizing planet. Proceedings of the National Academy of Sciences 114(34), 8935–8938
Sidhwani P (2015) Spatial inequalities in big Indian cities. Econ Pol Wkly 50:55–62
Simone A (2020) Cities of the global south. Ann Rev Sociol 46:603–622
Sircar N (2021) Assets and spatial inequality. In: Chakravorty S, Sircar N (eds) Colossus: the anatomy of Delhi, vol 15. Cambridge University Press, pp 59–87
United Nations, Department of Economic and Social Affairs, Population Division (2018). World Urbanization Prospects: The 2018 Revision, Online Edition
United Nations, Department of Economic and Social Affairs, Population Division (2022). World Population Prospects 2022, Online Edition
Wang YP, Kintrea K (2019) Sustainable, healthy and learning cities and neighbourhoods. Environ Urban ASIA 10(2):146–150

11 Neighbourhood Inequalities and Challenges to Urban Planning in Dhaka

Shilpi Roy

Abstract

Fast-growing Southern cities accommodate diverse and intricate neighbourhood clusters that exhibit disparities in opportunities. This chapter delves into the socio-spatial elements of spatial inequality among neighbourhood types in Dhaka and entrenched political and economic factors. A multi-stage sequential quantitative analysis using a wealth-based neighbourhood categorisation plus data from various sources reveals deficiencies and differences in services, facilities, and the built environment across five neighbourhood types. Inadequate resources, weak institutional capacity, market-driven resource distribution, and elitist planning contribute to Dhaka's status as an unequal city. This chapter advocates for innovative, people-centred neighbourhood planning that recognises the internal differentiation of neighbourhoods both in Dhaka and similar contexts.

Keywords

Inequalities · Deprivation · Neighbourhood · Urban planning · Dhaka

S. Roy (✉)
Urban and Rural Planning Discipline, Khulna University, Khulna, Bangladesh
e-mail: shilpi.roy@urp.ku.ac.bd

11.1 Introduction

Within the reality of unprecedented rates of urbanisation (UN Habitat 2022), neighbourhoods of Southern cities, in general, struggle to meet the growing demand for services and facilities, infrastructure and housing (Wang and Kintrea 2019). Overall, higher economic growth has improved living conditions in large cities; still, poverty and neighbourhood inequality in basic infrastructure, housing, education, and health services are explicit (Moshi et al. 2018; Kundu et al. 2018). In response, public policies, deprived of neighbourhood focus, have appeared ineffective in mitigating urban socio-spatial inequalities (Sowgat and Roy 2022b; Delos Reyes et al. 2018).

Neighbourhood environment, the accessibility of services and facilities, and housing influence an individual's health and well-being (Roy and Sowgat 2024; Maddock et al. 2021), educational attainment (Jensen and Seltzer 2000), and economic prosperity (Dodini and Fantini 2006). Historically, neighbourhood inequalities remain an important driver of deprivation and higher-order urban segregation (Sampson 2019). The prevailing socio-spatial inequality and differentiation of urban neighbourhoods in the developing South are widely acknowledged as a key barrier to achieving several sustainable development goals, specifically SDG 1, SDG 3, SDG 4, and SDG

Y. P. Wang et al. (eds.), *New Drivers of Division*, Sustainable Development Goals Series,
https://doi.org/10.1007/978-981-96-5629-5_11

11 (Wang and Kintrea 2019). Thus, neighbourhood studies are essential in understanding urban spatial inequalities regarding policy implications for determining the overall progress and sustainability of cities (Sampson 2017).

Inequality studies of neighbourhoods generally focus on livelihoods, housing, service gaps, health and education inequalities in informal settlements (Hosseini et al. 2023; Lubeck-Schricker et al. 2023; Shrum et al. 2023; Bharathi et al. 2022). Communities, particularly the affluent ones, have begun to create an emerging literature examining gated communities, health and health-related behaviour, safety, and access to and the quality of local facilities, including public green spaces, the built environment and quality of life (Bhandari 2023; Dang et al. 2023; Ehwi 2023; Ji et al. 2023; Heroy et al. 2023; Robb Larkins and Durão 2023). However, there remains a significant knowledge gap concerning diverse neighbourhood agglomerations and their effect on inequalities (Lynge et al. 2022) in the large, fast-growing Southern cities where the rapid urbanisation process contributes to diverse and complex neighbourhood clusters with disparities in opportunities (Kundu et al. 2020).

As the fifth fastest-growing megacity and sixth densest city in the world (Roy 2023; Statista 2023), Dhaka, the capital of Bangladesh, confronts critical sustainability challenges, including failure to meet the rising need for housing, infrastructure, and urban services, widespread poverty, and acute environmental pollution (Roy 2021). Despite the introduction of formal planning practices predominantly during the 1950s, its neighbourhoods have developed spontaneously and unplanned across the city (Baffoe and Roy 2023). More importantly, the organic nature of urban transformation in Dhaka has resulted in the emergence and entrenchment of diverse and complex neighbourhood structures (Roy and Sowgat 2020) that suffer from an unequal distribution of resources and life opportunities while simultaneously being rife with social discrimination as low-wealth communities, and urban villages are at a heightened risk of facing injustice (Sowgat and Roy 2022a). A recent study by Sowgat and Roy (2022b) shows that the city faces acute socio-spatial division concerning access to urban services, housing, employment, and religion. They find clusters of poorly served areas at the periphery and slums as isolated communities.

This chapter aims to analyse and shed light on spatial inequality in Dhaka using the neighbourhood as the analytical unit. It answers the questions of (1) What are the diverse socio-spatial elements of spatial inequality and differentiation between the neighbourhood classes in Dhaka, and (2) What are the deep-rooted factors of inequalities in this city that contribute to inequality?

The chapter contributes to the intellectual and policy arena in several areas. A neighbourhood wealth-based approach to developing a typology for Dhaka reveals an advanced understanding of the diverse residential environment in large, fast-growing, South Asian cities. The systematic analysis uses a multi-dimensional approach to reveal internal characteristics of these types from the latest data in various natures that may offer urban scholars a better understanding of the internal structure of these classes and create a baseline for tracking future changes in the neighbourhood groups. A relative comparison of these neighbourhood types concerning natural and built environments, infrastructure, services and facilities, and the socio-economic conditions of the residents would provide new evidence of and insights into inequalities and deficiencies in urban neighbourhoods in the Southern context and promote more sustainable urban development in similar settings.

The next section defines the neighbourhoods in Dhaka, followed by the methodology applied to classify them and the different data sources that have contributed to explaining the internal structure and differences of the communities in this city. The following section offers insight into the disparities in different neighbourhood classes in the city. The final two sections discuss the underlying entrenched forces associated with these inequalities and conclude with remarks on reducing these differences.

11.2 Defining Neighbourhoods: The Context of Dhaka

According to the 2011 national census, Dhaka has 945 neighbourhoods (NH), locally called Mahalla. Among these, 863 are residential communities; the remaining are primarily restricted areas or those with less than 35% residential land use. The rapid urban reconstruction process has reconfigured the attributes of these Mahallas, including the emergence and growth of non-residential land uses, and redefined their boundaries over time. Some Mahallas may serve as physical neighbourhood boundaries for some, where they live, share infrastructures, services, and facilities, and interact with others (Alkan and Maksudyan 2020). For others, their perceived neighbourhood "pseudo environment" may differ from these boundaries (Lippmann 1991, in Wong et al. 2012). Instead of the subjective perception of neighbourhoods, which are fluid and often multiple and dynamic (Alkan and Maksudyan 2020), the objective approach has laid the foundation of defining neighbourhoods for this study as the means to have policy implications. Although the smallest census tracts may not resemble the perceptual reality of individuals in all cases, these units, in most cases, demonstrate the nature of neighbourhoods in Dhaka where groups of residents are associated with a collection of spatially grounded attributes. This study uses 863 Mahallas to represent neighbourhoods in Dhaka for which census data are available.

Out of the 863 residential neighbourhoods, a small share is developed through plans laid out by public and private sectors, namely Gulshan, Uttara, Banani, Wari, and Bashundhara Residential Areas. Individuals and real estate companies lead the organic transformation of most of the neighbourhoods in Dhaka. These planned and unplanned neighbourhoods are distributed across the city core and periphery; some are old, while others are very young. To explain the internal structure and differences of the communities in Dhaka, it is crucial to classify the neighbourhoods that would reflect the variations in development patterns, locations, and age.

11.3 Neighbourhood Classification and Data Sources

Most macroscale (intra and inter-city) quantitative studies use limited variables and outdated census data to understand fast-growing and complex neighbourhood dynamics and inequality, be it income, wealth, or urban services (Lynge et al. 2022). Those who tend to use diverse attributes for studying socio-spatial segregation, isolation, and clustering patterns can inform sectoral policy solutions to redress the housing and service gaps in the neighbourhoods (Sowgat and Roy 2022a), yet these are not suitable for offering more profound insights into diverse neighbourhood classes which are necessary for targeted interventions. Hence, investigating prevailing residential wealth-based neighbourhood types through systematic analysis and revealing their distinct environmental, infrastructural, and socio-economic characteristics are crucial for demonstrating the persistent and multi-dimensional nature of residential inequalities in large, fast-growing, Southern cities such as Dhaka.

Dhaka's 863 neighbourhoods were initially categorised as a part of a large study following a multi-stage sequential quantitative analysis of selected variables (services, facilities, and socio-economic) of the 2011 national census that produced five clusters, each representing a neighbourhood category. Before clustering, important variables were initially identified through principal component analysis. Once clustering was done, Exploratory data analysis (EDA), with the help of scatter plots of these important variables, helped to define each cluster's neighbourhood characteristics. For EDA, the values of the variables were divided into five categories: low (0–20%), lower-moderate (21–40%), moderate (41–60%), higher-moderate (61–80%), and high (81–100%). Thus, for assigning the wealth class against each cluster, these figures were scored on a scale of 1 to 5, where 1 indicates the best status, and 5 is the worst status against each variable used for EDA. The average score of all 19 variables closer to 1 indicates the wealthiest neighbourhoods, and closer to 5 represents low-wealth neighbourhoods.

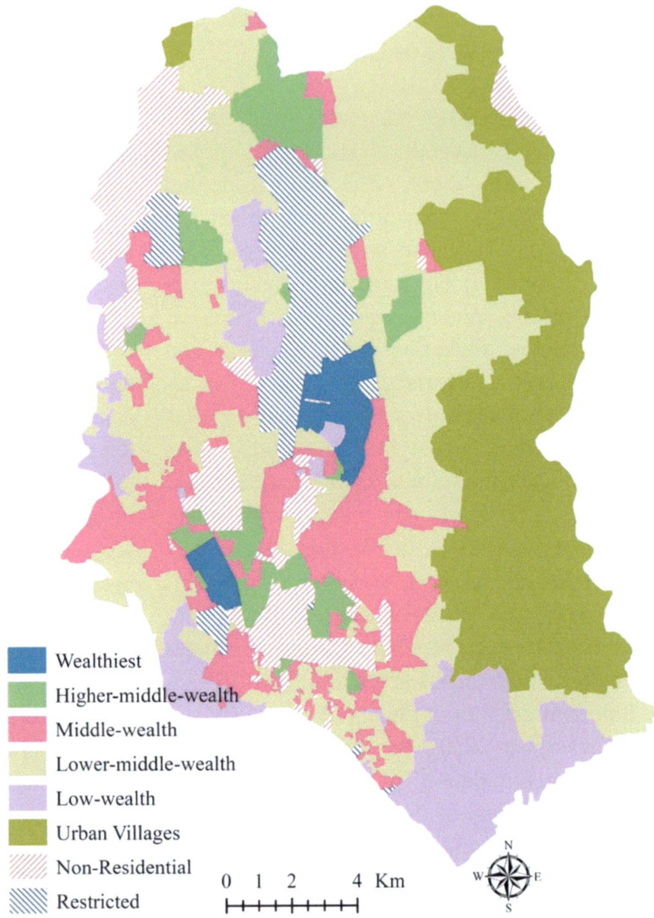

Fig. 11.1 Neighbourhood clusters categorised by wealth in Dhaka. *Source* Author, 2025

Then, screening was performed to identify if any neighbourhood belonged to an inappropriate category with the help of EDA and neighbourhood sustainability auditing (NSA). NSA was conducted between 2018 and 2019 and was aimed at achieving an overall idea about the 863 neighbourhoods in Dhaka through field auditing, photographic analysis, interviews, and GIS analysis of the spatial data on land use. For refining the neighbourhood category, data on various socio-economic, built and living environment-related indicators, including access to and coverage of urban services and facilities, were gathered from multiple sources, including online platforms, spatial data used for the Dhaka Detailed Area Plan and direct field observations. These data largely resembled the wealth of neighbourhoods, approximating the economic status of their residents.

Consequently, following scoring, screening, and judgement, several neighbourhoods shifted from old clusters to new ones. Finally, following various iterations, applying intuition and specialist opinion regarding the resources of neighbourhoods and the wealth of the residents, the 863 neighbourhoods have been classified into five wealth-based categories that best represent the various neighbourhoods of Dhaka: (1) wealthiest neighbourhoods (2) higher-middle-wealth neighbourhoods (3) middle-wealth neighbourhoods (4) lower-middle-wealth neighbourhoods and (5) low-wealth neighbourhoods and urban villages (Fig. 11.1). These neighbourhood classes incorporate all possible varieties in the city concerning development patterns, locations, and age.

Neighbourhood, wealth-based categorisation applied a long list of variables gathered through various data sources specified above. For this chapter, the variables of socio-economic status, demographic characteristics, ethnicity, housing characteristics, the quality of the built environment, and the urban service situations of the neighbourhoods that are used worldwide for measuring socio-economic segregation in cities (see Randon-Furling et al. 2020) informed the neighbourhood, structure-based differentiation.

There are differences within the same wealth categories in terms of age, location, pattern of development, and so on. It is noteworthy that neighbourhoods are dynamic spaces that are constantly changing; two neighbourhoods within the same wealth class often do not exhibit the same characteristics. Even due to the dominance of unplanned development of a significant share of neighbourhoods in Dhaka, neighbourhoods are home to mixed-income groups, the differential quality of the built environment, and urban amenities. To better understand the nature and differentiation of neighbourhoods, 11 neighbourhoods from the five wealth-based categories that reflect the variations in development patterns, locations, and age were further studied with the help of a walkthrough analysis, a study of urban amenities, health and education institutional surveys, physical feature surveys, photographic analysis, interviews (conducted between 2019 and 2022), 11 focus group discussions (FGDs) (conducted in 2022) and a GIS analysis of the recent spatial data prepared for the Dhaka Detailed Area Plan, 2016–2035. The use of several tools and the latest quantitative and qualitative data offered a deeper understanding of the diversity, deficiencies, and differences in Dhaka at the neighbourhood scale that are discussed in the next section.

11.4 Neighbourhood Differentiation in Dhaka

A significant share of the wealthiest neighbourhoods in Dhaka, except Uttara, are planned and located in the city core. The other four types have predominantly developed organically. A small section of upper middle-wealth neighbourhoods is planned, mainly in the city core. In contrast, large pockets of slums are distributed throughout the city, and the urban villages are in the peripheral areas. This section broadly features housing, the built environment characteristics, urban amenities, services and facilities, and access to employment opportunities prevailing in different neighbourhood categories in Dhaka, all of which explain the internal differentiation among the neighbourhoods.

11.4.1 The Wealthiest

Dhaka hosts a small number of the wealthiest neighbourhoods (21 out of 863) (Table 11.1), where the quality of the built environment, services, and amenities are significantly better than those in other categories (Fig. 11.2). Being public sector-led planned residential areas (e.g. Gulshan, Dhanmondi, Bonani, and Uttora), the city's best available service coverage and amenities, good road networks with sidewalks, natural bodies of water, parks, high security and high-end retail, restaurants, and banks have made these areas the most desired places for living in the city for the rich and super-rich, such as in the case of Gulshan (NSA 2018–2019). The EDA of the 2011 national census data shows that all households use water seal sanitation facilities and have full electricity coverage. The share of owner-occupied homes/flats in the wealthiest neighbourhoods is 21–40%, which is not as impressive as the urban villages. These neighbourhoods offer the possibility of getting the highest rental value in the city and a good return on investment in property. Most owners rent out their properties; 61–80% of residents live in rented properties.

Residents in this category have a high literacy rate (81–100%). In reality, the wealthiest neighbourhoods enjoy much better quality education facilities than the rest. Residents in these areas have access to the country's top-quality private English medium schools and universities, which are within close reach (NSA 2018–2019; Education Institution Survey 2019; FGD 2022).

Table 11.1 Distribution of neighbourhoods in the wealth categories

Wealth-based neighbourhood category	Number of NH	Number of NH in %	Area in Sq. Km.	Area in %	Population as of 2011 national census	Population per Sq. Km. as of 2011
Wealthiest	21	2	7.09	3	108,787	15,344
Higher-middle-wealth	65	8	15.68	6	580,224	37,004
Middle-wealth	287	33	36.61	14	2,656,509	72,562
Lower-middle-wealth	367	43	103.7	40	3,695,710	35,638
Low-wealth neighbourhoods	80	9	32.59	13	1,309,285	40,174
Urban villages	43	5	61.52	24	219,791	3573

Source Author, 2025

Fig. 11.2 Upscale residences for affluent residents in Gulshan. Photo by author

Due to higher levels of security and close access to the diplomatic zone and other conveniences, foreign nationals prefer to live in the Gulshan, Banani, and Baridhara areas (NSA 2018–2019; FGD 2022). Their presence and interest in these areas have created exclusive zones with premium-quality housing, convenience shops and gated, highly secured areas (NSA 2018–2019). Most planned residential neighbourhoods like Gulshan, Banani, and Dhanmondi tend to accommodate a higher share of non-residential land uses like offices and commercial properties at the cost of liveability (NSA 2018–2019). Ultimately, the concentration of white-collar employees (both men and women) is more noticeable in these neighbourhoods than in any other area of Dhaka.

11.4.2 The Higher-Middle-Wealth Neighbourhoods

Like the wealthiest neighbourhoods, poor-quality sanitation facilities are almost absent in the higher-middle-wealth neighbourhoods (EDA of the 2011 National Census data). Households predominantly have access to tap water (81–100%), and the presence of any low-quality housing structure is almost zero (0–20%). In some areas, there is a majority of house ownership, yet residents generally live in rented properties (81–100%) (EDA of 2011 National Census data; NSA 2018–2019).

However, unlike the wealthiest, out of 65 higher-middle-wealth neighbourhoods, a significant share of neighbourhoods that are developed

Fig. 11.3 New residential development in a young, planned, upper-middle-wealth neighbourhood, Bashundhara. Photo by author

organically suffer from high-density, chaotic environments, increased noise pollution due to excessive non-residential land use, and a severe deficiency of parks, playgrounds, open spaces, and quality walkable roads (NSA 2018–2019; walk-through analysis and study of urban amenities, 2019–2022). Even the city's top private sector-led planned residential area, Bashundhora (Fig. 11.3), which drew interest to the educated service sector population and families with school-going children, faces the issues of poor drainage, water-logging, the prominence of mosquitoes, and a deficiency of public spaces and sidewalks (Walk-through analysis 2019–2022; GIS analysis of the spatial data; FGD 2022). Financing to improve poor road and drainage conditions is an added burden for the property owners. Residents feel the area is regimented, isolated, and lacks spontaneity due to high security (FGD 2022). There are, however, exclusive enclaves in organically developed neighbourhoods where residents of relatively new properties benefit from increased safety and cleanliness through active resident-led management associations (FGD 2022).

As the EDA of the 2011 national census data suggests, working in the service industries is the primary income source for males and females living in the neighbourhoods of this category (81–100%). As in the wealthiest neighbourhoods, residents in this category also have a high literacy rate (81–100%). Residents have easy access to some of the city's top private and public schools and colleges that they can afford (FGD 2022).

In contrast, high-end private health services remain inaccessible for a significant share of tenants. A few neighbourhoods in this category host some of the most eminent private clinics and hospitals (FGD, 2022). For example, the Central Road neighbourhood (Fig. 11.4), an organically developed neighbourhood, has a healthcare density of 9.37 per thousand population, of which 97% comprises private clinics and hospitals (population projected on 2011 national census; Health Institution Survey 2019–2022).

11.4.3 The Middle-Wealth

The 287 middle-wealth neighbourhoods are comparatively disadvantaged, where residents compete to access quality urban services. The organically grown neighbourhoods in the city core are more open to accommodating new populations but struggle to provide essential urban services and infrastructure. Due to the incompatible land uses and dominance of commercial uses, these neighbourhoods are crowded, dense, and congested (NSA 2018–2019). The city corporation recently widened the roads and, where possible, included sidewalks in the Tajmahal

Fig. 11.4 Central Road area boasts the highest concentration of private clinics and hospitals in the city. Photo by author

Road area (Walkthrough analysis 2019–2022) (Fig. 11.5). In general, narrow streets, limited open spaces, and play areas (which cover less than 2% of the neighbourhood areas in Tajmahal Road and Purba Kazipara) are typical in the middle-wealth communities (NSA 2018–2019; GIS analysis of the spatial data). In Purba Kazipara, an organically developed neighbourhood, household density is around 17,000 per km^2 (projection of the 2011 national census data); residents here are frustrated about ill-maintained narrow roads, poor surveillance, the lack of open spaces, limited public education and health facilities, and poor cleanliness. Drug addiction among the youth is an acute cause of concern here (FGD 2022). However, these neighbourhoods grow and accommodate changes to respond to the needs of the changing population (Sowgat and Roy 2022a). The rental population is high (81–100%), and there is low homeownership (0–20%) (EDA of the 2011 National Census data).

As a significant share of the communities in the middle-wealth category are in the city core, they benefit from a dominant and centrally located service sector along with commercial and business sector employment (NSA 2018–2019; Sowgat and Roy 2022b). There are traces of slum-like settlements in this category, yet their presence is not as prominent as in the lower-middle-wealth and low-wealth neighbourhoods (NSA 2018–2019). A good share of middle-wealth communities has jobs in the industrial sector (EDA of the 2011 National Census data). The Mirpur area has many garment industries, and 80% of the population of Mirpur Section 11 are industrial workers. Likewise, the neighbourhoods near the planned Tejgaon industrial area have a high concentration of people involved in the industry.

Due to limited investment in public schools, private schools have mushroomed in Dhaka over the last 30 years (Roy et al. 2018). Although middle-wealth neighbourhoods have access to

Fig. 11.5 Tajmahal Road, a densely developed planned middle-wealth neighbourhood. Photo by author

primary-level educational institutions, most of these are private schools with high fees compared to public primary schools. For example, in the Tajmahal Road neighbourhood, the ratio of public, private, and non-governmental organisation (NGO)-/charity-led education institutions is 2:92:6 (Education Institution Survey 2019–2022).

11.4.4 The Lower-Middle-Wealth Neighbourhoods

Lower-middle-wealth neighbourhoods are primarily organically developed settlements demonstrating good community cohesion yet comprising high-density, congested residential quarters (NSA 2018–2019). Most people in 367 neighbourhoods live in rented properties (EDA of 2011 National Census data). For example, more than 80% of households in Khilkhet, Pirerbagh, Rampura, Banasree, Jatrabari, and Azimpur live in rented properties. Both brick-built and partially brick-built structures are the dominant types of structures here. Almost half of the households have no water seal sanitation facilities. Access to tap water (81–100%) and electricity (81–100%) are not as bad as in the low-wealth neighbourhoods and urban villages.

Some of the oldest neighbourhoods in Dhaka belong to this category, where a high intensity of one-story buildings is evident (EDA of 2011 National Census data). Other such concentrations outside the old part of the city include Dhaur, Ashitia, Diabari, and its adjacent area. About 98% of the buildings in these areas are old one-story buildings (structural analysis of the GIS database prepared for Dhaka Detailed Area Plan 2016–2035). Because of congested living spaces and narrow roads, the demolition or construction of high-rise buildings has been challenging

Fig. 11.6 New residential developments alongside the old buildings in the historic city core. Photo by author

Fig. 11.7 Improved roads between buildings in Mirpur are now social spaces for the residents. Photo by author

here (NSA 2018–2019) (Fig. 11.6). Most of the neighbourhoods in the city's historical core have become the most hazardous (fire hazard) areas in the city due to non-conforming land uses, new vertical expansion against extremely poor accessibility, and high density (Roy 2021). Still, many heritage buildings within the historical core are unique features of these neighbourhoods (NSA 2018–2019; FGD 2022).

According to a Walkthrough analysis, physical feature surveys, photographic analyses, and interviews conducted between 2019 and 2022, under the strong leadership of the former Mayor, old narrow roads in several blocks were upgraded for improved accessibility, promoting new high-rise buildings in certain aged public sector-led planned residential areas. New sidewalks and urban green also improved walkability in this area. In the hands of neighbourhood associations, neighbourhood security is ensured through gates and night patrols (Fig. 11.7). Without community space and children's play areas, neighbourhood roads become vibrant in the afternoon, where children play, and neighbours meet and greet. Community gardening along one road has improved community cohesion, belongingness, and walkability. However, the areas still suffer from poor waste management and inadequate health and education facilities.

The EDA of the 2011 national census data shows that female work participation in the service sector is comparatively lower in this neighbourhood category than in other types (41–60%). Yet, their concentration in the industrial sector (41–60%) is higher than the wealthiest (0–20%) and upper-middle-wealth neighbourhoods (21–40%). Although most males from this category are engaged in the service sector (61–80%), a high clustering of industrial workers can be found in Old Dhaka. More than 80% of people living in Nawab Yousuf Road are involved in the industrial sector because the area has housed many home-based traditional small industries for many years.

The literacy rate for this category is not satisfactory (61–80%), lower than the wealthiest and higher-middle-wealth neighbourhoods (81–100%). School attendance is similar in low-wealth communities and urban villages (61–80%). Primary school attendance of females is relatively poor compared to other categories. Also, their attendance in higher education is nearly zero (EDA of 2011 National Census data). These neighbourhoods have access to primary-level educational institutions, yet there is an inadequate number of secondary schools (NSA 2018–2019: FGD 2022).

Consequently, the neighbourhoods struggle to provide education for a large part of their population. Regardless of wealth categories, the existing education system in Dhaka is highly dependent on private schools, colleges, and universities. Most private education institutions charge higher fees than public institutions. Access to government institutions is highly competitive. Parents from lower-middle-wealth neighbourhoods choose to send their children to moderately low-cost private institutions that lack the standard requirements of education facilities and struggle to deliver quality education (FGD 2019–2022).

Interestingly, some of the oldest communities in Dhaka that belong to this category have some of the city's oldest and best educational institutions, initially set up by the British for the affluent people who lived in these neighbourhoods during the nineteenth and early twentieth centuries (Education Institution Survey 2019–2022). These neighbourhoods suffer from inadequate, poor-quality, and unaffordable healthcare facilities. Healthcare, in most cases, is limited to primary health care (Health Institution Survey 2019–2022).

11.4.5 Low-Wealth Neighbourhoods and Urban Villages

Eighty neighbourhoods fall within the category of low-wealth communities. Slum and squatter settlements dominate this category, where tenure insecurity, serious lack of basic amenities and poor and unhealthy living environments remain vital aspects of deprivation (NSA 2018–2019; FGD 2022). Settlers have had to rely on illegal connections for water and energy at high costs. These areas also suffer from congested living environments, poor housing conditions and circulation space, and the absence of necessary urban amenities. In Karail, with the largest squatter population in Dhaka, 40,000 people live in every km^2 area, only 1.8% of the neighbourhood area is dedicated to services and facilities, and 93% of households live in temporary shacks (GIS analysis of the recent spatial data prepared for Dhaka Detailed Area Plan 2016–2035) (Fig. 11.8). Among squatter settlements, the concentration of temporary shacks is twice the city average (BBS 2015). Poor migrants who come for low-paid jobs live in these neighbourhoods for affordability and to benefit from good social cohesion. Apart from slums and squatters settlements, peripheral neighbourhoods like Hasan Nagar are poverty-stricken. Residents in all these areas survive by following a minimalist approach and informal arrangements (FGD 2022).

Urban villages (43) are unplanned, usually left out of basic service coverage, and suffer most from poor infrastructural development demonstrated through narrow and unpaved roads, an impoverished water supply (21–40% have access

Fig. 11.8 Life in the largest squatted area (Korail) in Dhaka. Photo by author

to tap water for drinking), and poor drainage systems and sanitation facilities (low water seal toilets with few exceptions and lower-moderate non-sanitary toilets) (NSA 2018–2019). These neighbourhoods have recently become part of the city and are yet to be brought under broader development and service coverage by the city authorities (Fig. 11.9). Electricity coverage is not much better in urban villages (61–80%) (EDA of the 2011 National Census data). Newly emerged urban villages along the city's edge are still less dense than all other categories of neighbourhoods. Although house ownership is the highest in urban villages (81–100%), a good share of households live in temporary structures (41–60%) (EDA of the 2011 National Census data). Still, a mixture of temporary and partly brick-built structures is evident in some urbanising neighbourhoods. These neighbourhoods accommodate people in transition, mainly low-income groups (NSA 2018–2019; FGD 2022).

Although urban villages experience gradual change and eventually become urban, unimproved road conditions, poor urban services, and low urban growth make these areas unsuitable for economic investment (Sowgat and Roy 2022b). The population here lack the skills and education required to access white-collar office jobs. In the presence of agricultural lands, over 14% of people residing in peripheral neighbourhoods are engaged in agriculture. Despite limited service provision and poor economic opportunities, low-income communities along the city's edge are still drawing new people and contributing to urbanisation by replacing natural agricultural and marshy land and affecting ecological settings (NSA 2018–2019; FGD 2022).

Fig. 11.9 Urbanising urban village, Mausaid. Photo by author

The limited concentration of healthcare and education facilities in low-wealth neighbourhoods, slums and urban villages (NSA 2018–2019; FGD 2022) remain the fundamental aspects of inequality. Available education service delivery by the public sector often does not match the social and economic context of the residents in these areas. In Hasan Nagar, a low-wealth neighbourhood, religion-based elementary education opportunities dominate the area due to deficient public sector facilities (Education Institutional Survey 2019–2022). Hasan Nagar provides primary health care, although the residents of ultra-poor neighbourhoods generally lack easy access to public and formalised health care (Health Institutional Surveys 2019–2022). The poorest communities lack indoor, specialised and quality health services; for example, 75% of health care provision in parts of the Karail slum is semi-formal without indoor and specialised facilities (FGD 2022). Residents commute to other areas for primary healthcare needs and access limited government healthcare. Although the city offers free medical care for all, the existing system is overburdened because of the lack of hospitals (FGD 2022; NSA 2018–2019). Mausaid, an urban village, has no health facilities (Health Institutional Surveys 2019–2022).

As per the EDA of the 2011 national census data, like low-wealth neighbourhoods and slums, the literacy rate in urban villages is significantly low (41–60%); it is the worst among the five types of neighbourhoods. School attendance is also the weakest (21–40%) in low-wealth neighbourhoods, slums and urban villages compared to other groups. Although female members' primary school attendance is high (81–100%), secondary school attendance is very low (0–20%). A good share of these females is engaged in industrial jobs (21–60% among those employed), yet the percentage of employed females remains low (0–20%).

11.5 The Economic and Political Reality of Inequality

The previous section reveals the distinctive features of spatial and neighbourhood inequalities in Dhaka (Table 11.2). This section discusses the deep-rooted political and economic reasons that have created and maintained neighbourhood inequalities in Dhaka.

When Dhaka became a municipality and the divisional headquarters in 1864, it was a small urban area of about 21 sq. km. (Mamun 2013, cited in Roy et al. 2019). Currently, this megacity hosts 10.36 million people in its 307 sq km city area (Worldometers 2023). Despite going through major political and economic shifts for most of its lifetime, Dhaka has managed to remain the financial growth hub and political centre of the

Table 11.2 Key features of inequality in the wealth-based neighbourhood categories

Wealth-based neighbourhood category	Key features of inequality
Wealthiest	Planned; best possible built environment; active neighbourhood associations play a crucial role in the provision of safety, cleanliness, and management of rickshaws; services and amenities in the country; highly secure; 100% of households use water seal sanitation facilities and have electricity; lower-moderate owner-occupied homes/flats; high literacy rate; access to the country's top-quality private English medium schools and universities; concentration of white-collar employees
Higher-middle-wealth	Predominantly developed organically; low home/flat ownership; increased safety and cleanliness through active resident-led management associations; high access to tap water; poor-quality sanitation facilities are almost absent; high-density chaotic environments; excessive non-residential land use and deficiency of parks, playgrounds, and open spaces; a few neighbourhoods host some of the most eminent private clinics and hospitals in the city but high-end private health services remain inaccessible for a significant share of tenants; high literacy rate; access to some of the city's top private and public schools and colleges; the service industry is the primary income source
Middle-wealth	Predominantly developed organically with low homeownership; residents compete to access quality urban services; ill-maintained narrow roads; poor surveillance; scarce open spaces; limited public education and health facilities; poor cleanliness; the city core has service, commercial and business sector employment, but some residents have jobs in the industrial sector
Lower-middle-wealth	Developed organically; low number of owner-occupied homes/flats; high-density, congested residential quarters; half of the households have no water seal sanitation facilities; high access to tap water and electricity; poor waste management and inadequate health and education facilities; moderate school attendance and literacy rate; there is a deficit in the adequate number of secondary schools; struggle to provide education to a large part of their population
Low-wealth neighbourhoods	Developed organically; slum and squatter settlements dominate; tenure insecurity; severe lack of basic amenities and poor and unhealthy living environments; congested living; poor housing conditions and circulation space; limited concentration of healthcare and education facilities; significantly low literacy rate; low school attendance
Urban villages	At the edge, developed organically; left out of basic service coverage; suffers most from poor infrastructural development; low number of water seal toilets and lower number of moderate non-sanitary toilets; moderate electricity coverage; high level of home ownership; the dominance of temporary structures; limited concentration of healthcare and education facilities; significantly low literacy rate; weakest school attendance; very low female secondary school attendance

Source Author, 2025

region and country for an extended period (for details, see Roy et al. 2019). The concentration of political power and resources in the capital has colonial roots, a key reason for the uneven urbanisation in Bangladesh (Baffoe and Roy 2023). Consequently, in its journey, the city has had to accommodate an influx of population, almost without any preparation, yet the situation remains the same.

The inequalities in the five neighbourhood categories are reflected in the quality of the built environment, availability of urban services and amenities, health and education facilities and management. In all these aspects, planned, up-market neighbourhoods outperform unplanned areas. The public sector in Dhaka attempted to bring order primarily through site and service schemes for a marginal share of middle-income people; sadly, these areas were converted into wealthy enclaves over time when Dhaka began to demonstrate its economic potential in the global economy during the 80s (FGD 2021). Piecemeal and ad-hoc neighbourhood-level interventions primarily improved accessibility and drainage situations in the public sector-led residential areas. While there have been improvements in access to essential services for most communities, development control measures have failed to balance the booming residential growth against the limited services and facilities in most areas that have been expanding haphazardly. The city boundary has been extended a few times for guided transformation, often to include emerging peripheral neighbourhoods formed through sprawling. These neighbourhood agglomerations are vastly preferred places for low-earning farmers and workers. City plans for Dhaka could not follow the fast rate of urbanisation coupled with weak institutional capacity and inadequate resources which have led to poor implementation of policies and limited investment in infrastructure. Ineffective national urban policies on migration and growth management and poor application of development control regulations have contributed to the rapid, haphazard, and unregulated growth of its neighbourhoods (Sowgat and Roy 2022b).

In addition, political and economic elites reside in the most affluent neighbourhoods of Dhaka. They capitalise on their wealth, social status, and connections to access individuals who can influence policies and interventions to bring quality services to their areas. More precisely, reputed citizens who lead community-based organisations in these areas have access to power and decision-making. The city's diplomatic zone resides in wealthy neighbourhoods; foreign nationals live in adjacent Gulshan and Baridhara. Therefore, efficient and quality service delivery remains a government priority for these neighbourhoods (Roy 2023). To some extent, these have contributed to significantly better services and facilities in the wealthiest neighbourhoods than the rest of the city.

Emerging leadership has begun to lead in addressing the persisting inequalities that have long been neglected. For example, accessibility and public spaces have been upgraded in a few middle and lower-middle-wealth public sector-led planned residential areas under new political leadership. Still, low-income communities and organically developed neighbourhoods are predominantly excluded from targeted public sector interventions concerning services, facilities, and infrastructure. Instead, informal settlements are denied access to formal urban services. Welfare programmes for poverty alleviation in the form of safety nets, maternal and child health, universal access to primary education, and slum upgradation have been diverse attempts to address multiple forms of poverty. Over the last 30 years, non-governmental organisations have been working on water supply, sanitation, health, capacity building, education, and employment, which has brought tangible benefits to the people in poverty. However, supply remains inadequate against demand, and project-based interventions often fail to have lasting benefits beyond the project period (Sowgat and Roy 2013). Also, the education and health sectors depend heavily on private sector financing because of the state's persistent low allocation in these sectors, affecting low-income people. For example, public sector investment is minimal for educating children and adults in the low-wealth localities. Therefore, the uneven distribution of resources and lack of distributive justice in policies and interventions are explicit in the case of Dhaka.

The high proportion of the population in rental accommodation across all neighbourhoods, including the wealthy ones, clearly reflects the country's transition from a traditional and rural-based economy towards an urban industrial-based economy. An equal society requires overall national and local financial wealth accumulation. Although, economically, Bangladesh is making progress, overall, it is at the early stage of industrialisation. Dhaka as the most important economic engine of the country, has received disproportionately higher national investment. Still, resources for meeting its developmental needs remain inadequate.

Neoliberal planning encourages the market to play an active role in transforming neighbourhoods in the absence of resources. Because of the weak political power, development regulators find it challenging to control violations of height, setback, and land use regulations, especially in unplanned residential areas. This violation affects population density and access to air and light and exhausts the available urban services in these neighbourhoods.

There is limited scope for regulating the built environment features in the planned private sector-led residential areas. The urge to maximise profit drives them to ignore education, health facilities, and open spaces within these neighbourhoods. Market freedom has even created isolated enclaves of planned residential areas like the Basundhora Residential Area, a large, gated community where the public sector can do little to influence any decisions related to the built environment (FGD 2021).

Private land ownership seems to have played an important role in transforming existing urban areas and suburban expansion in Dhaka. If the prevailing population density remains unchanged, Dhaka will require a significant amount of developable land to meet the affordable housing challenge. Large property investors are limited in the city as the economic development and accumulation have not reached the appropriate level for that. Even after the extension of the city boundary, it is very difficult for the government and commercial developers to acquire large pieces of land for planned development in the extended city area due to its cost.

Planning and development practices in Dhaka are yet to realise how neighbourhoods shape people's life opportunities and embrace the value of neighbourhood planning. Although the Detailed Area Plan created scopes for addressing immediate needs, the scale of spatial units in this plan is inappropriate for capturing the unique features, specific necessities, and possibilities that have special significance in each neighbourhood. The ward level needs assessment and detailed area planning units are too big to understand the neighbourhood-level issues. Consequently, neighbourhood-level planning and targeted interventions are limited. These findings draw attention to the theoretical debate in the political economy that can shape how we could engage in creating sustainable neighbourhoods in the developing South.

11.6 Conclusion

Neighbourhoods in Dhaka are dynamic, and a significant share of neighbourhoods have evolved organically. These neighbourhoods, being flexible to change, welcome migrants. Yet, the city is rapidly becoming unequal because of organic and uncontrolled changes. Urban policies and interventions are struggling to effectively address the inequalities in services, facilities, and the quality of the built environment, which has led to crowded and polluted living conditions and communities with deficient urban services and facilities, and an extreme shortage of public open spaces in a very high share of neighbourhoods. In contrast, planned areas are rigid in welcoming changes and reject social mixing. They are safe and well-served and intend to create an ideal and stable image of functional space.

The distribution of health and education facilities and service provision varies considerably across Dhaka's neighbourhoods and results in spatial inequalities. To some extent, institutions, partisan politics, power, and the market are responsible for the widening gap. Inadequate resources for meeting developmental needs, weak institutional capacity for guided urban growth, disproportionate distribution of the available resources due to market forces, and elitist planning are the prime forces making Dhaka an unequal city. The prevailing nature of colonial governance, ineffective urban policies and the neo-liberalisation of planning and governance have all contributed to sustained neighbourhood differentiation in this economic powerhouse of Bangladesh.

If neighbourhood inequalities are not managed carefully, Dhaka will fail to offer its full economic and social potential. There is an apparent urgency to promote the distribution of resources in a way that would contribute to tackling the spatial disparities in the city. Innovative approaches to people-centred neighbourhood planning by acknowledging the internal structure and differentiation of neighbourhoods would make Dhaka a relatively equal city. The neighbourhood lens in the planning process and policies would improve the scope of addressing neighbourhood-based needs and the inequality of opportunities.

References

Alkan H, Maksudyan N (2020) Introduction. In: Alkan H, Maksudyan N (eds) Urban neighbourhood formations, boundaries, narrations and intimacies. Routledge, pp 1–14

Baffoe G, Roy S (2023) Colonial legacies and contemporary urban planning practices in Dhaka, Bangladesh. Plan Perspect 38(1):173–196

Bhandari P (2023) Lived reality of elite neighbourhoods: geographies of inequality in Delhi. Contemp South Asia 31(1):36–50

Bharathi N, Malghan D, Mishra S, Rahman A (2022) Residential segregation and public services in urban India. Urban Stud 59(14):2912–2932

Dang Y, Zhan D, Qiu L, Wu S, Cui Y (2023) Effects of the built environment on residents' subjective well-being and behaviours: a case of Hangzhou, China. J Housing Built Environ 38(1):497–514

Delos Reyes MR, Gamboa MAM, Rivera RRB, Daguio KGL (2018) The Philippines: national urban policies and city profiles for Manila and Batangas. GCRF Centre for Sustainable, Healthy and Learning Cities and Neighbourhood (SHLC)

Dodini M, Fantini M (2006) The EU neighbourhood policy: implications for economic growth and stability. JCMS: J Common Mark Stud 44(3):507–532

Ehwi RJ (2023) Walls within walls: examining the variegated purposes for walling in Ghanaian gated communities. Hous Stud 38(4):527–551

Heroy S, Loaiza I, Pentland A, O'Clery N (2023) Are neighbourhood amenities associated with more walking and less driving? Yes, but predominantly for the wealth. Environ Plan B: Urban Anal City Sci 50(4):958–982

Hosseini A, Finn BM, Momeni A (2023) The complexities of urban informality: a multi-dimensional analysis of residents' perceptions of life, inequality, and access in an Iranian informal settlement. Cities 132:104099

Jensen B, Seltzer A (2000) Neighbourhood and family effects in educational progress. Aust Econ Rev 33(1):17–31

Ji Q, Yin M, Li Y, Zhou X (2023) Exploring the influence path of high-rise residential environment on the mental health of the elderly. Sustain Cities Soc 98:104808

Kundu D, Pandey A, Sharma P, Bhusan S, Mondal B, Lahiri B et al (2018) India: national urban policies and city profiles for Delhi and Madurai. GCRF Centre for Sustainable, Healthy and Learning Cities and Neighbourhood (SHLC)

Kundu D, Sharma P, Lahiri B, Sangtani C (2020) Delhi: a city of diversity and disparity. GCRF Centre for Sustainable, Healthy and Learning Cities and Neighbourhood (SHLC)

Lubeck-Schricker M, Patil-Deshmukh A, Murthy SL, Chaubey MD, Boomkar B, Shaikh N et al (2023) Divided infrastructure: legal exclusion and water inequality in an urban slum in Mumbai, India. Environ Urban 35(1):178–198

Lynge H, Visagie J, Scheba A, Turok I, Everatt D, Abrahams C (2022) Developing neighbourhood typologies and understanding urban inequality: a data-driven approach. Reg Stud Reg Sci 9(1):618–640

Maddock A, Blair C, Ean N, Best P (2021) Psychological and social interventions for mental health issues and disorders in Southeast Asia: a systematic review. Int J Ment Heal Syst 15(1):56

Moshi I, Msuya IR, Todd G (2018) Tanzania: national urban policies and city profiles for Dar es Salaam and Ifakara. GCRF Centre for Sustainable, Healthy and Learning Cities and Neighbourhood (SHLC)

Randon-Furling J, Olteanu M, Lucquiaud A (2020) From urban segregation to spatial structure detection. Environ Plan B: Urban Anal City Sci 47(4):645–661

Robb Larkins E, Durão S (2023) Guarding the urban elite: hospitality security in São Paulo. City Soc 35:27–37

Roy S (2021) Sustainable urbanisation in Bangladesh and Dhaka: challenges and way forward [Background Paper]. 23rd ASEF Summer University: Liveable Cities for a Sustainable Future. Asia-Europe Foundation. https://asef.org/wp-content/uploads/2021/11/ASEFSU23-Background-Paper_Sustainable-Urbanisation-in-Bangladesh-Dhaka.pdf

Roy S (2023) Growth management challenges for a rapidly urbanising megacity. In: Routledge handbook of Asian cities. Routledge, p 225

Roy S, Sowgat T (2020) Dhaka: diverse, dense, and damaged neighbourhoods and the impact of unplanned urbanisation. GCRF Centre for Sustainable, Healthy and Learning Cities and Neighbourhood (SHLC)

Roy S, Sowgat T, Ahmed MU, Islam ST, Anjum N, Mondal J, Rahman MM (2018) Bangladesh: national urban policies and city profiles for Dhaka and Khulna. GCRF Centre for Sustainable, Healthy and Learning Cities and Neighborhood (SHLC). https://www.centreforsustainablecities.ac.uk/wp-content/uploads/2018/06/Executive-Summary-Bangladesh-National-Urban-Policies-and-City-Profiles-for-Dhaka-and-Khulna.pdf6

Roy S, Sowgat T, Mondal J (2019) City profile: Dhaka, Bangladesh. Environ Urban ASIA 10(2):216–232
Sampson RJ (2017) Urban sustainability in an age of enduring inequalities: advancing theory and ecometrics for the 21st-century city. Proc Natl Acad Sci 114(34):8957–8962
Sampson RJ (2019) Neighbourhood effects and beyond: explaining the paradoxes of inequality in the changing American metropolis. Urban Stud 56(1):3–32
Shrum W, Walker M, Fan J, Mbatia PN, Miller P, Harsh M, et al (2023, March) Tough ties and rough networks: inequality and exploitation in African slums. Sociol Forum 38:375–394
Sowgat T, Roy S (2013) Pro-poor development: an assessment of the national level policies and programs in Bangladesh. Plan Plus 6:43–61
Sowgat T, Roy S (2022a) Throwntogetherness in Dhaka: rethinking urban planning: a visual essay. City 26(2–3):422–432
Sowgat T, Roy S (2022b) Neighbourhood segregation in Dhaka. J Urban Aff 46:1371–1391
Statista (2023, September 7) Cities with the highest population density worldwide in 2022. www.statista.com. https://www.statista.com/statistics/1237290/cities-highest-population-density/
UN Habitat (2022) World Cities Report 2022: envisaging the future of cities. United Nations Human Settlements Programme: Nairobi, Kenya, pp 41–44
Wang YP, Kintrea K (2019) Sustainable, healthy and learning cities and neighbourhoods. Environ Urban ASIA 10(2):146–150
Wong C, Bowers J, Williams T, Simmons KD (2012) Bringing the person back in: boundaries, perceptions, and the measurement of racial context. J Polit 74(4):1153–1170
Worldometers (2023, September 7) Main cities by population in Bangladesh. https://www.worldometers.info/world-population/bangladesh-population/

Urbanisation and Neighbourhood Diversification in Chongqing

12

Yunxia Liu, Chenguang Li, Yue Yin, Yucheng Zhan, Xuan Sun, Lei Zhai, and Tao Sun

Abstract

Chongqing, a megacity in southwest China, has undergone a unique urbanisation process, which can be likened to 'one tiny horse pulling a huge cart'. That is, one small area of the municipality defined as urban (5% of land area and 31% of population) is driving the development of the much more extensive area defined as rural (95% of land area and 69% of population). Chongqing's former socialist organisation of residence and the work-unit system has been transformed through housing privatisation and commercial property development into a diversified and complex urban neighbourhood arrangement. Different types of neighbourhoods have emerged, each exhibiting diversity in residents' socio-economic characteristics, housing conditions and the neighbourhoods' internal spatial arrangements, commercial amenities and access to public services.

Keywords

Urbanisation · Neighbourhood division · Neighbourhood diversity · Housing reform · Work-unit neighbourhood

12.1 Introduction

Chongqing, the largest city in inland China, has undergone rapid urbanisation, as it has benefited from being elevated to a 'national-level city' directly administered by the central government since 1997. The neighbourhoods created by the danweizhi or work unit system from 1950 to early 1990s have gradually disintegrated due to the separation of work and living, as well as the privatisation of housing. Large-scale urban expansion also brought about thousands of new-type planned residences, including commercial housing as well as government-subsidised housing and resettlement housing, ranging from luxury homes to ordinary apartments. Neighbourhoods have been created through these real estate developments, and have become differentiated socio-economically through the urban housing

Y. Liu
School of Management, Lanzhou University, Lanzhou, P.R. China
e-mail: liuyunxia@lzu.edu.cn

C. Li · T. Sun (✉)
Harbin Institute of Technology, Shenzhen, P.R. China
e-mail: suntao2020@hit.edu.cn

C. Li
e-mail: lichenguang@nankai.edu.cn

Y. Yin · Y. Zhan · X. Sun · L. Zhai
Zhou Enlai School of Government, Nankai University, Tianjin, P.R. China

Y. P. Wang et al. (eds.), *New Drivers of Division*, Sustainable Development Goals Series,
https://doi.org/10.1007/978-981-96-5629-5_12

market. This means that political-oriented divisions based on the work unit system have been replaced by outcomes that follow more closely from the individual's social and economic status.

This chapter will focus on how the state and the market have interacted to drive Chongqing's urbanisation and examine the transformation process of neighbourhood diversification. The chapter is organised as follows: Sect. 12.2 analyses Chongqing's urbanisation and its driving forces. Specifically, it summarises the challenges to Chongqing's urbanisation, reviews the strategies and measures to push it forward, and identifies the state and market driving factors that facilitate urbanisation, as well as the challenges ahead to further sustainable development. Section 12.3 illustrates the spatial configuration of Chongqing's residential neighbourhoods and reasons for it. It highlights the transformation from production-residence neighbourhoods based on a socialist working-unit system to diversified neighbourhoods; typical neighbourhood types are then introduced. Finally, the diversity of neighbourhood is elaborated. Section 12.6 concludes the findings.

12.2 Challenges of Urbanisation and Development Strategies

Chongqing is located in the southwestern part of inland China, covering an area of 82,400 km^2. In 2022, the permanent resident population reached 32.1 million, with an urbanisation rate of 71%.[1] In 1997, by virtue of its unique strategic position and locational advantages, Chongqing became the fourth municipality directly under the central government after Beijing, Tianjin, and Shanghai.

The meaning of 'city' in China differs from that in other countries. The administrative divisions in China are as follows: the whole country is divided into provinces (autonomous regions or municipalities directly under the central government). Larger cities are further divided into urban districts and rural counties. Counties are then subdivided into townships and towns. Most large cities may administer some rural counties. In the case of Chongqing, apart from the urban districts, the city has 12 counties.

As a direct-administered municipality like Beijing, Shanghai, and Tianjin, Chongqing has the same hierarchic level as a province. Not only does it include administrative regions with a high level of urbanisation, but it also encompasses counties where rural areas make up a significant proportion. For instance, in 1997, out of the 28.7 million permanent residents in Chongqing, a substantial 19.8 million were rural residents. Even in 2022, with a total permanent resident population of 32.1 million, there are still 9.3 million rural residents.

Chongqing went through a rapid urbanisation process during the past 30 years or so. Its permanent urban resident population increased from 8.9 million in 1997 to 22.8 million in 2022, and the urbanisation rate jumped from 29.5% in 1996 to 71% in 2022 (Fig. 12.1).

However, Chongqing's total population didn't increase as quickly as other large cities in China, with a growth rate of 12% from 1996 to 2022. First, it is historically a manufacturing hub in western China, so, driven by industry development, it already had a huge population base before the 1990s. Second, its topography restricts the growth of liveable areas and makes the construction of infrastructure and facilities costly. Third, there is competition between Chengdu and Chongqing in terms of city attractiveness (Wang et al. 2022). Chengdu is adjacent to Chongqing. Its rural and peri-urban areas are more affluent and comfortable to live in. For those who move from other cities, the first place to settle in a new city is more likely to be in the peri-urban areas in Chengdu, which is well recognised for its leisurely lifestyle and is a more liveable city.

[1] Chongqing Municipal Bureau of Statistics, 2022 Chongqing National Economic and Social Development Statistical Bulletin, https://tjj.cq.gov.cn/zwgk_233/fdzdgknr/tjxx/sjzl_55471/tjgb_55472/202303/t20230317_11775723.html?eqid=eaaa1c73000bdcbf0000000664675477.

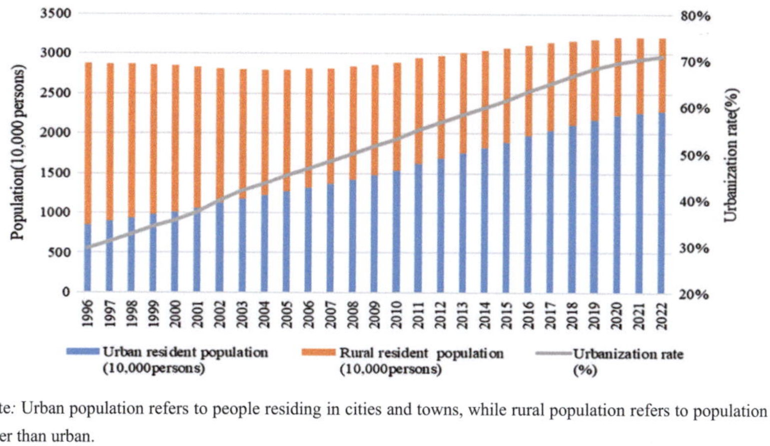

Fig. 12.1 Urbanisation rate and urban rural population in Chongqing. *Note* Urban population refers to people residing in cities and towns, while rural population refers to population other than urban. *Source* Chongqing Statistical Yearbook (1997–2023)

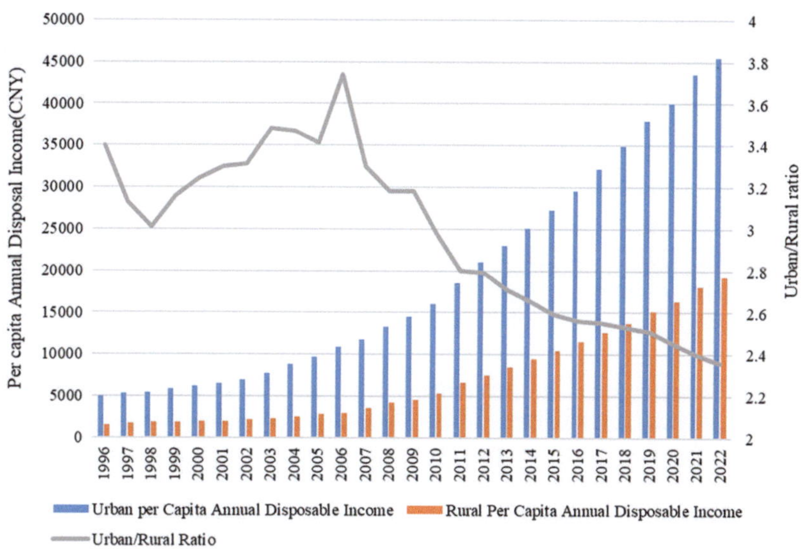

Fig. 12.2 Urban and rural residents' per capita disposable incomes in Chongqing (1996–2022). *Sources* 2022 Statistical Bulletin on Economic and Social Development of Chongqing City, and Chongqing Statistical Yearbook

12.2.1 Challenges of Urbanisation

Chongqing faced challenges in urbanisation. The city exhibits a typical Chinese dual economic structure of 'a large city coexisting with large rural areas', where there exist substantial urban–rural gaps. In 1996, the disposable income per capita of urban residents in Chongqing was 5023 yuan, while that of rural residents was only 1479 yuan, so the urban income was more than three times that of the rural income (Fig. 12.2). In recent years, the ratio of disposable income per capita between urban and rural residents in Chongqing has shown an overall downward trend, but the gap remains significant. Rural areas also lag far behind urban areas in terms of infrastructure conditions, industrial development, and production and operation methods.

12.2.2 Development Strategies and Measures

Chongqing adopted an innovative urbanisation strategy, metaphorically described as 'a small horse pulling a big cart'. The 'small horse' represents the central urban district, which accounts for a smaller proportion of the total land, population, and other resources, while the 'big cart' symbolises the entire Chongqing city. The essence of this strategy is to leverage the limited resources of the central urban district to drive economic growth and social progress throughout the entire municipal area.

From the 1990s to early twenty-first century, Chongqing vigorously promoted the construction of urban infrastructure, such as transportation, public services, and security, to enhance the city's competitiveness and attract foreign investment. At the same time, relying on the industrial foundation of the original urban area, Chongqing upgraded traditional industries and actively developed strategic emerging industries to expand the industrial cluster effect and attract world-class enterprises. During this process, on the one hand, the development and expansion of Chongqing's urban area attracted the rural population from surrounding areas. On the other hand, the Chongqing government also promoted the development of surrounding areas by expanding the scope of public services and building infrastructure, achieving urbanisation in place. After 2017, these newly built urban areas have gradually become the main force driving the urbanisation development of Chongqing (Zong and Yu 2024).

Moreover, with the implementation of this strategy, the inequality between urban and rural areas in Chongqing has been alleviated to some extent. From 2011 to 2020, the average annual growth rate of income for rural residents in Chongqing was 147.7%, while that for urban residents was 116%, and the urban–rural income ratio also gradually decreased from 2.8 to 2.45, indicating that rural areas have also achieved significant development during the urbanisation process (Aik and Zhang 2023). The measures to achieve this include:

1. **Providing Public Services and Job Opportunities to Migrant Populations**

An important strategy of urbanisation is to transform the rural population into urban. In 2010, Chongqing launched a reform of the integrated planning of the urban and rural registered residence systems, primarily targeting rural–urban migrant workers with agricultural household registration but already engaged in urban employment. The reform encourages them to switch to an urban registered residence while allowing them to retain their contracted land, homestead land (zhaijidi or rural land reserved for housing construction), and forest land in the rural areas. Within just over a year of this policy, the number of rural residents in the city converting their registered residence exceeded 3.2 million. At the same time, efforts have been made to promote equal access to public services, ensuring agricultural-labour-turned-urbanites can enjoy the same social security and public services as other urban citizens.

The municipal government has also taken a series of measures to eliminate disparities in public services between rural and urban areas. Following its new status as a centrally administered municipality, Chongqing initiated the unification of the urban and rural household registration systems in various districts and counties. In 2007, the city took the lead nationwide in introducing pension insurance for migrant workers, integrating it into the urban employee pension insurance system, which alleviated the thorny problem of transferring and continuing pension insurance for migrant workers when moving across provinces, districts or counties.[2] In 2016, the city government began to gradually abolish the policy standards bonded to registered residence identities, reducing the obstacles encountered by rural migrant workers in accessing urban public services.[3] In 2022, the government explicitly proposed in its municipal planning to establish

[2] 重庆市农民工和城镇职工养老保险实现"并轨" . The merging of pension insurance for migrant workers and urban employees has been achieved in Chongqing. https://www.gov.cn/jrzg/2010-07/04/content_1645073.htm.

[3]《2016年重庆市推进新型城镇化工作要点》. 'Key Points for Advancing New Type of Urbanisation in

a basic service supply mechanism for all permanent residents which is marked by citizens' resident identity card numbers and bonded to their lengths of residence. The goal is to achieve full coverage of basic public services in urban areas.[4]

Another measure is to invest in infrastructure and improve the liveability of newly transformed urban areas in the administrative sense of what used to be rural areas. The municipal government, by means of investing in infrastructure, planning and constructing industrial parks, and taking full advantage of national strategies, harnesses the invisible hand of market forces to drive the transformation of rural residents into urban citizens. There is significant investment in infrastructure. From 2020 to 2022, the city invested a total of 398.3 billion yuan (about 55 billion US dollars) in constructing infrastructure such as late-model networks, intelligent computing, and information security, building a state-of-the-art new generation of information infrastructure as a support system.[5] On the one hand, these construction projects have directly created an abundance of job opportunities, many of which target low skilled labourers. On the other hand, the improvement of infrastructure has stimulated local industrial development and generated additional indirect employment opportunities.

2. **Planning and Expansion**

The population and the built-up area of the city have grown continuously since the 1980s with urban planning playing a crucial guiding role. In 1983, the Master Plan of Chongqing City (1981–2000) was approved by the State Council. In the initial stage of the planning, Chongqing had an urban population of 1.52 million and a built-up urban area of 73.4 km^2. The goal was to reach a population of 1.7 million and expand the built-up area to 102 km^2 by the year 2000. In 1997, Chongqing became the fourth centrally administered municipality in China, marking the onset of a phase of blockbuster growth in urbanisation. To adapt to the new circumstances, a new master plan was drafted, Master Plan of Chongqing Municipality (1996–2020), which was approved by the State Council in 1998. The goal of this plan was to reach a population of 3.7 million and expand the built-up area of the main city to 300 km^2 by the year 2020. This plan proposed to overcome the geographical barriers posed by Zhongliang Mountains and Tongluo Mountains, and expand urban development space with 11 clusters on the outskirts of the main city. Additionally, in terms of urban road traffic, the plan proposed for the first time to build a layered road network configuration by constructing a fast road network system, significantly increasing river-crossing bridges, and strengthening connections between clusters to enhance the unity of the city.

After 2000, governments at all levels in China began to attach importance to the coordinated development of urban and rural areas, when Chongqing became a key pilot city. In October 2007, the first nationally urban and rural integrated master plan was approved.

By 2005, the population of the central urban area of the city reached 5.4 million, with a built-up area of 364 km^2. The 2007 Master Plan proposed that, by 2020, the urban population in the main city area reach 9.3 million, and the urban construction land reach 865 km^2.

In 2011, in order to facilitate the implementation of the national 'Belt and Road Initiative', a revision was made to the Urban and Rural Master Planning of Chongqing (2007–2020), again, which was approved by the State Council for implementation in the same year. The planned urban population in the main city in 2020 was adjusted from 9.3 million to 12 million people, and the planned urban construction land in the main city was increased from 865 km^2 to 1188 km^2.

Chongqing in 2016' https://www.cq.gov.cn/zwgk/zfxxgkml/szfwj/qtgw/201605/t20160506_8613882.html.

[4]《重庆市新型城镇化规划(2021-2035年)》. 'Chongqing's New Type of Urbanisation Planning (2021-2035)'. https://wap.cq.gov.cn/zwgk/zfxxgkzl/fdzdgknr/ghxx/zxgh/202202/t20220208_10373560.html.

[5] 重庆市人民政府《重庆市新型基础设施重大项目建设行动方案(2020—2022年)》. Chongqing Municipal Government, 'Action Plan for Key Project Construction of New Type Infrastructure (2020–2022)'. https://admin.cq.gov.cn/zwgk/zfxxgkml/szfwj/xzgfxwj/szf/202006/W020230221542091384555.pdf.

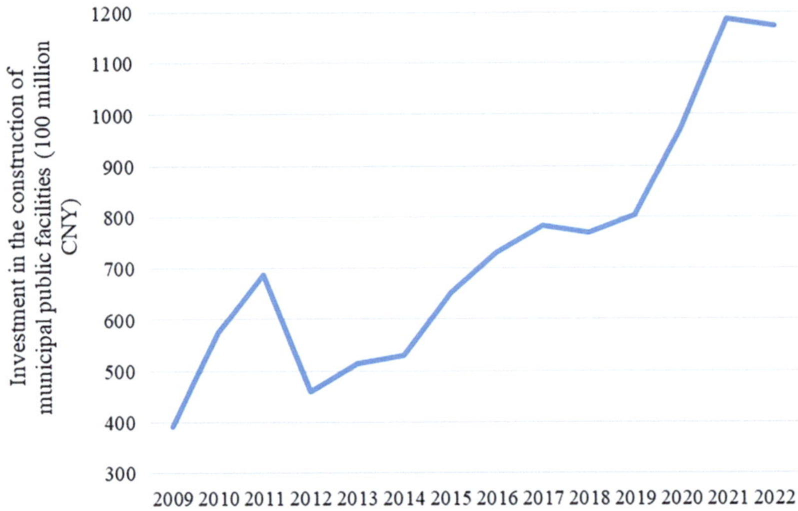

Fig. 12.3 Investment in the construction of municipal public facilities of Chongqing. *Source* Authors, 2021

3. **Housing Development and Public Amenities**

In addition, the government has propelled the construction of large residential areas through master planning. During the 13th Five-Year Plan, the Chongqing municipal government took the lead in planning 21 large residential areas, each having a population between 200,000 and 300,000 people and covering an area of 10–30 km^2 (Wang et al. 2012). When laying out these large residential areas, full consideration was taken in terms of the basic conditions such as the distribution of industrial space, liveable environmental quality, and supporting infrastructure for transportation and municipal facilities in the main city. The selection of sites was based on the key areas for the future development of the main city. The establishment of these residential areas primarily served the development of regional industries. In the planning phase, based on indicators such as land areas of various industries and job coefficients, the government calculated the number of residents required to be accommodated in the residential areas and accordingly laid out their populations and land scales, ensuring the coordinated development of industrial and residential spaces. Moreover, overall annual investment in the construction of municipal public facilities has also been on the rise, from 39.1 billion yuan in 2009 to 117.4 billion yuan in 2022 (Fig. 12.3).

12.3 Major Causes of Fast Urbanisation

Chosen as the major urban growth pole in the west of China, elevation in the administrative hierarchic level is the major external factor driving the rapid urbanisation of Chongqing, while economic development is the most important internal driving force.

12.3.1 Administrative Level Elevation and Preferential Input of Resources

Benefiting from its elevation from an ordinary prefecture level city to a directly administered municipality, Chongqing has received more support in terms of human resources, fiscal policies, and industrial policies during the last 30 years. In terms of human capital, demand for various types of talent increased after the city became a directly administered municipality in 1997. Taking public servants as an example: In 1997, Chongqing publically recruited and selected public service personnel from across the whole country, which not only improved the overall quality of urban management personnel and enhanced urban governance capabilities, but also attracted a large number of talented persons and their families to settle in Chongqing. This provided human capital for urban development and enhanced the vitality of the city; the increase in population also stimulated the demand for housing, medical, and other public services, as well as commercial facilities, driving economic development.

On the other hand, the administrative level and authority of a directly administered municipality have also provided Chongqing with the convenience of conducting policy experiments. By undertaking various pilot projects and their accompanying initiatives, the city has been able to pioneer the adoption of new ideas, technologies and modes of management, invigorating its innovative spirit and fueling its urban development. For example, Chongqing fully utilised its advantageous position in national strategies such as the Yangtze River Economic Belt Construction, and the Belt and Road Initiative to boost its economy. In 2010, the State Council approved the establishment of the Liangjiang Xinqu or Two Rivers New Area in Chongqing to undertake the major strategic task for national development. Spanning three administrative districts including Jiangbei, Yubei and Beibei, it set up national-level economic and technological development and high-tech zones. Supported by the central government, this new area has attracted introduced a number of high-tech manufacturing giants such as BOE (Best On Earth), Selis Automotive and Corning, fostering two pillar industries in automotive and electronic information, each worth 200 billion yuan, and significantly enhancing the attraction of Chongqing to skilled labour. The government's substantial investment resulted in industrial prosperity and economic growth, bringing in high-paying jobs and high-return business opportunities, attracting an influx of residents from Chongqing's surrounding areas, and significantly improving the city's urbanisation level.

12.3.2 Economic Development and Industrial Restructuring

Once an old and traditional industrial base, Chongqing has now transformed into a modern metropolis, where the secondary and tertiary industries thrive. The market economy has unleashed its developmental vitality. Following China's reforms era, Chongqing was given more powers as one of the 'cities specifically designated in the state plan' (jihua danlie shi) in 1984 and accordingly reforms were initiated related to economic transformation. Chongqing's economic dynamism was stimulated, and its industrial structure began to show a diversified trend. Not only has the industrial sector achieved further development through restructuring, integration, and attraction of foreign investment, but sectors previously suppressed during the planned era, such as commerce and finance, have also made significant progress in tandem with the rapid expansion of private economy and township enterprises. With a series of infrastructure projects such as railways, highways, and airports, Chongqing's construction and tourism industries witnessed unprecedented development. Meanwhile, the manufacturing sector struggled to compete with the east coastal areas, leading to continuous losses, and shaking its previously dominant position among Chongqing's three pillar industries.

In 2007, the city was positioned as a key growth pole in western China, endowing it with the significant responsibility of driving economic development in this region and the upper reaches of the Yangtze River, thereby achieving the national developmental objective of xiaokang (to deliver a life of moderate prosperity to the people in an all-around way). During this phase, the establishment of the Liangjiang New Area, Lianglu Cuntan Free Trade Port Area and Xiyong Comprehensive Free Trade Zone further improved the degree of Chongqing's opening up to the outside world. With the support of the central government, the city attracted numerous high-tech manufacturing enterprises, leading to the emergence of a modern manufacturing sector, anchored by two pillars of electronic information and automobile manufacturing. The substantial growth of the manufacturing industry directly spurred the development of the productive service sector, leading to rapid growth in Chongqing's finance, design, logistics and commerce industries. Furthermore, the expansion of the manufacturing sector brought about income growth and urban expansion, which in turn generated robust demand, indirectly driving the growth of the non-productive service sector. By 2018, the tertiary industry in Chongqing accounted for 52.3% of its GDP, surpassing the 40.9% contribution of the secondary industry.

In recent years, the scale of Chongqing's manufacturing sector has further expanded. Beyond the existing automotive and electronics industries, it has further developed strategic emerging manufacturing sectors such as cutting-edge equipment, biopharmaceuticals and advanced materials leading to a steady increase in the production of high-tech industrial products. The service sector in Chongqing has also entered a new phase of growth, bolstered by the digital economy. In 2020, the service sector significantly enhanced its contribution to the city's overall GDP growth to 72.9%. This was primarily driven by the rapid growth of other service industries represented by ICT and software sectors, as well as scientific and technical service industries. In 2022, Chongqing's cultural and tourism industries achieved an added value of 106.326 billion yuan, while the e-commerce and comprehensive health service sectors began to take shape alongside the continuous emergence of innovative business models such as internet celebrity incubation, online health care and smart elderly care. The constant optimisation of Chongqing's industrial structure is conducive to achieving stable economic growth and sustained urban development.

12.4 Neighbourhood Diversity and Social-Spatial Division

The definitional struggle of the concept of neighbourhood is well documented. Neighbourhood can be a place, a community and a unit of policy design. In this study, neighbourhood refers to the place where people reside and spend their time. For urban residents, neighbourhoods are places that determine their quality of life and economic standing.

Neighbourhood here equals the Residence Committee areas (or Shequ), the lowest tier of China's four-tier administration system in cities, under the Street (Jiedao) Committee, district governments and the municipality.

12.4.1 Distribution of Neighbourhoods

Shaped by its predominantly mountainous topography, Chongqing has formed a polycentric and clustered spatial configuration of residential areas. Unlike the urban sprawl mode which may be metaphorically likened to spreading a big cake on a flat pan, Chongqing's residential areas are composed of multiple relatively autonomous clusters. These clusters maintain a degree of spatial isolation from each other (Fig. 12.4), and internally each cluster has its own centre and road system.

The formation of this clustered structure results from the combined effect of geographical conditions and historical changes. Geographically, Chongqing is characterised as a 'mountain + river city', with hills and mountains as the dominant landforms, of which mountains occupy 76% of its terrain. The main city encompasses diverse topographical features such as parallel mountain ridges, isolated high hills, urban mountains and cliff lines. Districts segmented by mountains, hills and rivers exhibit significant variations in natural conditions and resource endowments. The complex mountainous topography naturally blocks the connections between different districts within the city. Additionally, Chongqing is situated at the confluence of the Yangtze and Jialing rivers, which flow through the city from southwest to northeast, forming natural barriers that divide the city and further weaken the connections between residential settlements within the city.

From the perspective of historical changes, Chongqing once briefly assumed the role of a national capital during wartime in the 1940s,

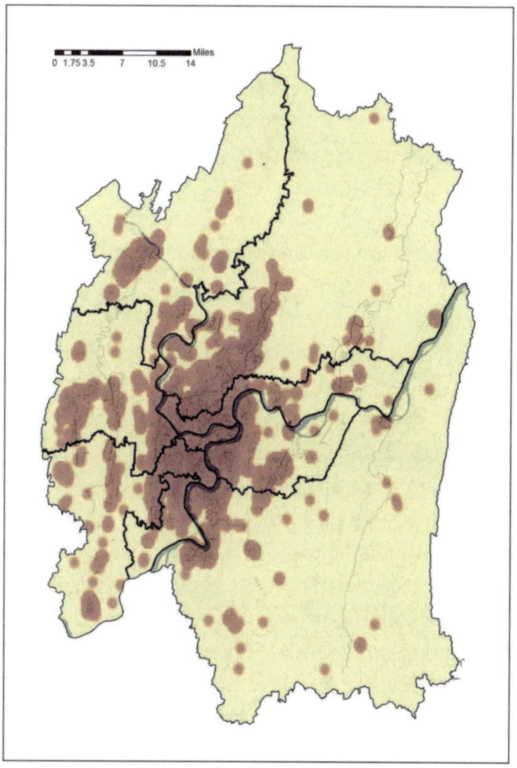

Fig. 12.4 Spatial distribution of residential neighbourhoods in Chongqing. *Source* Authors, 2021

resulting in a significant increase in its residential population. To avoid air raids, residential areas were dispersed to the outskirts, leading to an initially broadly dispersed yet locally concentrated arrangement in the distribution of residential clusters. Afterward, during the planned economy era from the 1950s to the late 1980s, Chongqing, as a large industrial base, gathered a diverse category of factories, both large and small. To enhance production efficiency, numerous housing areas were built around those factories. As workers and their families moved in, production-residence areas centred around work units took shape, further consolidating the clustered urban structure.

Chongqing was once a heavy industry city, characterised by a large number of former work-unit neighbourhoods due to historical reasons. From the 1960s to 1980s, large industrial and mining enterprises (such as the Chongqing Steel Plant, Jialing Plant, Special Steel Plant, Construction Plant, and Chang'an Plant) in the main urban area of Chongqing had good production efficiency and a high level of profit. A large number of working staff residential areas were built around these plant areas to solve housing problems of factory workers, showing a strong work-unit characteristic. Later, with the relocation or closure of plants, these unit-based old residential areas were incorporated into the city's overall management mechanism. Compared with the decentralised layout of small and medium-sized enterprise residential areas in the city, these residential areas were larger in scale and more concentrated in distribution. According to a national survey on housing conducted in 1985, municipal public housing comprised only 24.1% of all housing, and work-unit housing accounted for 58.1% (Logan et al. 1997). Few still-intact work-unit residential neighbourhoods are concentrated in government, public institutions and state-owned enterprises.

After becoming a city directly administered by the central government, Chongqing has undergone a shift in terms of city positioning, and its urban development has been significantly accelerated. During this period, driven by the demands of industrial development and an increase in population size, its clustered structure has become more diverse. On the one hand, this is evident in the continuous emergence of new clusters as the city expands. Small towns and industrial parks located outside the metropolitan area have merged and gradually evolved into multifunctional new urban clusters. On the other hand, existing clusters have undergone reorganisation and differentiation. Some industrial clusters, initially centred around specific work units, disintegrated due to factory relocations or bankruptcies, while some other clusters have developed new functions in line with industrial restructuring. Moreover, the evolution of Chongqing's cluster structure has exhibited some common features. First, the spatial scale of its urban clusters has generally expanded due to population growth, which gradually blurs the spatial boundaries between these clusters. Second, the enhancement of infrastructure has significantly facilitated urban transportation. Coupled with the greater mobility brought about by economic and social transformations, this has gradually disrupted the previously balanced work-residence structure within the clusters. While facilitating functional changes in these clusters, this development also strengthens connections between them. Third, it has encouraged further integration of clusters within designated areas, resulting in an increasingly diverse and enriched spectrum of neighbourhood forms within these areas.

In terms of residential space allocation, relying solely on market mechanisms often leads to resources favouring higher income groups, while lower-income groups face housing difficulties. China's housing construction policies stipulate that newly built commercial residential areas must include a certain proportion of affordable housing to ensure fair distribution of public resources and social justice, and to ensure that different social classes can enjoy basic housing rights. The government is responsible for the subsequent operation and management of the affordable housing, ensuring high standards and quality planning and construction, balanced layout, convenient transportation, complete supporting facilities, suitable living environment and shared community environment and professional property management services. This ensures that middle- and low-income groups have the same living environment as commercial housing residents, reducing the sense of disparity between the rich and the poor.

In recent years, Chongqing has promoted the construction of affordable housing by adopting mixed construction with commercial housing and public rental housing, using a 'small concentration, large dispersion' model to prevent concentrated areas of affordable housing from becoming urban ghettos, allowing residents to live better, more comfortably, and with more dignity. As a typical representative of affordable housing communities, the Minxin Jiayuan community allocates 10% of its total construction area for public rental housing. Public rental housing is designed simply according to the standard of 400 yuan per square metre, with unit sizes ranging from 35 m^2 to 80 m^2, with more than 85% being below 60 m^2. These systematic considerations enable middle- and low-income groups living in public rental housing to enjoy the living standards of mid-range commercial housing.

12.4.2 Transformation of Neighbourhoods

Like other Chinese cities, Chongqing's production-residence neighbourhoods originally based on the work unit system have changed into a diversified neighbourhood system (Zhang and Chai 2014; Deng 2018; Logan et al. 1997).

Under the work unit system during the planned economy era from the 1950s to the late 1980s, the production functions and living of employees and their families were linked together forming a 'work unit'. This work unit was more than just a place where employees engaged in production, but also the provider of residential services, including kindergartens, primary and secondary schools. Large work units typically had facilities such as public bathhouses, canteens and recreational spaces for staff members to watch movies, exercise, swim, or engage in other leisure activities. In one word, all residential services were basically provided by work units. Meanwhile, staff members of the same work unit generally resided in affiliated housing compounds (Bray 2005).

During the work unit organisation period, residents in neighbourhoods had a high degree of homogeneity in terms of occupational composition. Moreover, as housing was provided to staff members at low rent as social welfare, residents did not possess full property rights (Yang et al. 2017). Residential property rights belonged to the work units, resulting in minimal residential mobility during this time (Xiao et al. 2020). As long as residents did not change jobs or were not allocated new housing in different locations by their work units, they usually lived in these work-unit communities for their entire lives (Li et al. 2019).

The economic reforms in the 1990s loosened the strict social management institutions of the planned era, breaking down the foundation on which work-unit communities relied for its maintenance, leading to the disintegration of the work unit system in urban areas, and resulting in large-scale dismantling or disappearance of work-unit communities. On the one hand, as the market began to play a decisive role in resource allocation, some work units lacking in market competitiveness that had previously survived under the protection of government administrative directives went bankrupt. The collapse of these entities, which used to provide housing, education, medical, and sanitation services under the work unit system, severed the networks and relational ties that sustained communities. On the other hand, in order to adapt to the marketised environment, work units that used to provide almost everything for their staff members began transitioning from administrative organisations to commercial enterprises, stripping themselves of their previously held social welfare functions. Work units ceased to be the sole channel through which individuals could access scarce survival resources such as material products, development opportunities, and social recognition. This has ended the individuals' organizational dependence on their work units, leading to the gradual disintegration of the previous work-residence model centred around the work unit. Consequently, individuals gained substantial autonomy and freedom by detaching themselves from their work units and started to move around under the driving force of the market. The static-natured work-unit communities are destined to be unsustainable against the backdrop of ever-increasing residential mobility.

In the late 1990s, the system of urban commercial housing provision was largely established, and housing was primarily constructed and sold by private property developers. Urban residents obtained property rights through purchases, enabling them to buy and sell houses. During this period, the real estate industry experienced rapid growth. Real estate developers began constructing a large number of residential communities, which varied in terms of location, layout, size, property services and amenities. These included large-scale high-rise residential buildings and low-rise villas, catering to a range of segmented markets from ordinary apartments to luxury residences. Following market logic, developers adopted a series of commercial strategies to attract homebuyers, resulting in significant differences between various grades of commercial housing, while commercial housing communities with similar pricing tended to offer comparable living conditions. Concurrently, with the diversified development of the national economy, the social stratification of urban residents deepened, with more pronounced differences emerging between social groups defined by income, occupation, class, and other dimensions. The intertwining of demographic diversification and housing variety, coupled with the establishment of private housing ownership, made it possible to buy, sell, and re-select housing properties. For instance, residents who previously lived in work-unit communities, under the market economy system, experienced income differentiation. Wealthy residents chose to sell their deteriorating housing in work-unit communities and buy housing with higher-quality living conditions and amenities. Residents who lacked the financial means to improve their living conditions continued to live in these communities. In this process, there is a turnover in community demographics. As the physical conditions of work-unit communities decline, their residents gradually become those with lower economic capabilities. Meanwhile, with the process of urbanisation, housing with poorer living conditions is often rented by migrant workers relocating from other regions.

In the wake of activation of residential mobility within the city, there has been a reshuffling of the residential layouts. Such spatial reorganisation is primarily driven by housing prices and the economic capabilities of the residents. Under the influence of market forces, the previous residential pattern centred around work units has been replaced, giving rise to a more diversified demographic of community residents.

Apart from market forces, the government has also played a more important role in shaping residential spatial configuration. To advance the industrial development in specific areas within the city, some local governments have planned and constructed large residential areas within the city. These residential areas adopt the planning concept of mixing different social classes, constructing various types of housing such as government-subsidised housing and commercial housing in designated proportions, and allocating corresponding public service resources such as education and health care facilities. These large residential areas have a unified planning and design that cater to specific industries or urban development strategies. As a result, compared to similar types of communities in other areas, communities within these large residential areas exhibit certain differences in terms of resident demographics, housing forms, and property services.

In terms of residential space allocation, relying solely on market mechanisms often leads to resources favoring higher income groups, while lower-income groups face housing difficulties. China's housing construction policies stipulate that newly built commercial residential areas must include a certain proportion of affordable housing, to ensure fair distribution of public resources and social justice, and to ensure that different social classes can enjoy basic housing rights. The government is responsible for the subsequent operation and management of the affordable housings, ensuring high standards and quality planning and construction, balanced layout, convenient transportation, complete supporting facilities, suitable living environment, and shared community environment and professional property management services. This ensures that middle and low-income groups have the same living environment as commercial housing residents, reducing the sense of disparity between the rich and the poor.

In recent years, Chongqing has promoted the construction of affordable housing by adopting mixed construction with commercial housing, public rental housing, and affordable housing, using a 'small concentration, large dispersion' model to prevent concentrated areas of affordable housing from becoming urban 'ghettos', allowing residents to live better, more comfortable, and with more dignity. As a typical representative of affordable housing communities, the Minxin Jiayuan community allocates 10% of its total construction area for public rental housing. Public rental housing is decorated simply according to the standard of 400 yuan per square meter, with unit sizes ranging from 35 m^2 to 80 m^2, with more than 85% being below 60 m^2. These systematic considerations enable middle and low-income groups living in public rental housing to enjoy the living standards of mid-range commercial housing.

12.5 Major Types of Neighbourhood and Inequality

In the last 30 years, a variety of new neighbourhood types have emerged. Each community type exhibits differences in terms of residents' socio-economic characteristics, housing conditions, spatial arrangement within the community, accessibility to public services, as well as commercial amenities.

To gain a better understanding of the diversity of neighbourhoods in Chongqing, a variety neighbourhoods located in the central urban districts of Chongqing in late 2022 and early 2023 were surveyed. The central districts have the highest proportion of urban population (93.3% in 2022). The five administrative districts covered in the fieldwork are Yuzhong, Shapingba, Yubei, Nan'an and Jiulongpo. These five districts are core urban areas of Chongqing Municipality, encompassing all types of neighbourhoods. However, there are certain differences in the proportionate distribution of different neighbourhoods types among different areas.

Among these, Yuzhong District, known as the 'mother town' of Chongqing, is relatively small in size and yet serves as the economic, political, and cultural centre of Chongqing, predominantly characterised by the tertiary industry. Shapingba District is renowned for its high-quality educational resources, while its urban area is relatively old and worn-out. Jiulongpo District is part of the heavy industrial base of Chongqing. Because they were suitable for industrial development, these three districts once gathered a large number of state-owned enterprises or units. As a result, they have a greater number of work-unit neighbourhoods and urban villages compared to other districts.

Yubei District, known as the largest contributor to Chongqing's GDP, is a hub for affluent residents, with a strong capacity to realise industrial development plans and a pleasant surrounding environment. Among its neighbourhood types, the proportion of high-end commercial housing is higher than that in the other four districts. Nan'an District, located at the unique 'two rivers and four banks' in Chongqing, is one of the districts with areas having the longest history and the richest cultural heritage. Although its terrain is not very suitable for industrial development, it is highly conducive to living. Commercial housing communities currently account for the highest proportion in Nan'an District.

Consistent with the surveys conducted in Datong and other cities prior to this project, four categories of community types were classified: work-unit neighbourhoods, commercial housing neighbourhoods, government-subsidised housing neighbourhoods, and urban village neighbourhoods. The distribution is as follows: work-unit neighbourhoods constituted 24.14%, commercial housing neighbourhoods dominated with 53.4%, government-subsidised housing neighbourhoods comprised 19.01%, and urban villages accounted for a mere 3.45%. Mirroring trends in other Chinese mega-cities, the substantial share of commercial housing underscores the pivotal impact of land developers on urban sprawl. Conversely, the government's concerted efforts to tackle urban village challenges have made such extensive settlements a rarity in Chongqing's urban landscape.

Examining neighbourhood size, the average household count stands at 2100 for commercial housing neighbourhoods, 1331 for urban villages, 1480 for work-unit neighbourhoods, and an impressive 4298 for government-subsidised housing neighbourhoods. It is clear that commercial and government-subsidised neighbourhoods boast a more substantial scale compared to their urban village and work-unit neighbourhood counterparts. Characterised predominantly by high-rise buildings exceeding 20–30 storeys, these expansive neighbourhoods are capable of housing a greater number of residents within the same footprint. In a mountainous city like Chongqing, where land is at a premium, such compact living arrangements offer an undeniably prudent and space-efficient solution.

Drawing from survey feedback across various neighbourhoods in Chongqing, a distinct income distribution was identified. Residents from urban village neighbourhoods report a median monthly income ranging from 3000 to 6000 yuan over the last six months. Government-subsidised housing neighbourhoods fall within the 6000–9000 yuan bracket, work-unit neighbourhoods hover just below 9000 yuan, and commercial housing neighbourhoods surpass this threshold.

These figures speak of a pronounced income disparity within Chongqing's urban landscape. Commercial housing neighbourhoods, offering superior living conditions, draw in the city's middle- and upper-income demographics. Work-unit housing neighbourhoods, often linked to the government or state-owned entities, have residents with a stable income and

Fig. 12.5 Residential buildings in the Jialing Factory Employee Dormitory area. *Source* Taken by the author

employment. On the other hand, residents from government-subsidised neighbourhoods and urban village neighbourhoods exhibit substantially lower incomes, reflecting their role as hubs for the city's lower-income and migrant workers.

Based on the fieldwork findings, neighbourhoods in Chongqing are mainly categorised into the following types:

12.5.1 Former Work-Unit

Work-unit neighbourhoods, a legacy of the planned economy era, are a type of welfare housing whose construction is funded by work units and rented (or privatised) to employees at rates lower than market prices. Unlike commercial housing neighbourhoods, these generally do not have commercial property services but are managed and maintained by the work units or municipal authorities. Different work-unit neighbourhoods vary significantly in terms of environmental greening, housing quality, and maintenance services, primarily depending on the economic strength of the respective work units.

The old work-unit neighbourhoods built before the 1980s have now become outdated. These vary in size, but the majority are in a state of disrepair, lacking public spaces for activities, and many residences even lack independent bathrooms or kitchens. Moreover, due to the lack of anticipation for the surge in residential parking needs during the planning phase, vehicles are densely and haphazardly parked in these neighbourhoods, further encroaching upon the already limited public spaces (Fig. 12.5). Outdated design and insufficient maintenance are significant contributors to the decline of these neighbourhoods. Since the work units responsible for maintaining them may have long ceased to exist, these have been left unattended for a long time. Maintenance of public hygiene relies on residents or cleaners hired by local communities. Many public facilities are in disrepair. Apart from the elderly who choose not to move away due to social connections or economic reasons, the residents of these old work-unit neighbourhoods are primarily young manual labourers with medium or low income, given the rents are low in these neighbourhoods. In addition, these are mostly located in commercially vibrant

Fig. 12.6 Public space of a new type of work-unit neighbourhood. *Source* Taken by the author

and densely populated central urban areas of the city, providing ample employment opportunities for them.

Due to the fact that the reforms primarily took place in the economic sector, many residential communities built by production units such as factories from the past have gradually vanished. However, many units in the public sector continued to construct work-unit neighbourhoods to address the housing issues of their employees. Those late-model work-unit neighbourhoods were mainly completed after 2000, and were mostly constructed by government agencies or public institutions, represented by university faculty and staff family compounds. These late-model work-unit neighbourhoods are relatively superior in terms of building quality, supporting facilities, and management level (Fig. 12.6). In addition, late-model work-unit neighbourhoods also have ancillary facilities not commonly found in other types of neighbourhoods. For example, in a community belonging to a government department, there are not only a seniors' university and an activity centre dedicated to serving retired cadres, but also a public kindergarten that formerly schooled children of cadres. With these advantages, the selling price of late-model work-unit neighbourhoods can be comparable to that of high-end commercial housing neighbourhoods, attracting a large number of non-unit employees to become residents through rental or purchase. However, the residents of these neighbourhoods are still primarily work-unit employees and their families.

12.5.2 Commercial Housing

Commercial housing neighbourhoods built and sold by real estate developers have become the most common type of community in Chongqing's urban areas. As the housing supply model with the highest degree of marketisation, the evolution of commercial housing neighbourhoods has gone hand in hand with the overall development of the real estate industry during the same period.

The early commercial housing neighbourhoods in Chongqing were mostly built in the late 1990s when the real estate industry just got off the ground. Geographically, they are primarily concentrated in the densely populated old urban

Fig. 12.7 A commercial housing community built in the late 1990s. *Source* Taken by the author

areas, and are relatively small in scale. They are typically characterised by residential buildings of around ten floors, equipped with elevators, and public activity spaces decorated with trees and greenery. However, many of them now appear time-worn and face maintenance issues (Fig. 12.7).

In general, ordinary commercial housing neighbourhoods targeting medium-income groups prioritise affordability. These neighbourhoods are situated in somewhat less favourable locations and have higher plot ratios. They use standardised property services and landscape designs to lower housing prices (Fig. 12.8).

As the most common type of housing neighbourhood in Chinese cities, ordinary commercial housing neighbourhoods are widely distributed throughout the city from downtown to peripheral development zones. The living and service amenities and infrastructure in these neighbourhoods are supplied and developed to meet the basic living requirements of the residents, and to meet the overall average standard of the city. However, security, environmental, and sanitary conditions are inferior to those of high-end residential neighbourhoods. Some neighbourhoods face problems such as high building density, deteriorating green spaces, insufficient parking spaces/garages, and building quality deficiencies such as wall seepage and foundation settlement.

Though the ordinary commercial housing neighbourhoods generally have clear boundaries and are enclosed by surrounding walls, fences or commercial buildings, they are not as enclosed as high-end commercial housing neighbourhoods, and are less strict on access control for non-community residents. They may have no community access control, or, if any, they are normally completely or partly out of function.

In the past decade, as the real estate industry matures, the late-model commercial housing neighbourhoods constructed between 2010 and 2020 not only retain the enclosed and liveable attributes of earlier commercial housing neighbourhoods, but also develop new features:

Firstly, expanded community scales. Under the model of centralised development, the number of units in the late-model commercial housing neighbourhoods has rapidly increased from hundreds in early commercial housing to thousands, with high-rise residential towers of over 30 floors becoming one of the signature features.

Fig. 12.8 Exterior view of a sampled ordinary commercial housing community. *Source* Taken by the author

Secondly, significant increase in commercial facilities. The late-model commercial housing neighbourhoods are typically equipped with more facilities. Some larger neighbourhoods even include commercial complexes such as pedestrian shopping streets or large shopping centres. This new trend, to some extent, alleviates the inconveniences brought about by the expansion of community size. Moreover, constructing commercial facilities can increase the selling price and sales volume of the property, generating more profit for developers.

Thirdly, more pronounced differences between different tiers of neighbourhoods. High-end commercial housing neighbourhoods targeting high-income groups are distinguished by their superior geographical locations, strict access control, concierge-style property services, personalised garden landscaping, health centres, and other high-end ancillary facilities, offering even an exquisite and luxurious living experience (Fig. 12.9).

In general, primarily comprising villas and large-sized high-end residences, these neighbourhoods are gated to form an enclosed space surrounded by walls and fences. Electronic access control systems are installed and protective measures are taken at the entrances to keep out non-community residents. Monitoring covers the entire neighbourhood, guaranteeing high levels of security. Meanwhile, these neighbourhoods are constructed with upscale building materials to provide higher architectural quality and better architectural appearance.

Neighbourhoods possess meticulous construction planning, high-end property management, nice ecological environments, high greening rate, and all necessary infrastructure. Most of them are furnished with pavilions, central squares, ball fields and fountains. Some even have their own swimming pools, fitness centres, leisure clubs, and other high-end living and service amenities. Inside the neighbourhoods, there are sufficient parking spaces/garages or underground parking places. Most neighbourhoods have separate passageways for pedestrians and vehicles above the ground to pass through.

Targeting the high-income population, these neighbourhoods in Chongqing are mostly located downtown or in the inner suburbs to provide residents with convenient and high-quality living amenities, or accommodate residents' desire for

Fig. 12.9 Exterior view of a sampled high-end commercial housing community. *Source* Taken by the author

both modern metropolitan life and a beautiful natural environment. More importantly, the ability of resident families to migrate from an ordinary commercial housing community to a high-end commercial housing community is often seen as a symbol of elevated social identity or social status.

12.5.3 Government-Subsidised Housing

Government-subsidised housing neighbourhoods refer to those consisting of public rental housing, low-rental housing, affordable housing, and a few limited-price commercial apartments. Unlike commercial housing or work-unit housing neighbourhoods, subsidised housing neighbourhoods are developed mainly through governmental initiatives and financial funds. Fundamentally, these are neighbourhoods offered by the government at below-market rates to eligible buyers or renters who are unable to afford market-rate housing. Clearly, subsidised housing communities serve as a social welfare measure for residents in need.

These neighbourhoods inevitably have inferior building quality and infrastructure provision compared to commercial housing neighbourhoods (Fig. 12.10). Furthermore, when occupied, they may gradually encounter issues related to environmental sanitation, accessibility to public services, and property management, such as green space deterioration and access control failure.

To minimise development costs, subsidised housing neighbourhoods are usually located in the suburban areas of a city, which poses a significant challenge to the well-being of residents due to housing–work segregation. Furthermore, due to the high heterogeneity of residents in these neighbourhoods and the young working population making up a large proportion, their low sense of belonging to the community and low desire to participate in community activities have become key hindrances to community development.

Despite all these deficiencies, the provision of subsidised housing neighbourhoods has enabled new urban dwellers and a new working population to settle down temporarily and adapt to a new urban lifestyle, and also enabled the lower income population to access basic housing security. In addition to the community functions described above, the 'blending' and 'merging' of residents

Fig. 12.10 Interior view of a subsidised housing neighbourhood. *Source* Taken by the author

within a shared community who have migrated from rural areas or small towns into a megacity, has become the major topic for the development of these neighbourhoods.

12.5.4 Urban Villages

The phenomenon of urban village neighbourhoods is a product of the urban expansion process that absorbs rural villages. With Chongqing's substantial investment in rebuilding urban villages in recent years, urban village neighbourhoods are becoming rare in the city. Largely retaining the characteristics of rural villages, urban villages feature open physical spaces and relatively basic established living circles. Mostly privately constructed by residents, housing in urban village neighbourhoods was often built before the villages underwent urbanisation, so their architectural qualities vary significantly, and their architectural forms are diverse, ranging from six-storey small buildings to rough shanties. But they are mostly defective with poor structural integrity and insufficient living functionality. Due to the lack of planning, housing distribution within urban village neighbourhoods is haphazard and disorderly, resulting in many cul-de-sacs. Affected by urbanisation, many young people from urban villages move out for work, leaving behind elderly residents or renting their houses to others. In contrast to hollowed-out villages resulting from population loss, urban villages on the city's peripheries have absorbed a significant number of low-income individuals, both local and from outside, providing them with accommodation. These residents operate various daily services within urban villages, ranging from food catering to medical care, forming a multipurpose and relatively autonomous small living circle (Fig. 12.11).

12.5.5 Disparity of Public Services and Amenities in Different Types of Neighbourhood

The differences between various types of neighbourhoods are also prominently reflected in the accessibility of public services. In addition to basic needs for personal survival, factors such

Fig. 12.11 Interior view of an urban village community. *Source* Taken by the author

as education, health care, leisure, culture, entertainment and transportation have become increasingly important in affecting residents' quality of life. As direct carriers of social public services, the spatial arrangement of relevant infrastructure enables citizens residing in different areas to actually enjoy different opportunities for social services. Despite a series of minimum standards outlined in urban planning documents for the construction and configuration of infrastructure, various types of public service facilities are often unevenly distributed across different types of urban neighbourhoods.

Accessibility is typically employed to reveal the level of availability of facilities under certain constraints. Collectively determined by factors such as time, distance and energy costs, it effectively reflects the extent to which individuals can access public services in the physical world. In order to quantitatively analyse and compare the living conditions of neighbourhoods within the main city of Chongqing, the Cumulative Accessibility Index is used to describe the accessibility of different types of social infrastructure within a 15-min community living circle.

Research findings of Chongqing indicate five results in terms of inequality: Firstly, there is relatively inadequate allocation of social public services in government-subsidised housing neighbourhoods and urban village neighbourhoods, where, except for shopping and primary education, there are noticeable deficiencies in leisure, entertainment, transportation, and many other facilities.

Secondly, there is limited allocation of social public services in newly constructed commercial housing neighbourhoods, whether high-end or ordinary. Particularly, the development of more supporting facilities such as middle- and high-level education, culture, and health care has yet to be strengthened. This aligns broadly with other cities in China at present.

Thirdly, there are significant internal differences in terms of allocation of social public services among early commercial housing, work-unit housing, resettlement housing (huiqianfang, a form of subsidised housing provided by the government or developer to the original residents after they have been relocated for urban renovation, demolition and other projects), and housing reform neighbourhoods (fanggaifang, built-up public housing purchased by urban workers at cost price, a transitional policy in the process of transforming urban housing in China from the work-unit allocation system to the market economy system). Some neighbourhoods excel in conditions related to shopping, primary education, community health care, and transportation, while others exhibit noticeable deficiencies. The disparities within neighbourhoods of the same category stem from a complex array of factors. One plausible explanation lies in the use of policy instruments. China has a preference for establishing administrative regions with specific policy objectives in urban development (called 'Te qu') such as economic development zones or historical-cultural districts, and allocates corresponding resources accordingly, including public services. These special zones receive varying levels of resources based on their policy goals, leading them to distinct developmental trajectories. Consequently, this affects the type, quantity, and accessibility of services available within the neighbourhoods located within these zones.

Fourthly, there are often rich social service facilities near neighbourhoods 'inside the system' (tizhinei, referring to those working for government organs and public sectors) represented by housing reform neighbourhoods, late-model work-unit neighbourhoods, and old-type work-unit neighbourhoods. Notably, the accessibility to high-quality service facilities such as department stores, high schools, museums, sports centres, and subways are several times or even higher compared to those 'outside the system' (tizhiwai), contributing to a superior community living environment. This may be an example of administrative power playing a significant role in resource allocation. Land resources and even future development plans are essentially in the hands of the government, so there is reason to believe that when the government constructs work-unit neighbourhoods to serve its own employees, it will prioritise allocating more valuable land parcels.

Fifthly, neighbourhoods 'outside the system' represented by resettlement housing neighbourhoods, government-subsidised housing neighbourhoods, and urban village neighbourhoods, primarily focus on the widespread availability of basic public services. There are few particularly prominent situations in terms of accessibility to social services, and yet the lack of high-quality public services poses a universal challenge.

12.6 Conclusion

This chapter provides a systematic analysis of the urbanisation, neighbourhood diversification and differentiation of social spaces in Chongqing.

Chongqing stands as a distinctive exemplar within the discourse of urban and regional studies. In contrast to the large cities of the economically prosperous southeastern coastal regions, where urbanisation is marked by both intra-regional rural-to-urban transitions and inter-regional migratory flows, Chongqing's urban development is primarily propelled by internal migration within its administrative confines. This phenomenon is characterised by a swift escalation in urbanisation rates while maintaining a relatively constant overall population: over a period of a quarter century from 1997 to 2022, the proportion of urban residents in Chongqing leaped from 29.5% to 71%, while the total population increased from 28.73 million to just 32.13 million.

Several multifarious elements contribute to these variances. Initially, as a city situated in western China, Chongqing's urban development commenced subsequent to that of cities in the east, thereby facing more acute competition for both inhabitants and professional talent. Moreover, the city's growth has been significantly influenced by policy measures from the central government, including its elevation to a directly administered municipal status and initiatives such as the 'Develop the West' strategy. These policies, though prompt in delivering benefits, tend to produce a pulsatile pattern of development due to their ephemeral nature. Lastly, the formidable topography of Chongqing, defined by its mountainous and riverine landscape, has engendered a robust local culture and customs. The climatic conditions, particularly characterised by severe heat in summer and cold, damp winters, colloquially known as the 'City of Fog', serve to further exacerbate the challenges associated with inter-regional migration, thus uniquely shaping its pattern of urbanisation.

Moreover, Chongqing's development follows a typical mode of urbanisation which may be metaphorically likened to 'a tiny horse pulling a huge cart', which merely relies on the 5% urban area and the 31% urban population to drive the development of the 95% large rural areas and the 69% rural population, facilitating urbanisation throughout the entire municipality. Specific policies facilitating urbanisation include:

(1) Measures such as reform of the registered residence system to be implemented to encourage the transformation of the rural population into urban residents. Meanwhile, equal social insurance and public services to be provided to the incoming population to address their concerns, facilitating their settlement in the city. (2) Planned expansion of built-up urban areas to be executed to accommodate a larger population. (3) The construction of housing and municipal public facilities to be promoted through the reform of the housing system, by leveraging the combined forces of the government and real estate developers in the capacity of market entities.

Overall speaking, the driving forces behind Chongqing's urbanisation include both the visible hand of the government and the invisible hand of the market. On the one hand, administrative power is indispensable, as the central government's will and coordinated planning are essential drivers of Chongqing's urbanisation. The elevation to a directly administered municipality was a crucial opportunity, providing Chongqing with the factor resources and policy support for its economic and social development. On the other hand, the reform of the economic system characterised by the transition from a planned economy to a market-oriented economy unleashed the vitality of market forces, significantly propelling Chongqing's industrial development. In the process of reform and urban innovation, the manufactory industry has been consolidated, and has brought higher income and increased demand, driving the growth of the service industry.

In sum, the dynamic of Chongqing's urbanisation has resulted from the coordinated efforts of both the government and the market. Government planning and support provide a foundation and direction for the city, while market dynamism injects vitality into economic diversification and sustainable development.

The typology of neighbourhoods within Chongqing manifests a pronounced diversity. Predicated upon both topographical and historical determinants, the city's residential districts are characterised by a polycentric and clustered spatial configuration. Distinct residential clusters preserve a degree of spatial segregation, concurrently facilitating relatively autonomous employment opportunities and public amenities within their respective precincts. Consequently, disparities have emerged amongst these clusters, contingent upon variances in educational institutions, health care facilities, income brackets and property valuations. Meanwhile, the advancement of infrastructure—including motorways, underground railways and light rail systems—alongside escalating property and land values, has catalysed the development of previously intractable mountainous and riverine sites into novel residential zones. This development has precipitated the gradual obfuscation and eventual confluence of certain cluster boundaries, engendering a continuum of urban fabric. Such dynamics have given rise to a multiplicity of neighbourhood forms within the city.

Additionally, akin to other major Chinese conurbations, the production-residence neighbourhoods in Chongqing, initially based upon the work-unit system, have evolved towards a more heterogeneous residential typology. This metamorphosis has been accelerated by the transition from a command to a market-driven economy, precipitating the decline of the work-unit system.

Concurrently, reforms towards housing commercialisation have stimulated increased residential mobility within the city. Furthermore, the large-scale resettlement housing for the Three Gorges migrants, along with the development of substantial affordable housing projects catering to up to 100,000 individuals from the urban low-to-moderate income strata, constitute residential neighbourhoods seldom observed in other urban centres.

In conclusion, during the urban transformation process, the typology of urban neighbourhoods has emerged, including work-unit neighbourhoods, commercial housing neighbourhoods, government-subsidised housing neighbourhoods, as well as urban villages. Each type of neighbourhood exhibits differences in terms of residents' socio-economic characteristics, housing conditions, internal spatial arrangements, accessibility to public services, and commercial amenities. As a result, two major kinds of neighbourhood, namely, 'generative neighbourhood' and 'constitutive neighbourhood' have come into place.

References

Aik NC, Zhang Q (2023) Use of Theil for a specific duality economy: assessing the impact of digital inclusive finance on urban-rural income gap in Chongqing. Fintech 2:668–679

Bray DB (2005) Social space and governance in urban China: the Danwei system from origins to reform. Stanford University Press

Deng F (2018) Work unit and private community in the evolution of urban planning in contemporary China. Plan Theory 17:533–550

Li X, Kleinhans R, Van Ham M (2019) Ambivalence in place attachment: the lived experiences of residents in Danwei communities facing demolition in Shenyang, China. Hous Stud 34:997–1020

Logan JR, Bian YJ, Guan Y, Lu HL, Pan YK (1997) "Work units" and the commodification of housing: observations on the transition to a market economy with Chinese characteristics. J Soc Sci China:28–35+194

Wang FC, Zhang CL, Yu Y, Fan Y (2012) Livability planning strategies for large residential agglomerations in the main city of Chongqing. Planner's Forum 28:24–27

Wang W, Wang SQ, Li YS, Feng C (2022) Assessing the sustainability and competitiveness of tourism economies in China's Chengdu-Chongqing metropolitan area. Environ Sci Pollut Res 29:66960–66978

Xiao ZP, Liu TB, Chai YW, Zhang MK (2020) Corporate-run society: the practice of the Danwei system in Beijing during the planned economy period. Sustainability 12

Yang Y, Zhang D, Meng Q, Yu W, Yuan L (2017) Stratified evolution of urban residential spatial structure in China through the transitional period: a case study of five categories of housings in Chengdu. Habitat Int 69:78–93

Zhang C, Chai Y (2014) Un-gated and integrated work unit communities in post-socialist urban China: a case study from Beijing. Habitat Int 43:79–89

Zong HM, Yu XX (2024) Urbanization of Chongqing municipality: regional contributions and influencing mechanisms. Sustainability 16

Spatial Inequality in Manila: A Case Study of (Un)Caring Urban Ethics in Jaime Cardinal Sin Village and the Manila North Cemetery

13

Ryan Randle B. Rivera, Mark Anthony M. Gamboa, Mario R. Delos Reyes, Rhay Daniel R. Racoma, and Irish M. Manlapas

Abstract

This chapter explores the lived manifestations of a particular spatial inequality in two Manila neighbourhoods. Taking the notion of the ethics of (un)care as a spatially reinforced phenomenon, the chapter argues that the living conditions in Jaime Cardinal Sin Village and the Manila North Cemetery tell a tale of an urban area that cultivates, on one hand, a caring environment, and on the other, the forgetfulness of a global city. The chapter concludes by reflecting on the potential of a bottom-up, solidarity-based urban citizenship in navigating spatial inequalities. With state recognition and support, this enduring social capital offers a pathway towards a kinder form of sustainability—one that is shaped and co-owned by the residents, cognisant of inherent interconnectedness.

Keywords

City of Manila · 'Planning of care' · Care ethics · Urban inequalities · Global South

13.1 Introduction

The City of Manila is among the world's densest and most overcrowded cities. A highly urbanised city in the westernmost portion of the metropolitan Manila region, the city hosted more than 1.84 million inhabitants in 2020. It yielded a gross population density of 73,920 persons per square kilometre, which is over 200 times the density of the Philippine archipelago.

The persistent growth trend in an already burdened city begs the question of its capacity to facilitate and maintain inclusive urban living conditions. These are said to take form as the 'interspersion of slums and enclaves', where affluent and impoverished neighbourhoods are located close to each other, but often transacting in unequal and stigmatised terms (Garrido 2020).

R. R. B. Rivera (✉)
Department of Community and Environmental Resource Planning, College of Human Ecology, University of the Philippines Los Baños, Los Baños, Philippines
e-mail: rbrivera7@up.edu.ph

M. R. D. Reyes · R. D. R. Racoma · I. M. Manlapas
Centre for Neighbourhood Studies Philippines (CeNS), Quezon City, Philippines
e-mail: mrdelosreyes@up.edu.ph

R. D. R. Racoma
e-mail: rrracoma@up.edu.ph

I. M. Manlapas
e-mail: immanlapas@up.edu.ph

M. A. M. Gamboa
University of the Philippines School of Urban and Regional Planning (UP SURP), Quezon City, Philippines
e-mail: mmgamboa@up.edu.ph

Y. P. Wang et al. (eds.), *New Drivers of Division*, Sustainable Development Goals Series,
https://doi.org/10.1007/978-981-96-5629-5_13

The contrasting coexistence of decrepit slums and gated communities and high-rises illustrates what Florida (2017) describes as the *new urban crisis*—where urbanisation has accelerated a growing political and economic divide—albeit at a more local, intra-urban scale. Nijman and Wei noted a similar trend, arguing that recent information-based urban economies tend to create 'new and deepening inequalities across multiple dimensions' in the form of wealth inequality, residential segregation, housing quality differentiations, and inequitable access to social services (2020: 1). These urban trends had already gained prominence even a decade earlier, culminating in Fainstein's (2010) framework of a good and just city. There she called for urban policies and processes predicated on transcendent values of equity, diversity, and democracy. She underscored the imperative for an urban environment that integrates the economically and socio-culturally disadvantaged groups of the city (Fainstein 2014). Finally, in its most recent iteration, the United Nations' New Urban Agenda (2017) has, once again, called upon cities to guarantee the right of all its inhabitants to inclusive and non-discriminatory enjoyment and utilisation of the fruits of urban development.

Against this backdrop, this chapter explores the lived implications of inequalities in Manila, using as analytical lens the spatialities of *care* and *uncare* in the city. Drawing from the work of Gabauer et al. (2022) it emphasises the fact that urban areas are a salient site of the tension between caring and uncaring practices among residents and state actors. It is in urban environments where the need for care 'emerges forcefully', and where the institutional capacity to fulfil caring practices is at the highest (Cohen and Knierbein 2022). Similarly, it is in the urban space where geographies can be extremely uneven, emanating from deeply entrenched differentiations along wealth, class, race, age, or gender lines. It is also in the urban context where there is high susceptibility to an 'entrepreneurial script' of space production, in which the welfare provisions of the government tend to retreat to make way for 'neo-liberal capitalist ventures'—privatising and thereby commodifying care (Chatzidakis et al. 2020; Gabauer et al. 2022; Madanipour 2022).

The first part of this chapter briefly describes the notions of socio-spatial inequalities in Manila that have long been a problem; the second part deals with previous conceptualisations of *care* and *uncare* as a spatially produced and negotiated phenomenon; the third part presents the manifestations of caring and uncaring practices observed in two case study neighbourhoods within the city of Manila. The chapter concludes with an assertion that local, solidarity-based, caring citizenship—which draws comparison with indigenous notions of *kapwa* and *pakikipagkapwa (mutuality and companionship)*—could offer a pathway to humanise and communally enable sustainability emerging from the smallest units of urban management.

13.2 Inequalities in the City of Manila

In broad terms, economic inequality in cities delves into the asymmetric distribution of some economic variables among individuals, households, or groups. Often, the debates as to how this could be addressed separate into two perspectives (Núñez and Tartakowsky 2007; Lefranc et al. 2008). One is *inequality of outcomes*, which grapples with the issues of resulting material wealth and living standards, including the differences in the levels of income, nutritional status, or educational attainment; and the other is *inequality of opportunity*, which tackles outcome-defining circumstances beyond one's doing, such as availability of employment opportunities, customs and traditions, and access to social services (United Nations Development Programme [UNDP] 2013).

The former tries to alleviate inequality mainly through wealth redistribution measures, enabling disadvantaged households to acquire goods and services that uplift their well-being (UNDP 2013). Inequality of outcomes are, thus, measured through asset-based indices of wealth from national surveys, as demonstrated in Molina et al. (2013), Palafox et al. (2016), and Palipudi et al. (2012). The latter, meanwhile, takes from Roemer's (1998: 4) principle of 'levelling the playing field' by compensating for exogenous circumstances that tend to cripple disadvantaged individuals. It echoes Sen's (2000) view of material wealth as the means to pursue the kind of life that people value. Inequality of opportunities, thus, tends to be measured using variables such as access to utilities, years of schooling, and levels of literacy, as done in McDoom et al. (2019) and Reyes et al. (2017). Nonetheless, in practice, the two views are inseparable, as outcomes are inextricably linked with opportunities, each one reinforcing the other (UNDP 2013). Contending economic inequality, therefore, should encompass these two conceptualisations.

The works of Garrido (2020), Porio (2016), and Shatkin (2008) offer a nuanced understanding of inequalities observed in Metro Manila. Drawing from ethnographic approaches, Garrido (2020) observed that, unlike the American model of segregation (comprising isolated areas with concentration of minority groups), the one in the region evokes an 'interspersion of slums and enclaves' (p. 3), where affluent and impoverished neighbourhoods are located within the same vicinity, and where interactions among groups exist but in an 'unequal' (p. 8) and 'stigmatised' (p. 11) manner. He also concluded that—having been reinforced by the segregating practices of residents from either side—a sort of 'classed spaces' have emerged, making the residents mindful of the abstract boundaries in the very *barangays* (i.e. villages, the smallest administrative unit in the Philippines) they live in (Garrido 2018: 444).

Porio (2016) illustrated Metro Manila's persisting inequalities in terms of access to housing. She noted a deepening divide post the social reform gains in the early 1990s, attributing it to the intensifying real estate development and a weak state. She noted that like many Asian cities, Metro Manila hosts 'prosperous spaces' that sit close to impoverished 'brown spaces', with the former having the state's sanction on the premise of public–private partnerships (Porio 2016: 193). This is akin to Ortega's (2016) observation of a gentrification trend in Metro Manila, where upscale development initiatives appear to have been predicated on the displacement of informal households, making cities more exclusive with appreciating land values. Porio (2016) also noted the region's susceptibility to natural hazards as a confounding variable in the inequality conversation, since these carry uneven consequences, especially for the disadvantaged and vulnerable communities.

Shatkin's (2008) explanation of the roots of Metro Manila's uneven urban development is close to the arguments put forward by Porio (2016) and Ortega (2016). He argued that there is a prevalence of 'planned privatisation of urban and regional planning' in Metro Manila, evidenced by the range of urban infrastructure projects initiated by the private sector and sanctioned through partnerships by the government. Large property and infrastructure developers have a de facto power over how the region's landscape would turn out—and this is anchored primarily on building spaces for consumption and production (Shatkin 2008: 384). Hence, many households are often displaced and relocated to remote provinces, as they happen to live under informal tenure along the project's right-of-way or within an estate's perimeter fence. An unfortunate outcome of this neo-liberal turn is the tendency for the residual residential spaces—those that do not yield the *highest and best use of land*—to become 'forgotten places' (Shatkin 2004). These are usually interstitial areas of residence barely recognised by local planning agencies who leave them out of the formal spatial planning agenda. A participatory map of neighbourhoods in Manila, depicted in (Fig. 13.1), is an indication of this peculiar distribution of settlements.

13.2.1 (Un)Caring Practices as a Manifestation of Urban Inequalities

(Un)caring practices in urban areas can be seen as a physical and lived manifestation of a city's state

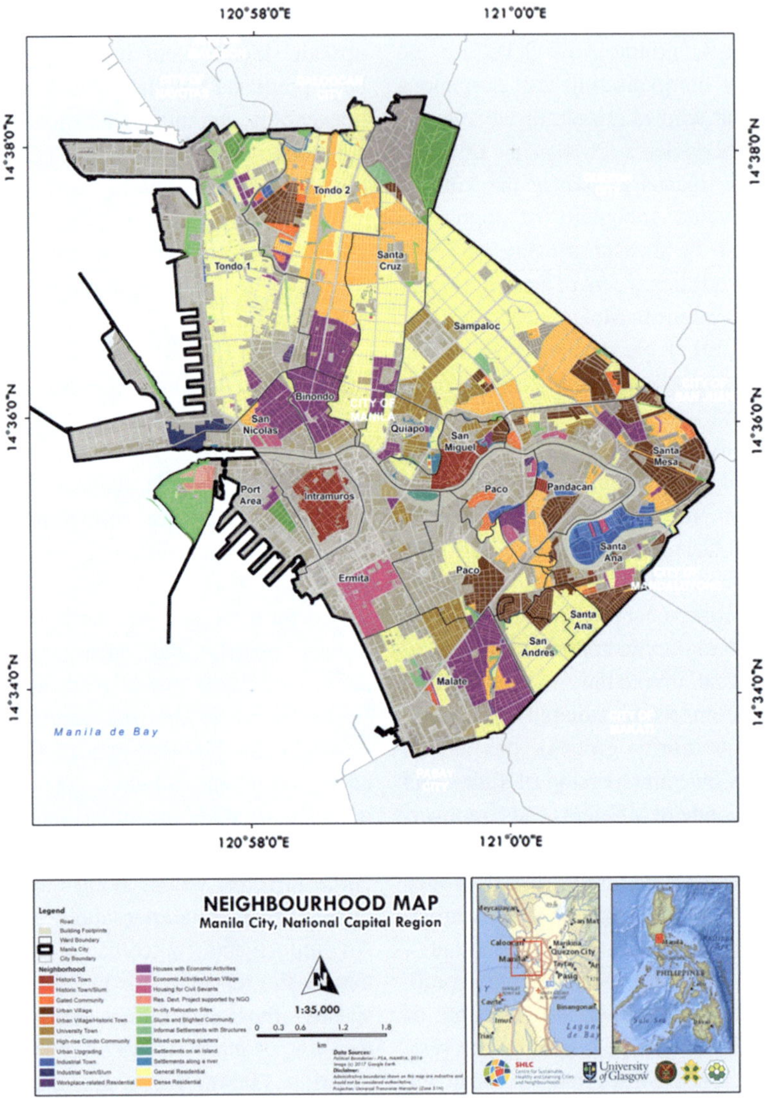

Fig. 13.1 Neighbourhoods in Manila based on a participatory mapping. *Source* Authors

of (in)equality. As Tronto (2015) has remarked, only in a 'truly equal and inclusive society' could residents be 'well cared for' and, at the same time, be able to 'engage in [meaningful] and caring relationships'.

13.2.1.1 Caring and Social Citizenship

Caring, in its barest sense, can refer to one's provision of the needs of another, which normally emanates from a degree of concern or proximity. It manifests most commonly at an intimate level within and among one's closer ties, where acts of kindness can be readily fostered out of sympathy. Despite its personal nature, the concept of caring could also transcend to the level of communities and be seen as 'a public and political responsibility' (Gabauer et al. 2022). In her seminal work, Tronto (1993) presented a collective notion of *care* as a 'species activity'— something that makes people distinctively *human*—which encompasses 'everything that [people] do to maintain, continue, and repair our world'. Such acts envision an ideal quality of life, where all beings could 'live as well as possible'. Any pluralistic society to endure and thrive, she argued, must embrace the notion of *interdependence*, in which people recognise their mutual responsibilities for each other (Tronto 1993). With this, they go beyond inward and rational choices and strive to become relational and moral agents who willingly give and receive care—fostering social bonds that could sustain communities (Trogal and Viderman 2022). When people are interdependent, they embrace a worldview of 'being-in-common', where they see themselves as reliant and heavily dependent on their social relations (in common) with others (Gabauer et al. 2022; McKinnon et al. 2022).

To operationalise the concept, Tronto (2013) put forward four phases underlying the practice of caring. It begins with '*caring about*', which involves a sensing of needs—of being attentive to the demands of those who require care. This entails recognising and intently listening to both 'articulated and unspoken needs' among stakeholders. Secondly, '*caring for*' takes place, which encompasses the decision to respond to a caring need. At this phase, there must be a mobilisation of all resources and competencies required to perform the act of caring. Eventually, '*care giving*' is carried out, during which the tangible act of caring is performed on the ground. All initiatives are implemented, and needs are acted upon during this stage. Finally, being a mutual act, caring ends with '*care receiving*', where the stakeholder-cum-care receiver responds to the act. This giving of feedback is two-pronged: first, it must affirm if the needs have been adequately met, and second, it must also indicate whether a new array of demands took form and must be addressed once again.

Caring, therefore, could be summarised as taking three different but interrelated forms. One, it is a kind of 'labour' associated with 'physical care giving' that could be done by one person for another. Two, it could refer to the interactions and the 'reciprocal relationship' between people who have given and received care. And three, it may account for an '*ethic-of-care*', where individuals recognise a shared sense of vulnerability and consequently embrace a mutual responsibility to act in a caring manner.

13.2.1.2 An Uncaring Turn?

With all the positive connotations of *care*, the lack of it could be detrimental to a city. 'Uncaring conditions', conceptualised as neglectful social practices that result in a deficient common environment, could manifest in various forms. It could be in a degradation of free and shared spaces, such as when the state 'disinvests' in social infrastructure like public libraries and open parks, or it could be in the persistence of a lack of adequate and affordable in-city housing upon which many vulnerable residents have come to rely (Gabauer et al. 2022; Katz 2008). The material conditions within an area, thus, are a strong manifestation of an area's prevailing (un)caring ethic. The resulting environment could correct, re-produce, or exacerbate the level of (un)care expressed on space, impacting how residents and state actors would relate and interact with each other.

This *uncaring* turn arises when the value that underpins caring—a mutual recognition of *shared vulnerabilities* and *interconnectedness*—comes to be ignored by most people. Here, the 'care of the self' becomes the primary mode of navigating urban life, with the state 'dismantling' its provisions in favour of a 'neo-liberal and entrepreneurial script' of dealing with societal needs (Dilts 2011; Gabauer et al. 2022). This leads to an 'uncaring climate' felt most intensely by those who depend on the government's welfare provisions (Gabauer et al. 2022). Those who fail to fend for themselves are then accused of free riding for excessively relying on social safety nets without much contribution. This notion, then, reinforces a level of callousness to others' plight, rendering an *uncaring attitude* acceptable and even quite reasonable for residents who are relatively more self-sustaining (Chatzidakis et al. 2020).

13.3 Methodology

To better understand the manifestations of (un)equal care in urban areas, this chapter focused on two neighbourhoods within the city of Manila: the Jaime Cardinal Sin Village (JCSV) in the district of Santa Ana, and the settlements within Manila North Cemetery (MNC) in Santa Cruz. The two areas were purposively selected following a series of neighbourhood focus groups among a sample of Manila's *barangays* conducted between 2019 and 2020. The following characteristics were considered: transition to tenurial security (or the lack of it), presence (or absence) of a recognised neighbourhood association, history of internal migration and neighbourhood development, varying physical conditions of the built environment, formal access (and lack of access) to basic utilities, and the diversity and perceived stigmatisation felt in the neighbourhoods. With the recommendation of focal persons and respondents at both city and sub-city levels, validated by the field observations of researchers, the two areas were found to illustrate the contrasting and varying tendencies for caring and uncaring urban practices. They are seen as *particular* cases of spatial inequality in Manila, where, on one hand, the city could be reasonably compassionate, humane, and caring, and on the other, tends to exhibit 'forgetful' and uncaring practices.

Such a granular approach is guided by Gabauer et al.'s (2022) position that the 'micro-research of lived space' is best suited in characterising the quality of '(un)caring encounters' among strangers in a city (Cohen and Knierbein 2022). It is at this scale where residents would generally develop a *sense of place*—where they get to form personal meanings about the city—as they experience the interface between domestic affairs and shared affordances and live with the relational dimensions of (in)equality cultivated in everyday interactions.

The audit of neighbourhood conditions was carried out in 2019. This is part of a four-year research carried out under the Centre for Sustainable, Healthy, and Learning Cities and Neighbourhoods (SHLC). The work for the Philippines covered twelve neighbourhoods in the city of Manila and ten neighbourhoods in Batangas city. The primary aim was to uncover differentiations on the level of sustainability and satisfaction experienced within and among the neighbourhoods in both cities. Aspects such as local demography, economic opportunities, housing and tenure, health and education services, condition of local utilities and infrastructure, and the role of governance actors were highlighted as indications of an area's general liveability. It is argued here that the similar yet diverging urban living conditions between the two case study neighbourhoods are strong illustrations of caring and uncaring attitudes within a city.

A series of qualitative data collection approaches were conducted for the study, namely, a participatory mapping of neighbourhoods, resident-led walking interviews, photo-elicitation focus groups, site observations and field works, and a review of land use plan and other local documents. The data was mostly transcripts of interviews and focus groups, videos and photographs shared by locals (related to their idea of their neighbourhood's sustainability), spatial field notes in the *.kmz* format, and segments of official documents deemed as salient to the research aim. These were analysed thematically through three stages of coding, beginning with *structural coding* to determine data segments for analysis, *concept coding* to process broad themes as they appear, and *in vivo coding* to capture neighbourhood nuances expressed by the residents. To guide the interpretation of results, the framework of caring and uncaring cities as well as the notions of inequality and segregation observed in Metro Manila were used and operationalised.

13.4 Neighbourhoods of (Un)Care: Living in Jaime Cardinal Sin Village and Manila North Cemetery

Taking the notion of the *ethic of (un)care* as a spatially produced and reinforced phenomenon, this work argues that the contrasting yet profoundly similar neighbourhood conditions in JCSV and MNC tell a tale of a capital city that cultivates, on one hand, a compassionate, humane, and caring environment, and on the other, a highly urban, *global city* that could be forgetful, restraining, and uncaring.

13.4.1 Jaime Cardinal Sin Village

Jaime Cardinal Sin Village (JCSV) is a 1.66-hectare medium-rise socialised housing estate, located at the lower eastern side of Manila, home to more than 980 households. The village comprises two adjoining residential complexes: one is a compound of three-storeyed buildings inhabited by about 250 families, built on their *sweat equity* with support of a church-based civil society organisation, Serviam Foundation; the other is a compound of five-storeyed residential tenements, hosting at least 728 families, built by a contractor for the government's relocation of informal dwellers along the banks of the Pasig River, through then Pasig River Rehabilitation Commission (PRRC). The two compounds sit on the property acquired by the Serviam Foundation, which is utilised following the terms of a 50-year usufruct agreement signed in 2005 with the PRRC. The estate is a case of in-city socialised housing initiative for lower middle to low-income families conceived, financed, and enabled by a church-based civil society organisation and an agency of the national government.

The tenure of residents living in the village can be considered formal. They pay a mortgage with minimal amortisation to the PRRC. Fees and charges (*butaw*, a term for membership fee) are also collected to facilitate access to utilities such as electricity, water supply, and wastewater treatment, and to undertake minor repairs and maintenance. The fund is being administered, in part, by the recognised homeowners' association in the village, the JCSV Kapitbahayan Incorporated. Following the terms of the usufruct agreement, the responsibility for major repairs, improvements, and maintenance works lies with the PRRC. With its decommissioning in 2019, however, the role has been in limbo, with two national government agencies—the Department of Environment and Natural Resources and the Department of Human Settlements and Urban Development—discussing, as of writing, the turnover of state duties and responsibilities over JCSV (Figs. 13.2, 13.3 and 13.4).

While already showing signs of age, the village's built environment is decent relative to other affordable housing in the capital city. The units are mostly of studio configuration, with some families having to incrementally construct partition walls and frontage improvements. The floor area ranges from 21 to 28 m^2, which is larger than the country's legal minimum floor area requirement for socialised housing.

Being an in-city relocation site, the area has ample access to neighbourhood and metropolitan facilities. The village is 3.3 kms from the Manila City Hall and is located close to secondary and tertiary health care facilities. Within the estate, there are also commercial units with services that range from a *carinderia* (eateries), a *sari-sari* store (sundry store), a computer and printing shop, a pharmacy, a laundromat, and a water refilling station. The site also hosts a basketball court, doubling as a multi-purpose hall for all public events, and a materials recovery facility meant for the village's solid waste management. Access to the property is regulated, with volunteer security staff manning the gate.

In general, households in the village can be grouped into two: those who were relocated by PRRC, and those who are beneficiaries endorsed by Serviam Foundation. The former belong to the families displaced by the rehabilitation initiative of the banks of Pasig River, while the latter are mostly indigent church members who were given dwelling units on the condition that they volunteer in the construction, and they maintain ties with the church. Both groups, nonetheless, are internal migrants, as most of them have lived outside Santa Ana, and even of Manila (Table 13.1).

Occasionally, conflicts arise between the two groups. A strong indication of this is the coexistence of two—at times, competing—neighbourhood associations in the village: (a) the JCSV Kapitbahayan Incorporated, the recognised village association formed following the agreement between Serviam Foundation and PRRC, and (b) the Dike Side Organisation of Punta

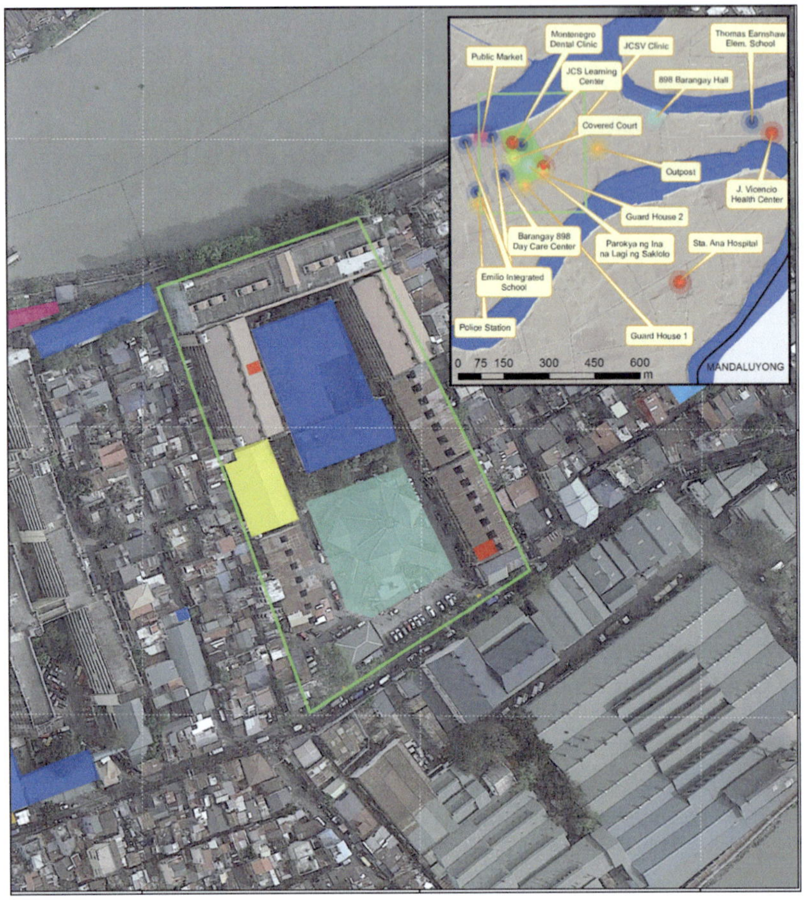

Fig. 13.2 A neighbourhood map of Jaime Cardinal Sin Village. *Source* Authors

(DSOP), the previous association of the beneficiaries living along Pasig River before their relocation to JCSV. Common issues include the cleanliness of unit frontage and communal areas, payment of dues, neighbourly conduct of residents, allegation of petty crimes, and overall concern on safety and security. There are also instances where members are stigmatised by other groups (*mga taga-likod* or those who live at the back, referring to PRRC's residential tenement beneficiaries), labelling some as having a peculiar way of living and relating with others (*iba ang ugali*). Nonetheless, the differences are often resolved amicably through formal and informal grievance mechanisms and, at times, with the intervention of the estate management of the church.

13.4.2 Manila North Cemetery

The Manila North Cemetery is a 54-hectare public burial ground located at the very northeastern portion of the city of Manila. It cuts across several *barangays* and is managed and operated by the city government of Manila through the Manila Health Department and the Manila North Cemetery Administration.

Many of today's Manila North Cemetery inhabitants trace their residency to makeshift and modular homes inherited from two to three earlier generations. The earliest wave of settlers, it was told, emerged in the late 1950s, following the stabilisation of the capital city after the Second World War (Billing 2018). When affairs went back to normal, the cemetery had its usual regional

Fig. 13.3 A view of the JCSV tenements supported by the PRRC and the Serviam Foundation

Fig. 13.4 A view of the *Paroky ng Ina ng Laging Saklolo* at the centre of JCSV. Photo by: Rhay Daniel R. Racoma

Table 13.1 (Un)caring urban conditions in Jaime Cardinal Sin Village

Phases of caring (Tronto 2013)	Jaime Cardinal Sin Village
Caring about 'Paying attention to the need for caring, which requires listening to articulated needs, recognising unspoken needs, and distinguishing among and deciding which needs to care about'	• Residents, within each complex, have developed good rapport (*palagayang-loob*) for having been together before and after their relocation; some have become close friends and have fostered older sibling ties and elderly reverence by affinity (*si nanay, mga kuya at ate*) • Building coordinators, who are also residents, have become the immediate mediator, voice for complaints and interests among unit dwellers, and regulation enforcers • Regular general assemblies are held among residents where major decisions and disputes are discussed and resolved
Caring for 'Someone assumes the responsibility to meet a need, to organise and marshal resources and personnel for the act of care giving'	• By a usufruct agreement, the Pasig River Rehabilitation Commission (PRRC), an agency created in 1999, assumed responsibility for the relocation of dwellers and the construction and maintenance of the socialised tenements; the PRRC, however, was decommissioned in 2019 and its responsibilities are the subject of an ongoing turnover discussion between two government agencies • The recognised neighbourhood association, JCSV Kapitbahayan Inc., serves as the official means of carrying out neighbourhood initiatives and regulation; they collect fees (*butaw*, a term for membership fee) for minor repairs, amenity improvements (such as the installation of CCTVs), and maintenance of the cleanliness of the estate • Serviam Foundation, through its JCSV estate manager, provides, in its initiative, moral and spiritual counselling, conflict mediation, and household development support
Care giving Deployment, the actual act of caring, with all the 'knowledge and moral competencies' required, 'individuals and organisations performing the caring tasks'	
Care receiving 'Response by the care receiver is given, indicating whether the needs have been met or not', and suggesting whether the addressed needs have produced a new set of demands to look after	• Residents acknowledge care by abiding by the regulations set by the administration, clinging to the notion that those would lead to the village's common good • In instances where the residents are divided on a regulation, certain groups can be slightly belligerent, opposing, for instance, by refusing to pay new fees and charges, by voicing out complaints to the building coordinators, or by outright criticism of the regulation during general assemblies

Source Authors, 2021

clientele. Hiring *sepulturero* (the private and independent contractors who construct, maintain, and secure tombs and mausoleums) increasingly became commonplace. Many of them resided in informal dwelling units close to, or along the peripheries of the cemetery.

Eventually, as trust was fostered, the families of the departed began allowing the regular caretakers to stay in the mausoleums. For the *sepulturero*, this had the advantage of living close to livelihood opportunities and metropolitan services with minimal to no rent. The owners, meanwhile, had a guarantee of localised security and regular upkeep of their departed's burial plots. As the

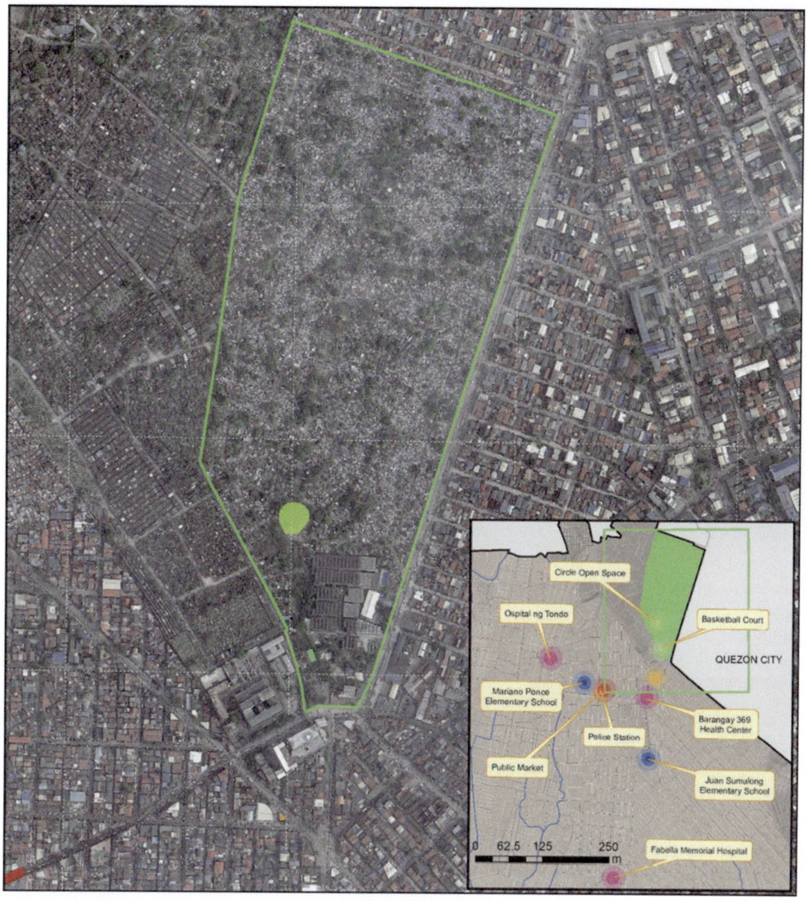

Fig. 13.5 A neighbourhood map of Manila North Cemetery. *Source*: Authors

families of the caretakers embraced the place they were given, they began making incremental housing improvements, normally with consent of the owner. They began installing improvised housing components made of makeshift materials, such as an extended canopy, a *do-it-yourself* clothesline, modular dining table extensions, and even foldable *sari-sari* store improvements—all to be dismantled annually in November during *undas*, the All-Saints' Day, as the owner's family visits the resting place.

Within the cemetery, one could observe a tendency among residents to stratify along four classes. Each class exhibits distinctive socio-economic characteristics, housing typologies, and migratory histories. For this chapter, they are described as follows: the 'generationals', born and raised inside the cemetery and whose residential identity dates back to ancestors who were early mausoleum caretakers; the 'formals', relatively well-to-do dwellers who have acquired subdivided titles of the plots of their departed and have decided to retrofit the mausoleums to include a living quarters; the 'renting migrants', those dwellers working in nearby establishments, paying rent for bed space to take advantage of the cemetery's urban proximity; and the 'relocatees', or those who have been displaced by government clearing operations and have opted to reside within cemetery grounds, usually referred to by residents as *mga dayo* (outsiders or newcomers) and often stigmatised as 'squatters within a squatters' area' (Figs. 13.5, 13.6 and 13.7).

An inevitable outcome of the cemetery's dwelling arrangement is the precarity of tenure.

Fig. 13.6 Frontage of Mausoleum-cum-residential strip

Fig. 13.7 A basketball court within the cemetery

Despite estimates of about 8000–10,000 inhabitants, the city government has always refused formal recognition of the dwellers in the cemetery (Dakudao 1998). Residents tend to experience perennial stigmatisation due to their address, being derogatorily referred to as *mga taga-loob* (those living inside a public cemetery). Nearby *barangays* have taken steps such as adopting villages, to provide primary support services, but these have remained intermittent and unofficial. Unlike the set-up in JCSV, there is also no formal organisation representing the interests of the residents. They are mostly self-correcting, with most minor conflicts having to be resolved among themselves, by familiarity and affinity, without formal mediation (Table 13.2).

Table 13.2 (Un)caring urban conditions in the settlements within Manila North Cemetery

<table>
<tr><th>Phases of caring
(Tronto 2013)</th><th>Manila North Cemetery</th></tr>
<tr><td>Caring about
'Paying attention to the need for caring, which requires listening to articulated needs, recognising unspoken needs, and distinguishing among and deciding which needs to care about'</td><td>• Residents take responsibility over their affairs, as there is no recognised neighbourhood association formally representing their interests
• They coalesce, self-regulate, and take shared initiatives among each other based on blood ties, affinity, and familiarity (e.g. living close to each other, or their migration is under the same circumstances), or by hustle networks (i.e. 'colleagues' in independent contracting, such as a circle of mason workers, pallbearers or buhat boys, and tomb caretakers)</td></tr>
<tr><td>Caring for
'Someone assumes the responsibility to meet a need, to organise and marshal resources and personnel for the act of care giving'</td><td rowspan="2">• Manila North Cemetery Administration has responsibility for the management and upkeep of cemetery grounds, which entails setting and promulgating cemetery regulations, performing maintenance works on road and utility infrastructure, and providing janitorial, solid waste management, and perimeter security services in the communal areas—all these being done without recognition of the settlements inside
• Chapters of people's organisations and socio-civic fraternities are active in clean-up operations and the promotion of public order inside the cemetery, particularly in inner areas where there is crowding of settlements
• Barangays near the cemetery have 'adopted' parts of the cemetery settlements, through which aid is given in the form of food and grocery packs, and services such as the conduct of outreach health and well-being programmes; major disputes can also be mediated at the level of the barangay through the Katarungang Pambarangay (Barangay Justice System)</td></tr>
<tr><td>Care giving
Deployment, the actual act of caring, with all the 'knowledge and moral competencies'. required, 'individuals and organisations performing the caring tasks'</td></tr>
<tr><td>Care receiving
'Response by the care receiver is given, indicating whether the needs have been met or not', and suggesting whether the addressed needs have produced a new set of demands to look after</td><td>• Residents welcome and receive any aid or assistance extended to them by the barangays, civil society, or the private sector, be it for the long term or only for immediate needs
• Informal self-help acts and solidarity-based practices, with residents building incrementally to address any gap in care</td></tr>
</table>

Source Authors, 2021

While some households are willing and able to pay, access to basic utilities has remained informal. For instance, to survive, most residents access electricity either through sub-metered consumption, self-generation (e.g. cheap, small-scale solar-powered lights), or by ingenious or even borderline extra-legal means (e.g. 'spiderman' connections, with dense wires resembling web slings). Water supply, meanwhile, is enabled through deep well pumps inside the cemetery, now prohibited in Manila, and through *do-it-yourself* taps on cemetery water lines. Outsiders and other residents get to deliver and sell the water, by gallon, on wooden carts.

As some residents have secured a de facto permission from plot owners to stay, the city government might find eviction politically challenging. The void in basic services, therefore, could be seen as a deliberate, yet subtle form of neglect intended to nudge the dwellers towards leaving their informal homes. With residents recognising the precarity of their tenure, they would rather build incrementally from their resources, to address any gap in care (e.g. modular improvements in the dwelling units, ingenious ways of accessing basic utilities and services) rather than expect and demand an institutionalised form of care from state agents. Many of them would navigate the precarity and uncertainty in a cautious, adaptive, and self-organising manner.

13.4.3 'Planning of Care' to Navigate Spatial Inequalities in Manila's Neighbourhoods

The foregoing discussion serves as a backdrop for planning cities with kindness at their core. The work of Jon (2020) is instructive in operationalising a *planning of care* towards a post-pandemic future. The most progressive iteration of planning, it suggests, is one that acknowledges the precarity of individualism. It embraces an 'outpouring of empathy' and recognises the 'self-*in*sufficiency' and shared vulnerability of each other. Here, spaces ought to be designed and cultivated on the logic of caring and kindness towards each other, without regard to one's class, race, age, or identity (i.e. acting with a 'veil of ignorance', being unmindful of one's background). More importantly, this logic of allocating and consuming spaces must also be cognisant of the well-being of 'nonhuman others' that coexist with human systems (Jon 2020).

Faint as it may seem, a bottom-up, solidarity-based, and empathy-enabled manner of urban citizenship was palpable in the two neighbourhoods—akin to Jon's (2020) planning of care. It draws comparison with indigenous notions of *kapwa* and *pakikipagkapwa*, which, in Filipino social psychology, refer to a 'oneness of self and others', having recognised a 'shared identity' (Enriquez 1978). These position one to empathise and toil with the other, extending acts of caring not on a pedestal but as equal beings. Bankoff (2007) has observed the same, arguing that, when faced with common hardship, the most vulnerable tend to exhibit a powerful sense of *social capital*. These self-help acts, he said, are morally enforced through reciprocation 'in due measure, at an appropriate time' (Bankoff 2007). This allows trust among informal ties to be fostered, cultivating an environment of *mutual reliance*. All this is taking place even without formal recognition or institutionalisation.

13.4.4 Preconditions of a Grassroots and Place-Based Caring Ethic

Reflecting on both neighbourhoods, four preconditions appear to lay the foundation for a grassroots and place-based caring ethic. The first precondition is of economy and practicality. The settlement must fend for the people's needs. It should afford basic services and facilities and must be within proximity to livelihood opportunities. Residents tend to stay, with little regard to the quality of spaces, so long as they have access to nearby public schools and health care facilities, reliable utility infrastructure, and places of work at a lower rent. Second, the community residents require social familiarity and place embeddedness. They need time to establish good rapport (*nakapalagayang-loob*) and familiarise themselves with the usual ways and routines (*gawi*) of their neighbours. They also require a period of acclimatisation with the physical space, allowing them to experience, incrementally personalise, and cultivate a strong attachment to the settlement. The third precondition is the imperative for shared toiling and thriving, an emergent outcome of the first two preconditions. Having spent decades together, residents tend to trust and mutually rely on each other, tempered by coping (*damayan*) with shared urban challenges and triumphs and labouring for the other in times of need (*bayanihan*). Finally, the presence of a continuously caring leadership remains integral to a caring neighbourhood ethic. Many residents revere a village leader (*pangulo*) as a selfless and proximate authority figure, one who empathises with their backyard toils. The *pangulo* is often influential enough to get things done, address needs, resolve disputes, and enable neighbourhood transformation that is tangible and proximate.

13.5 Conclusion

Manila was the focus of this work, where interstitial and relational segregation has long been observed, and where marginalised areas tend to become forgotten places. Like most global cities, it tends to prioritise the highest and best use of land, as it operates in an already overcrowded neo-liberal planning environment (Garrido 2020; Ortega 2016; Shatkin 2004).

Drawing from this broader notion of inequalities in Manila, this study dwells deeper into a *particular* form of spatial inequality that apparently transpires between and among sub-city areas. It is one where residents of different neighbourhoods tend to experience caring and uncaring urban practices in a differentiated and arguably unequal manner. On one hand, the city can afford a reasonably compassionate, humane, and caring environment, while on the other, the city can assume a global and neo-liberal position, where it could be forgetful, restraining, and ultimately uncaring especially towards informality.

Patterns and expressions of what Tronto (2013) describes as an 'ethic of care', and what these imply in how spatial inequality becomes manifest and diverges at micro-scale, were explored in two Manila neighbourhoods. The first case study is Jaime Cardinal Sin Village (JCSV), an in-city socialised housing initiative for the families affected by the clearing of the banks of Pasig River. JCSV was conceived, financed, and enabled by an agency of the national government and a church-based civil society organisation and is now maintained through a neighbourhood association assembled and governed by the residents themselves. While the arrangement is not without issues and conflicts, the estate provides a sound and doable model and could be described as an expression of an urban caring ethic. The other case explores the settlement in Manila North Cemetery, which, despite the absence of any formal state recognition, has grown and matured to host several thousand dwellers over at least six decades. To access basic utilities and services, the cemetery neighbourhood has relied on informal self-help acts and solidarity-based practices and the occasional aid rendered by *adopting barangays.* Worsening this are the petty but recurring acts of stigmatisation felt in daily encounters, reminding the residents of their precarious and meagre place in the highly urban capital city.

The above organic *caring ethic* manifestations—backed with a governance framework that embraces a community's 'being-in-common' and inherent interconnectedness—offers a mode of citizenship that could humanise urban living. An outcome of self-organising behaviour, it brings forward a *kinder* form of sustainability: one that is shaped and co-owned by residents at the local level. It has strong potential to inspire and transform practices at higher governance levels as cities confront a more uncertain and precarious future.

References

Apparicio P, Martori JC, Pearson AL, Fournier É, Apparicio D (2014) An open-source software for calculating indices of urban residential segregation. Soc Sci Comput Rev 32(1):117–128. https://doi.org/10.1177/0894439313504539

Bankoff G (2007) Dangers to going it alone: social capital and the origins of community resilience in the Philippines. Contin Chang 22(2):327–355. https://doi.org/10.1017/S0268416007006315

Billing L (2018) Graveyard living: inside the 'cemetery slums' of Manila. Guardian, March 21. http://www.theguardian.com/cities/2018/mar/21/cemetery-slums-life-manilas-graveyard-settlements-philippines

Chatzidakis A, Hakim J, Littler J (2020) The care manifesto: politics of interdependence

Cohen N, Knierbein S (2022) Social inequalities, uneven space, and care. In: Care and the city: encounters with urban studies. Routledge, pp 37–42

Dakudao ME (1998) The development of cemeteries in Manila before 1941. Philipp Q Cult Soc 26(1/2):254–271. http://www.jstor.org/stable/29792419

Dilts A (2011) From 'entrepreneur of the self' to 'care of the self': neo-liberal governmentality and Foucault's ethics. Foucault Stud:130–146. https://doi.org/10.22439/fs.v0i12.3338

Enriquez V (1978) Kapwa: a core concept in Filipino psychology. Philipp Soc Sci Hum Rev
Fainstein SS (2010) The just city, 1st edn. Cornell University Press
Fainstein SS (2014) The just city. Int J Urban Sci 18(1):1–18. https://doi.org/10.1080/12265934.2013.834643
Florida R (2017) The new urban crisis: how our cities are increasing inequality, deepening segregation, and failing the middle class—and what we can do about it, 1st edn. Perseus Books, LLC
Gabauer A, Knierbein S, Cohen N (2022) Care, uncare, and the city. In: Care and the city: encounters with urban studies. Routledge, pp 3–14
Garrido M (2018) How interspersion affects class relations. Int J Urban Reg Res 42(3):442–460. https://doi.org/10.1111/1468-2427.12632
Garrido M (2020, May) Reconceptualizing segregation from the Global South. City Community 1–19. https://doi.org/10.1111/cico.12504
Jon I (2020) A manifesto for planning after the coronavirus: towards planning of care. Plan Theory 19(3):329–345. https://doi.org/10.1177/1473095220931272
Katz C (2008) Bad elements: Katrina and the scoured landscape of social reproduction. Gend Place Cult 15(1):15–29. https://doi.org/10.1080/09663690701817485
Lefranc A, Pistolesi N, Trannoy A (2008) Inequality of opportunities vs. inequality of outcomes: are western societies all alike? Rev Income Wealth 54(4):513–546. https://doi.org/10.1111/j.1475-4991.2008.00289.x
Madanipour A (2022) Critical reflections on care. In: Care and the city: encounters with urban studies. Routledge, pp 15–23
McDoom OS, Reyes C, Mina C, Asis R (2019) Inequality between whom? Patterns, trends, and implications of horizontal inequality in the Philippines. Soc Indic Res 145(3):923–942. https://doi.org/10.1007/s11205-018-1867-6
McKinnon K, Healy S, Dombroski K (2022) Care from the beginning: birthing collective origins, interdependent cities, and new community economies. In: Care and the city: encounters with urban studies. Routledge, pp 24–33
Molina HF, Nakamura K, Kizuki M, Seino K (2013) Reduction in inequality in antenatalcare use and persistence of inequality in skilled birth attendance in the Philippines from 1993 to 2008. BMJ Open 3(6). https://doi.org/10.1136/bmjopen-2012-002507
Nijman J, Wei YD (2020, March) Urban inequalities in the 21st century economy. Appl Geogr 117:102188. https://doi.org/10.1016/j.apgeog.2020.102188
Núñez J, Tartakowsky A (2007) Inequality of outcomes vs. inequality of opportunities in a developing country. An exploratory analysis for Chile. Estud Econ 34(2):185–202. https://doi.org/10.4067/S0718-52862007000200004
Ortega AAC (2016) Manila's metropolitan landscape of gentrification: global urban development, accumulation by dispossession & neoliberal warfare against informality. Geoforum 70:35–50. https://doi.org/10.1016/j.geoforum.2016.02.002
Palafox B, McKee M, Balabanova D, Alhabib KF, Avezum A, Bahonar A, Ismail N, Chifamba J, Chow CK, Corsi DJ, Dagenais GR, Diaz R, Gupta R, Iqbal R, Kaur M, Khatib R, Kruger A, Kruger IM, Lanas F et al (2016) Wealth and cardiovascular health: a cross-sectional study of wealth-related inequalities in the awareness, treatment and control of hypertension in high-, middle- and low-income countries. Int J Equity Health 15(1):15–17. https://doi.org/10.1186/s12939-016-0478-6
Palipudi KM, Gupta PC, Sinha DN, Andes LJ, Asma S, McAfee T (2012) Social determinants of health and tobacco use in thirteen low- and middle-income countries: evidence from Global Adult Tobacco Survey. PLoS ONE 7(3). https://doi.org/10.1371/journal.pone.0033466
Porio E (2016) Prosperity and inequality in Metro Manila: reflections on housing the poor, climate risk, and governance of cities. In: Globalization and democracy in Southeast Asia. Palgrave Macmillan, pp 177–197. https://doi.org/10.1057/978-1-137-57654-5
Reyes CM, Mina CD, Asis RD (2017) Inequality of opportunities among ethnic groups in the Philippines, pp 1–66. https://pidswebs.pids.gov.ph/CDN/PUBLICATIONS/pidsdps1742.pdf
Roemer J (1998) Equality of opportunity. Harvard University Press
Sen A (2000) Development as freedom. Anchor Books
Shatkin G (2004) Planning to forget: informal settlements as "forgotten places" in globalising Metro Manila. Urban Studies 41(12):2469–2484. https://doi.org/10.1080/00420980412331297636
Shatkin G (2008) The city and the bottom line: URBAN megaprojects and the privatization of planning in Southeast Asia. Environ Plan A 40(2):383–401. https://doi.org/10.1068/a38439
Trogal K, Viderman T (2022) Everyday struggles and contestations around care. In: Care and the city: encounters with urban studies. Routledge, pp 103–108
Tronto JC (1993) Moral boundaries: a political argument for an ethic of care. Routledge. https://doi.org/10.4324/9781003070672
Tronto JC (2013) Caring democracy: markets, equality, and justice. NYU Press
Tronto JC (2015) Who cares? How to reshape a democratic politics. Cornell University Press. https://doi.org/10.7591/cornell/9781501702747.001.0001
United Nations (2017) New Urban Agenda. www.habitat3.org
United Nations Development Programme (2013) Humanity divided: confronting inequality in developing countries. 296. https://www.undp.org/content/dam/undp/library/Poverty-Reduction/Inclusive-development/Humanity-Divided/HumanityDivided_Full-Report.pdf

14 Income, Employment, and Housing Inequalities in Asian and African Cities

Sohail Ahmad, Tanjil Sowgat, and Ya Ping Wang

Abstract

This chapter analyses the income, employment, and housing conditions of residents covered in the SHLC Household Survey conducted across various neighbourhoods in 14 Asian and African cities. The analysis provides comparative insights into neighbourhood-level inequalities in the Global South. While continental or subcontinental differences are limited, significant variations in inequality are evident at the country and city levels. The findings reveal serious inequality across neighbourhoods, with neighbourhood wealth emerging as a critical catalyst. This intriguing micro-level data underscores the importance of neighbourhoods in understanding the various dynamics of rapidly growing cities in the Global South.

Keywords

Housing · Employment · Inequalities · Income · Neighbourhoods

S. Ahmad (✉)
School of Computing, Engineering and Physical Sciences, University of the West of Scotland, Paisley, UK
e-mail: Sohail.Ahmad@uws.ac.uk

T. Sowgat
Urban and Rural Planning Discipline, Khulna University, Khulna, Bangladesh
e-mail: tsowgat@urp.ku.ac.bd

Y. P. Wang
School of Geography and Tourism, Shaanxi Normal University, Xian, China

School of Social and Political Sciences, University of Glasgow, Glasgow, Scotland, UK

Y. P. Wang
e-mail: yaping.wang@snnu.edu.cn; yaping.wang@glasgow.ac.uk

14.1 Introduction

Earlier chapters focus on specific subjects of economic, social, or environmental significance that determine socio-spatial inequality in countries or cities of Africa and Asia. This final chapter builds on previous discussions which emphasised the impact of income and wealth distribution on urban poverty and inequality. It utilises the analysis of a large survey undertaken to focus on the interplay of income, employment, and housing inequalities in the Global South. It aims to comprehensively understand the intricate relationships between housing and income/employment inequalities at the neighbourhood level across various Asian and African cities. By delving into the disparities in different neighbourhoods, it seeks to uncover how these two forms of inequality (housing and income levels) intersect and reinforce each other. The findings from this study are instrumental in informing the develop-

Y. P. Wang et al. (eds.), *New Drivers of Division*, Sustainable Development Goals Series,
https://doi.org/10.1007/978-981-96-5629-5_14

ment of targeted strategies to foster inclusive and resilient urban environments. Understanding the specific dynamics of housing and income inequalities at a granular neighbourhood level directs the proposal of tailored interventions that address the unique challenges faced by different communities. These insights, on the one hand, contribute to the academic discourse on urban inequality and, on the other hand, offer guidance to policymakers, urban planners, and community leaders striving to build cities that focus on equitable and sustainable neighbourhoods.

The chapter commences with a literature review, establishing the analytical framework. Subsequently, case studies, data, and variables are studied, followed by initial inequality comparisons related to income and employment before shifting the focus to housing. The subsequent section explores the intricate relationship between these factors, revealing the multifaceted interplay of different inequalities. Finally, the conclusion brings the findings together and highlights the main differences between neighbourhoods, cities, and countries.

14.2 Inequality and Neighbourhoods: Conceptual Justification

Over the past 25 years, income inequality between countries has declined, yet within countries it has increased (United Nations 2020). Today, 71% of the world's population resides in countries grappling with widening inequalities. This alarming trend is underscored by the fact that only 26 of the world's wealthiest individuals hold wealth equivalent to half the global population (United Nations 2020). Fuelled by social phenomena, inequalities can become so pronounced that they catalyse citywide or national inequality (Khondker 2011; Otsuka and Yamano 2005). The most prominent disparities emerge in the Global South, where regional statistics reveal growing inequality concerns (UN-Habitat 2022; UNDESA 2020). In 2022, the share of the top 10% clusters in Africa, East Asia, and South Asia accounted for approximately 53%, 45%, and 54% of the total national income, respectively, according to the World Inequality Database. For the Global South, providing essential services—such as housing, employment opportunities, and reliable income—in an equitable manner becomes increasingly complex.

Income inequality and socio-economic segregation negatively impact health, well-being, and sustainability (Wilkinson 2020, Wilkinson and Pickett 2022; Patel et al. 2018). Numerous studies highlight their adverse effects on educational attainment, economic growth, social cohesion, and social mobility (Pryce 2021). Income inequality leads to spatial segregation between the rich and poor. As income inequality widens, those on the top opt for wealthy neighbourhoods with higher values of residence and better living conditions. Over time, properties appreciate, and the gap further deepens. On the contrary, low-income households tend to concentrate in the less desirable neighbourhoods. In this way, income inequality perpetuates segregation and inequality in housing and the built environment.

In the realm of policy discourse, the clarion call for cross-border collaboration reverberates prominently. However, amidst the chorus, a perplexing ambiguity persists regarding the spatial boundaries of inequality. Current equity planning literature urges a 'Southern' vocabulary to address these growing inequalities. Since inequalities are observed globally, regionally, and nationally, there are frequent calls in international policies to focus on different geographic levels. Some scholars advocate supra-national perspectives, weaving together continental threads of inequality, while others scrutinise national-level dynamics, revealing stark variations within countries (Khondker 2011; Otsuka and Yamano 2005; Spiegel 2012). However, understanding inequality at the sub-city level is essential for creating compelling, equitable, and sustainable urban policies. It allows the precise identification of needs, targeted resource allocation, and the development of tailored interventions that address the specific challenges faced by different communities and ultimately lead to inclusive and resilient cities.

The existing literature offers numerous examples demonstrating how neighbourhoods within cities experience unequal distribution of resources across different parts of the world (Duroudier 2014; Feitosa et al. 2007; Reardon and O'Sullivan 2004; Johnston et al. 2007). These divisions may result from power dynamics, political processes, or exclusionary planning and development (Talen 2018; Marcińczak et al. 2013). Income disparities significantly influence residents' choices, with financially advantaged citizens having preferences for specific neighbourhoods over low-earning families (Harvey 1989; Hulchansky 2010; Tammaru et al. 2020). Neighbourhoods are small but vital to understanding inequalities in fast-urbanising cities (Sowgat and Roy 2022). Although the neighbourhood is a fundamental unit of analysis for understanding this issue, only a few studies in the Southern context explicitly address the role of neighbourhood wealth in creating such inequality (Sowgat and Roy 2022). While scholarly works on European cities indicate increasing neighbourhood inequalities between high- and low-income groups (Fujita 2016; van Ham et al. 2020; Tammaru et al. 2014), similar investigations remain limited for cities in the Global South.

Urban inequality is multi-dimensional and complex, manifesting in economic, social, and spatial dimensions (UN-Habitat 2022). Inequality is characterised by unequal access to income and wealth, urban services and infrastructures, housing and environmental burdens. Current urban development patterns, driven by massive capital accumulation, commodification and privatisation of urban services, exacerbate these disparities (Bandauko et al. 2022). The specific nature of production and consumption of urban spaces (e.g. gated communities) contribute to inequalities massively as affluent households cluster in wealthy neighbourhoods while poor households are concentrated in others. The former group enjoys far better services compared to the latter. Therefore, inequality is closely associated with wealth. Income is widely used in Western and Southern literature to determine contrasting inequality (Alvaredo et al. 2018). Income disparities can lead to distinct residential patterns, intensifying differences in housing quality (Patel et al. 2018; Pryce 2021; Bhorat et al. 2014). In large American metropolitan areas, Dong (2018) provides empirical evidence that increasing income inequality has worsened rental affordability for low-income tenant households. Chen (2016) uses the case study of Guangzhou to demonstrate that income, among other factors, is positively correlated with home ownership among the 'native plebian' class—a sub-group positioned between urban elites and lower masses. So, within cities, certain socio-economic groups can achieve home ownership more readily than others, reflecting broader socio-economic stratifications. In Sweden, where housing wealth constitutes the most significant component of household wealth, Wind and Hedman (2018) reveal that high-income households experience higher capital gains from housing, while migrants see lower gains. Income inequality leads to spatial segregation between the rich and the poor and as it widens, those on the top opt for wealthy neighbourhoods with higher values of residence and better living conditions. Over time, properties appreciate, and the gap further deepens. On the other end, low-income households concentrate in the least desirable neighbourhoods. In this way, wealth inequality contributes to prevailing segregation and inequality in housing and the built environment. The findings in this SHLC study reveal that many people in the selected cities operate outside the formal income and tax systems. As a result, they often struggle to identify their exact income. With no national data on income at the household level, this fact could give a misleading measure of income data. Three alternative variables have therefore been used to address this problem. The usable income information includes data of perceived hardship or ability to maintain families with the income received. Employment patterns have been used to recognise current engagement in economic activities which reflect economic status. A random adult member of each household was interviewed about their employment status and since the primary earner was not targeted, this approach provides an approximate picture of the household's economic situation. Since engagement in formal employment is seen to be an indicator in recent inequality studies (Sassen 2013; Van Ham et al. 2020), this aspect has also been highlighted.

Household-level inequality often refers to disparities in access to quantity and quality of housing based on household economic status or ethnicity (Aizawa et al. 2020; Munir et al. 2022). In developed countries, particularly in the United States and Western Europe, research emphasises home ownership rates (Farahi and Jiao 2024). These studies often highlight how income and racial disparities affect the ability to purchase homes (Wu et al. 2019), leading to significant differences in wealth accumulation and housing stability. Studies in developing countries more often focus on physical aspects of housing, such as dwelling size, construction quality, and access to essential services and amenities (Ahmad 2015). Research by Bhan and Jana (2015) and Roy and Sowgat (2024) reveals housing construction quality and rental structure as inequality indicators in their South Asian studies. Housing satisfaction has also been a popular subjective measure to determine housing inequality (Dupont 2004; Nasrabadi et al. 2024).

This chapter follows the above thread of literature and investigates how wealth differences affect inequality in housing at the neighbourhood level. While inequality is a vast concept (Wilkinson 2020; Wilkinson and Pickett 2022), the focus here is only on income, employment and housing. This analysis draws from a survey conducted as part of the SHLC study. While the SHLC survey includes more than 200 questions, only select key parameters were strategically used to understand inequality. National statistics in the study cities fail to provide detailed income, employment, or housing data at the neighbourhood level. However, the cross-country data collected in the SHLC study presents a unique opportunity. For the first time, it is possible to make neighbourhood-level comparative inferences, shedding light on the intricate dynamics of income, employment, and housing inequality.

14.3 Data and Methods

This study uses all 14 SHLC project case study cities from seven Asian and African countries (see Table 1.2 in Chapter 1). These cities face different levels of poverty, growth, health outcomes, and access to urban services (Ahmad et al. 2021; Roy and Sowgat 2024). The larger cities have experienced a high rate of urbanisation over the last thirty years; they face different challenges in meeting the growing demands for urban services (Kundu et al. 2020; Roy et al. 2019; Sengupta et al. 2022). The smaller cities are relatively slower in population growth but face similar challenges (Kundu et al. 2020; Roy et al. 2018). While there are similarities in economic development and poverty among the study countries, they have significantly different social and political contexts for urban transformation (see Chapter 3).

This analysis is based on the household survey conducted as part of the SHLC study (see Chapter 1) of over 15,000 households and individuals from the 14 cities (Wang et al. 2023). In each city, neighbourhoods were categorised into five groups based on their perceived wealth and income levels—low, lower middle, middle, higher middle and high. The research team in each country carried out the income/wealth categorisations of the neighbourhoods., employing qualitative and quantitative tools. In each group, several typical neighbourhoods were selected for sample distribution. Interviews, mostly face to face, were conducted in 249 neighbourhoods with at least 10 households surveyed in each. The total sample (targeted for 1000) in each city was more or less evenly distributed among the five neighbourhood groups. Once a household was selected for the survey, a random adult was invited for the interview. Therefore, the survey questionnaire (adopted by all country research teams with only minor adjustments) includes variables on the household (majority questions) and the individuals.

Of many income and employment-related variables, household economic status and respondent employment status were selected to measure income inequality, following other similar studies (see Alvaredo et al. 2018; Van Ham et al. 2020). Regarding the household economic status, the questions asked were: 'Thinking about money in the last 12 months, would you say that your household: (1) Is doing very well with plenty of money left over after paying for essentials; (2) Is doing okay and has some money left after paying for essentials; (3) Receives just enough money to pay for foods, bills and other essentials; (4) Is occasionally short of money for essentials; (5) Is struggling to pay bills or to afford to feed everyone most of the time?' Although a few other questions were asked, this was considered most relevant to capture household economic status because it directly related to the household and pertained to a longer span of time. Another question about employment status should ideally have been asked from the primary earner of the household, but the survey addressed randomly selected adult respondents. The question was: 'Which of the following best describes your employment status: (1) Self-employed, no employees; (2) Self-employed, with employees; (3) Contributing family worker; (4) Formal employee; (5) Informal employee; (6) Member of a co-operative; (7) Any other?'

Three key aspects—overcrowding, housing tenure, and housing satisfaction—are used in the analysis of housing. Each aspect captures different dimensions of the housing experience that collectively reflect the overall quality and fairness of housing conditions. A dwelling is assumed to be overcrowded if the occupancy rate is three or more persons per room (or living space). This variable is constructed based on the number of rooms and household size reported in the survey. Regarding housing tenure, the question was asked: 'Please tell me about this place you live in, does your household: (1) Own it in full, outright; (2) Own it, paying off mortgage; (3) Rent it from a private landlord; (4) Rent it from a government/social organisation; (5) Is it an employer-provided housing; (6) Is it an informally occupied dwelling; (7) Any other?' For dwelling satisfaction, the question is re-categorised as: 'How satisfied or dissatisfied are you with the dwelling in which you are currently living: (1) Very satisfied; (2) Satisfied; (3) Neither satisfied nor dissatisfied; (4) Dissatisfied; and (5) Very dissatisfied?' Housing satisfaction is a multi-dimensional concept encompassing various aspects of residential living, including the first two variables: overcrowding and security of tenure. Therefore, it is considered for in-depth exploration.

Methodologically, the descriptive statistics of the variables were first presented and then cross-tabulations of the aforementioned variables were performed. Logistic regression was then employed to understand the factors influencing the likelihood of being satisfied with a dwelling. While doing so, the main idea was to identify how housing satisfaction is linked with income inequality. In order to do so the neighbourhood level's wealth, household level's economic status and individual level's employment status had to be understood. Logistic regression is a statistical method suitable for modelling binary outcome variables. It has been used here to predict the probability of living in a satisfied dwelling using the following equation:

$$\text{logit}(P) = \ln = \beta_0 + \beta_1 X_1 + \beta_2 X_2 + \cdots + \beta_k X_k$$

where P is the probability of living in a satisfied dwelling, β_0 is the intercept, and $\beta_1, \beta_2, \ldots, \beta_k$ are the coefficients of independent variables $X_1, X_2, \ldots, X_k$.

Maximum likelihood estimation (MLE) has been used to fit the logistic regression model. The fitting process involved iteratively adjusting the model parameters to maximise the likelihood of observing the given outcomes. The Hosmer–Lemeshow test was employed to assess the model's goodness of fit. A non-significant test result indicates a good fit. McFadden's R-square indicated the model's explanatory power. The estimated coefficients (log-odds) of the logistic regression model were interpreted to understand the impact of each independent variable on the likelihood of living in an overcrowded dwelling (or unsatisfied dwelling). The odds ratios were calculated to provide a more intuitive interpretation. Wald tests were conducted to assess the statistical significance of each coefficient. Variables with p-values less than 0.05 were considered significant predictors.

14.4 Results and Discussions

Initially, the comparative disparities concerning income, employment, and housing inequality and their interrelationships were emphasised across continents, countries, and cities. Within each subsection, variations in wealth within neighbourhoods were studied to gauge how they impact overall inequality. In the final subsection, the interrelationships have been examined between housing and other forms of inequality.

14.4.1 Income and Employment Inequality

On an average, only 42% of the households said they were 'doing okay' or 'very well' over the past 12 months with respect to household income and their economic status. This means that more than half of all households surveyed (58%) were experiencing economic difficulties. This percentage varies significantly across continents. In Asian countries, 49% of households reported being in a good economic state (doing okay or very well), while only 32% of households in African countries reported the same. The higher performance of Asian countries is, however, primarily driven by China, where an impressive 82% of households reported doing okay or very well. In contrast, the percentages are lower in Bangladesh (49%) and India (38%). The Philippines has the lowest at 26%. In African countries, the share of households doing okay or very well is more consistent, ranging narrowly between 31% (Tanzania) and 34% (Rwanda and South Africa). The situation across the countries of Africa shows a continental pattern, but for Asia, the income landscapes are very different.

There is a significant variation in household economic status across cities. As expected, most households in Chinese cities are doing okay or very well. Often within the country, secondary cities (e.g. Khulna in Bangladesh and Madurai in India) have a higher share of struggling households (Fig. 14.1). In particular, the Bangladeshi secondary city is far behind its capital as 62% of households in Dhaka reported doing okay or very well, as compared to only 31% in Khulna. Bangladesh is among the fastest-growing countries among all the study countries, and it is also ranked as one of the poorest among the study countries. Urbanisation and economic crises may have impacted the country's disparity. The two South African cities show a similar pattern, with Cape Town doing slightly better than Johannesburg.

This national picture is helpful for cross-country comparison; it, however, hides the critical differences between neighbourhoods. Figure 14.2 illustrates the distribution of household economic status across neighbourhoods by varying income/wealth levels. Unsurprisingly, within a city, low-income neighbourhoods have a relatively smaller proportion of households reporting that they are doing okay or very well. A positive, albeit modest correlation exists between household economic status and neighbourhood wealth groups (Pearson's correlation = ~0.30). This indicates that while the household economic status tends to improve with neighbourhood wealth, other factors also play a significant role. Notably, the disparity in economic status between a low-income and high-income household is quite pronounced in certain cities: households in high-income neighbourhoods in Madurai are 48 times more likely to be doing okay or very well compared to those in low-income households. In Delhi, this disparity is 13 times, in Khulna 8 times, and in Johannesburg 5 times. The wealth-wise variation of economic status indicates that neighbourhoods are gradually becoming segregated since those who have good affordability are more likely to live in wealthy neighbourhoods, while people who are in a hardship situation live in low-wealth neighbourhoods. This resonates with the early phases of the Western form of segregation, where there are elements of social exclusion within cities.

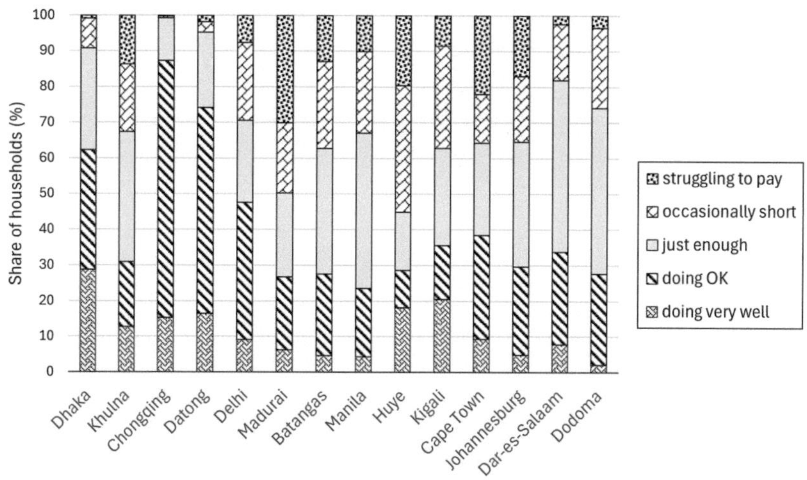

Fig. 14.1 Distribution of household economic status (based on the past 12 months) across case study cities, 2022. *Source* Primary Survey, 2022

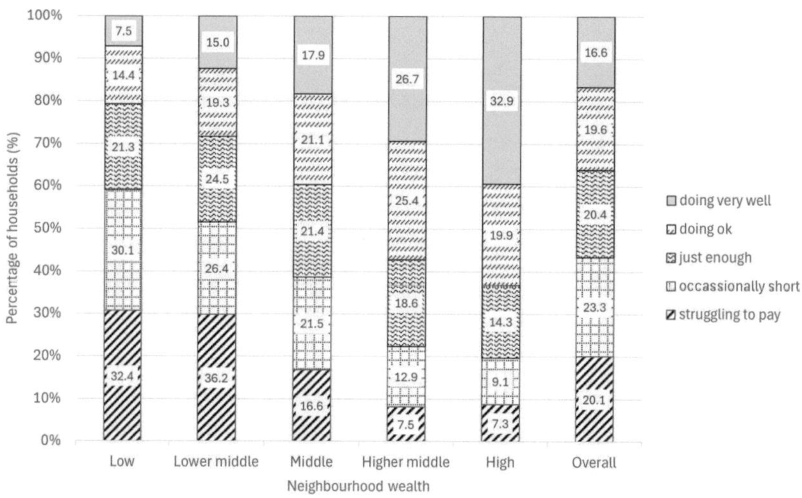

Fig. 14.2 Household economic status by neighbourhood wealth in case study cities, 2022. *Source* Primary Survey, 2022

However, there are also struggling households within the wealthy neighbourhoods, and within poor neighbourhoods there are relatively better-off households (Fig. 14.2). This suggests that these developing countries may be at an early stage of spatial and social exclusion compared to Western developed countries. The exclusions are yet to become extreme in the cities of the Global South. Related studies have shown that in developing countries, rapid urbanisation often leads to a mix of socio-economic groups within the same neighbourhood due to the lack of strict zoning laws and the informal housing market. This contrasts with more developed Western countries where spatial segregation is more pronounced due to established urban planning policies and socio-economic stratification (Tammaru et al. 2014; Musterd 2017).

Regarding the status of employment situation among the interviewees, around 20% of the respondents in the study cities rely on informal employment even though 40% have access to formal employment. Informal employment is widespread across both continents, though

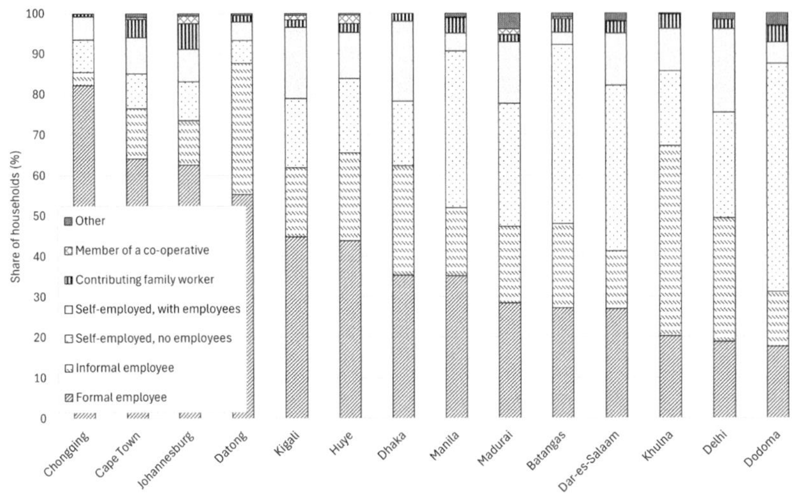

Fig. 14.3 Employment status of respondents by case study cities, 2022. *Source* Primary Survey, 2022

it is more prevalent in Asia (23%) than in Africa (15%). Bangladesh and India, both in the South Asian region, show high levels of informal employment, whereas other countries have similar patterns though not to such high levels. City-wise, Khulna has the highest level of informal employment (47%), followed by Datong (32%), Delhi (31%), Dhaka (27%), Batangas (21%), and Madurai (19%). However, a significant portion of respondents in Delhi (47%), Madurai (46%), Batangas (47%) and Manila (43%) reported being self-employed. Notably, over half of the respondents from Tanzanian cities reported being self-employed (with or without employees). The current lack of skilled manpower and scarcity of formal jobs in the study cities explains the reliance on informal jobs or self-employment (Fig. 14.3).

Informal employment is predominant in low- and lower middle-wealth neighbourhoods: 38% in low-income neighbourhoods, 29% in lower middle neighbourhoods, 21% in middle-income neighbourhoods, 14% in higher middle neighbourhoods, and 9% in high-income neighbourhoods (Fig. 14.4). Except in Batangas, across all cities, inhabitants of low-income neighbourhoods have a significantly larger share of informal employment than those of high-income neighbourhoods.

14.4.2 Housing Inequality

Overcrowding

Occupancy rate refers to the number of people living in a dwelling relative to the number of living rooms. A dwelling is considered overcrowded if the occupancy rate is three or more persons per room. In 70% of cases, the occupancy pattern is less than two persons per room and in 13% of cases more than three persons per room. Households in Asian cities are more crowded than in African cities. Country-wise, a high percentage of the housing has an occupancy of one or fewer people per room—in China (87%) and South Africa (67%). The Chinese strategy for small family size and investment in public housing may have resulted in low occupancy rates.

Overall, 13% of households live in overcrowded dwellings, with a higher share in Asia (17%) than in Africa (7%). At the country level, 32% of households in the Philippines live in overcrowded conditions, followed by 20% in India, 16% in Bangladesh and 9% in Tanzania. China has negligible overcrowding (less than 0.1%). At the city level, Manila is the most crowded, with 37%, followed by Batangas at 27%, and Khulna and Delhi at 21% each. African cities have less overcrowded dwellings, ranging from

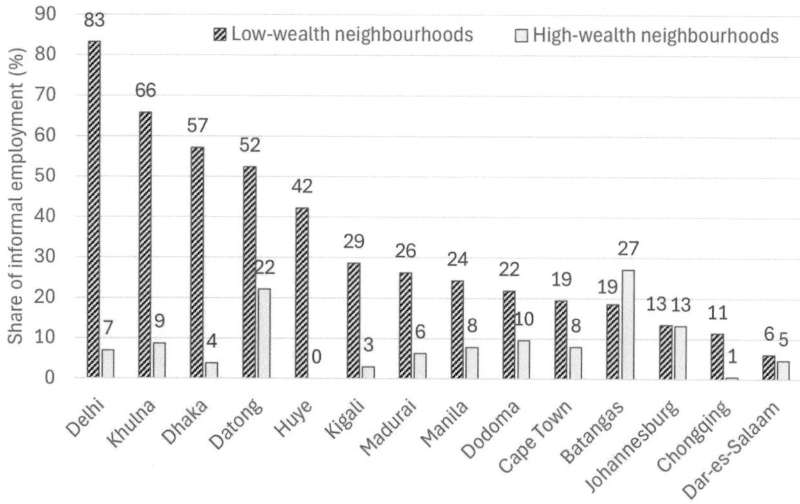

Fig. 14.4 Share of informal employment in low-wealth and high-wealth neighbourhoods in case study cities, 2022. *Source* Primary Survey, 2022

5 to 9%. As neighbourhood wealth increases, overcrowding decreases: 24% of households in low-income neighbourhoods live in overcrowded conditions, compared to just 5% in high-income neighbourhoods (Fig. 14.5). These huge variations reflect extreme housing inequality within and between cities. Widespread informal human settlements and high poverty rates contribute to overcrowding, especially in poor neighbourhoods in India, Bangladesh, and the Philippines (UN-Habitat 2020).

Housing Tenure

Housing ownership levels are almost similar in Africa (61%), and Asia (55%). Rental housing shows similar data across the continents, but more people in Asian cities reside in informal housing (5.35%) than in Africa (1.63%). Across the cities, India offers more ownership compared to the other cities in this study (79%), as the Indian city average is 57.56%. Rental accommodation is mostly found in Bangladesh (44%) and Rwanda (41%) because of the limited investments in public housing and the overall poverty situation in the urban areas. Both these countries have large slum populations. Again, 10.5% of the respondents in Bangladesh and 9% in the Philippines live in informal housing. City-wise, the two Chinese cities and Delhi have a high percentage of ownership. Khulna (9.5%) and Batangas (14%) have a high percentage of informal housing. Figure 14.6 reveals low-income neighbourhoods have a significantly large share of tenure in informally occupied dwellings (13%) compared to a much lower share (2.3% or less) in other neighbourhoods. Also, low-income neighbourhoods have the lowest share of rents received from government/social organisations (less than 1%) compared to others.

Satisfaction with housing

Housing satisfaction reflects the level of contentment individuals or households have with their living conditions and environment, making it a critical measure for further exploration.

On an average, 80% of households reported being satisfied with their dwellings. However, satisfaction levels vary significantly between Asian and African cities, with 86% of households in Asian cities expressing satisfaction compared to 70% in African cities. Within Africa, Huye exhibits the lowest housing satisfaction rate at 47%, followed by Dodoma at 62%, and Dar es Salaam at 73%. In Asian cities, housing satisfaction ranges from 76% in Datong to 91% in Madurai.

A notable trend is that households in wealthy neighbourhoods tend to report higher satisfaction

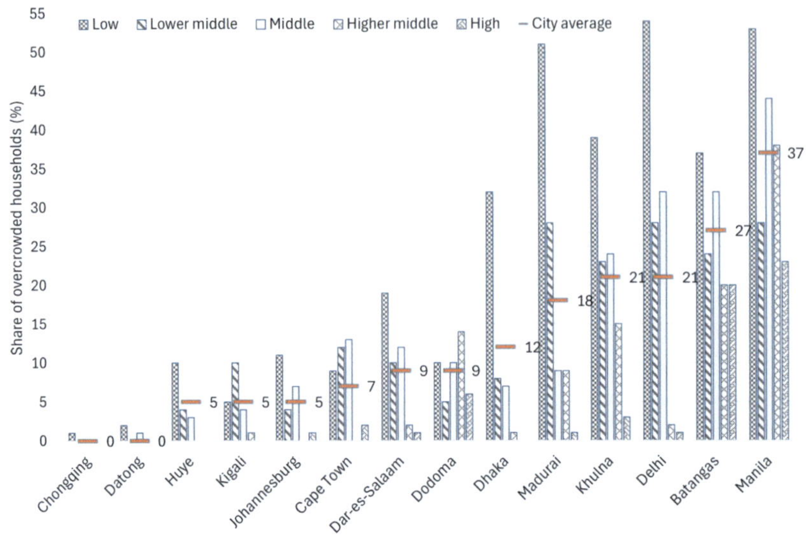

Fig. 14.5 Share of households living in overcrowded dwellings (3 or more persons per room) by neighbourhood wealth across case study cities, 2022. *Source* Primary Survey, 2022

with their dwellings compared to those in poorer neighbourhoods. For instance, while 93% of households in high-income neighbourhoods were satisfied with their residences, only 72% of those in low-income neighbourhoods reported the same. This disparity is evident across different cities. In low-income neighbourhoods, housing satisfaction ranges from 40% in Huye to as high as 93% in Madurai. Overall housing satisfaction increases with neighbourhood wealth, reflecting significant variations across different urban contexts.

14.4.3 Relationship Between Income and Housing Inequality

Three income-related factors are considered: household economic status, respondent employment status, and neighbourhood affluence. These factors collectively provide a comprehensive view of economic inequality and its impact on housing satisfaction. These dimensions uncover the interplay between economic status and housing conditions across different urban settings.

The household survey data reveals that if the housing satisfaction level is 0.8 ± 0.4, it indicates that eight out of ten households are content with their living arrangements. Madurai boasts the highest satisfaction level at 94%, while Huye demonstrates the lowest at 47%. It was observed that housing satisfaction positively correlates with economic well-being, with an increase from 72 to 93% in high-wealth neighbourhoods. Furthermore, households experiencing financial hardship exhibit lower satisfaction levels (68%) compared to those with more economic stability (91%). The analysis also uncovered that those respondents engaged in informal employment reported the lowest housing satisfaction (72%), while self-employed individuals with employees reported the highest (92%). Notably, the variation in housing satisfaction across employment statuses was found to be marginal.

Descriptive statistics reveal housing inequality is linked to income inequality. Patterns are clear from the descriptive statistics; however, regression analysis provides several benefits that evidence the robustness, accuracy, and interpretability of findings, and therefore logistic regression is executed. It quantifies the strength and nature

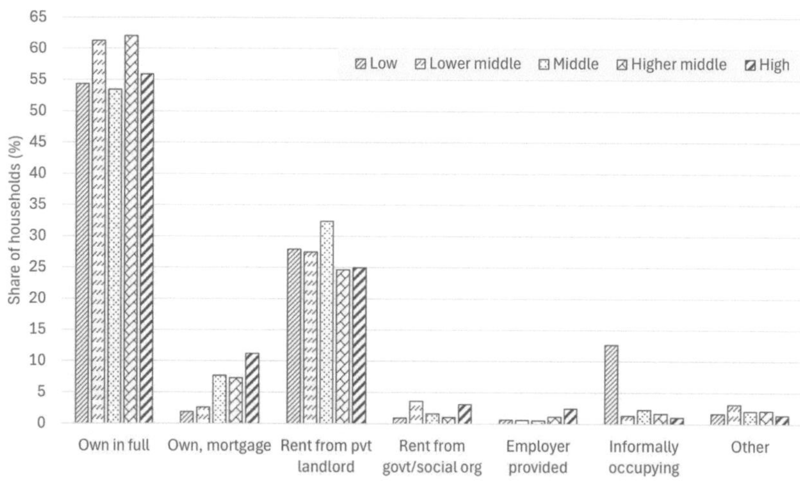

Fig. 14.6 Housing tenure by neighbourhood wealth across case study cities, 2022. *Source* Primary Survey, 2022

of relationships between variables. It allows for the inclusion of multiple independent variables, thereby able to isolate the effect of each variable. It also helps in building models that can generalise the findings. Therefore, logistic regression is performed to understand the socio-economic factors responsible for housing satisfaction. For this purpose, a household is assumed to be satisfied with its dwelling if it reports 'very satisfied' or 'satisfied', and dissatisfied if it reports 'neither satisfied nor dissatisfied', 'dissatisfied' or 'very dissatisfied' with its dwelling. The model was statistically significant, $\chi^2(33) = 2516.38$, $p < 0.001$, indicating a good fit. The pseudo R^2 was [e.g., 0.16], suggesting that approximately 16% of the variation in the outcome was explained by the model.

Results demonstrate that housing satisfaction increases with improvement in household economic status after controlling for city, neighbourhood wealth, housing tenure and education attainment (Table 14.1). For instance, the odds of a household doing very well with plenty of money left over after paying for essentials satisfied with their dwelling is 3.76 times higher than the odds of a household which is struggling to pay bills or being able to feed everyone in it. Therefore, bridging household economic inequality is crucial to reducing housing inequality (proxied by housing satisfaction).

Household economic status is closely aligned with neighbourhood wealth, typically with regard to residential segregation by income, local economic opportunities, social networks and resources, and property services and infrastructure, among other factors. In these contexts, the findings reveal that the odds of a household being satisfied with its dwelling in high-income neighbourhoods is 3.2 times higher than the odds in low-income neighbourhoods, after controlling for the city, household economic status, housing tenure, and education. Additionally, the corresponding odds for households in higher middle-income neighbourhoods is only 1.5 times higher. This implies that improvements in low-income neighbourhoods have significant potential to bridge the gap in housing inequality. However, improvements in low-income neighbourhoods have smaller effects than improvements in household economic status in an increase in housing satisfaction.

There is a significant variation in housing satisfaction across cities—Huye's 47% to Madurai's 95%. Results reveal that the odds that a household living in Huye (or Dar es Salaam) is satisfied with its dwelling are 0.15 times (or 0.31 times) lower than the odds of a household living in Dhaka, after controlling household economic status, neighbourhood wealth, housing tenure, and education. In general, African cities have significantly lower

Table 14.1 Descriptive statistics and logistic regression for housing satisfaction in case study cities

Variable	Category	Descriptive statistics		Logistic regression
		Number of observations (%)	% households satisfied with their dwelling	Housing satisfaction (odds ratio)
Neighbourhood wealth	Low-income	3131 (20.16)	72.09	Reference
	Lower middle-income	3621 (23.32)	72.66	0.81***
	Middle-income	3172 (20.43)	78.18	1.06
	Higher-middle-income	3042 (19.59)	87.27	1.51***
	High-income	2563 (16.5)	93.12	3.17***
City	Dhaka	1450 (9.34)	87.98	Reference
	Khulna	1136 (7.32)	77.11	0.69**
	Chongqing	1283 (8.26)	86.20	0.42***
	Datong	860 (5.54)	76.28	0.25***
	Delhi	1335 (8.6)	90.64	1.30
	Madurai	1043 (6.72)	95.39	3.82***
	Batangas	1079 (6.95)	88.88	1.24
	Manila	1000 (6.44)	86.50	1.05
	Huye	393 (2.53)	46.82	0.15***
	Kigali	1092 (7.03)	66.39	0.34***
	Cape Town	1001 (6.45)	81.00	0.58***
	Johannesburg	1000 (6.44)	78.86	0.52***
	Dar es Salaam	1616 (10.41)	73.40	0.31***
	Dodoma	1241 (7.99)	61.99	0.18***
Household economic status	Struggling	1459 (9.51)	67.56	Reference
	Occasionally short	2628 (17.14)	67.45	1.29**
	Just enough	4783 (31.19)	78.96	2.08***
	Doing okay	4713 (30.74)	87.86	3.45***
	Doing very well	1751 (11.42)	90.74	3.76***
Housing tenure	Own in full, outright	8930 (57.56)	83.39	Reference
	Own, paying off the mortgage	904 (5.83)	91.37	1.14
	Rent from private landlord	4282 (27.6)	73.37	0.51***
	Govt/social organisation	320 (2.06)	82.50	0.92
	Employer provided housing	155 (1)	83.23	0.64
	Informally occupying dwelling	594 (3.83)	62.06	0.31***
	Other	328 (2.11)	67.07	0.51***
Respondent education level	No school	950 (6.13)	68.14	Reference
	Some primary	1029 (6.64)	63.17	0.95
	Primary	2939 (18.96)	71.60	1.22*
	Secondary	4593 (29.62)	79.83	1.29**

(continued)

Table 14.1 (continued)

Variable	Category	Descriptive statistics		Logistic regression
		Number of observations (%)	% households satisfied with their dwelling	Housing satisfaction (odds ratio)
	Technical/vocational	465 (3.00)	86.45	1.45*
	College	1820 (11.74)	84.17	1.56***
	University	3708 (23.92)	91.29	1.59***
	Chi^2			2516.38
	p-value			$p < 0.001$
	Degree of freedom			33.00
	Pseudo R^2			0.16
	N			15,319

Note Housing satisfaction is the dependent variable, where 79.91% of households are satisfied with their dwelling ('very satisfied' or 'satisfied') and 20.09 are dissatisfied with their dwelling ('neither satisfied nor dissatisfied', 'dissatisfied' or 'very dissatisfied')
$^*p < 0.05$, $^{**}p < 0.01$, $^{***}p < 0.001$

housing satisfaction than their counterparts in Asian cities, after controlling household economic status, neighbourhood wealth, housing tenure, and education.

There is a significant variation in housing satisfaction by housing tenure—62% in informally occupied to 91% in own, paying off mortgage dwelling. After controlling for confounders, the odds that a household living in an informally occupied dwelling is satisfied with its dwelling is 0.31 times lower than that of a household living in owner-occupied housing. As expected elsewhere, data from this study reveals that the majority of households living in informally occupied dwellings are economically disadvantaged households (25% struggle to pay bills or can afford to feed everyone most of the time, and 32% are occasionally short of money for essentials).

Overall, the results reveal significant income and housing inequalities within and between cities, which are closely linked. Reducing economic inequality at both the household level (e.g. economically disadvantaged households) and the neighbourhood level (e.g. poor neighbourhoods) is crucial for mitigating housing inequality and improving well-being as envisaged in the Sustainable Development Goals.

The findings reveal a subtle contrast between different political systems, exemplified by comparisons between China and India, as well as South Africa and Rwanda. Households in China generally enjoy better economic and employment conditions. For instance, 74% of respondents work in the formal sector and 82% households are doing okay or very well. In contrast, in India, only 23% of respondents work in the formal sector and just 38% of households are doing okay or very well. Additionally, 11% of Indian households live in overcrowded dwellings, while less than 0.1% of households in China face similar conditions. These statistics suggest that individuals may sacrifice some personal freedom in exchange for relative equality in work and living conditions. China's unique combination of state capitalism and stringent governance has propelled rapid economic growth and significant improvements in living standards (Naughton 2018). Conversely, despite being a democracy, India's economic advancement is impeded by structural challenges and a large informal sector, resulting in notable disparities in living standards (Drèze and Sen 2013).

There are notable differences in household economic characteristics between South Africa and Rwanda. A larger percentage of South African households (63%) are employed in the formal sector compared to their Rwandan counterparts (45%). Additionally, home ownership rates are higher in South Africa (67%) than in Rwanda (57%). Furthermore, South African households report greater satisfaction with their dwellings (80%) compared to Rwandan households (61%). These disparities likely reflect the impact of more advanced economic structures and social policies in South Africa, which contribute to greater employment stability, higher home ownership rates, and increased housing satisfaction compared to Rwanda (Bhorat et al. 2014). This comparison prompts consideration of how different economic and social policies influence household outcomes and underscore the need for targeted approaches to addressing these inequalities.

14.5 Conclusions

The cities in the Global South face significant levels of inequality. Neighbourhood-level data indicates that income and housing disparities have led to categorisation of residents. Asian countries, particularly China, generally exhibit better income and employment stability compared to African countries. In addition, satisfaction with housing is higher in Asian countries, though a large number of households in these countries engage in informal employment and live in overcrowded dwellings compared to African countries. There are also noticeable differences among individual countries and their cities. For example, China and South Africa are making progress in reducing inequality, while Bangladesh and Rwanda continue to grapple with high levels of inequality. In China, a blend of state capitalism and stringent governance likely contributes to better economic and employment status, while in Bangladesh, issues such as high population pressure and poor economic growth are hindrances. South Africa's more developed economy and effective social policies have likely contributed to its progress in reducing inequality. The unique social and political contexts of different countries have significant impacts on inequality.

In several aspects, major cities have outperformed secondary cities within a country. For example, major cities like Dhaka and Delhi have lower rates of informal employment compared to secondary cities such as Khulna and Madurai. In five out of seven countries, major cities also exhibit higher levels of housing satisfaction than their secondary counterparts. Disparities in housing and income are prevalent across cities but become more pronounced at the sub-city level, where households in low-wealth neighbourhoods face significant disadvantages in both housing and income compared to those in affluent neighbourhoods. Prior studies have highlighted similar findings using household consumption and expenditure surveys but often fail to consider local spatial contexts or examine neighbourhoods at a smaller scale. These broader analyses can obscure the true extent of inequality and hinder the identification of extreme disparities at the neighbourhood level, hampering the development of targeted solutions. Moreover, it has been observed that there is a substantial presence of high-income households in poor neighbourhoods, while affluent neighbourhoods have a significant share of low-income households. This indicates that Asian and African cities are currently experiencing early stages of socio-spatial exclusion when compared to developed Western countries. Living in affluent neighbourhoods provides better housing and a superior built environment, contributing to overall well-being. Conversely, residents of less wealthy neighbourhoods face heightened exposure to economic inequality. Socio-economic data on income and employment patterns across neighbourhoods reveal that individuals are not randomly distributed; rather, their economic circumstances often confine them to reside in relatively low-wealth neighbourhoods. These are early indications of the emergence of segregated neighbourhoods and gated communities.

The findings from the micro-level data highlight the importance of neighbourhood data for understanding fast-growing cities in the Global South. Current statistical divisions in the study countries and, more generally, cities of the Global South have not collected census data on housing or income. Such data would enable inter and cross-city comparisons to critically identify the complex inequality situation.

Once again, a neighbourhood-focused strategy for reducing inequality proves to be highly effective in discerning micro-level differences in Global South cities. Despite varying levels of inequality, neighbourhood wealth is a shared challenge across all studied countries. Consequently, future discussions should focus on neighbourhood wealth, highlighting the importance of localised approaches to promoting fair urban development.

References

Ahmad S (2015) Housing poverty and inequality in urban India. In: Heshmati A, Maasoumi E, Wan G (eds) Poverty reduction policies and practices in developing Asia. Springer, Singapore, pp 107–122

Ahmad S, Baffoe G, Bhandri R, Young G, Osborne M (2021) Sustainable, healthy, learning cities and neighbourhoods. In: Venter M, Hattingh S (eds) Learning for a better future: perspectives on higher education, cities, business & civil society. Centre for Local Economic Development: Topics in Local Development, vol 1. AOSIS, Cape Town, pp 27–50

Aizawa T, Helble M, Lee KO (2020) Housing inequality in developing Asia and the United States. Cityscape 22(2):23–60

Alvaredo F, Chancel L, Piketty T, Saez E, Zucman G (2018) World Inequality Report 2018. World Inequality Lab. https://wir2018.wid.world

Bandauko E, Arku G, Nyantakyi-Frimpong H (2022) A systematic review of gated communities and the challenge of urban transformation in African cities. J Housing Built Environ 37(1):339–368

Bhan G, Jana A (2015) Reading spatial inequality in urban India. Econ Polit Wkly L(22):49–54

Bhorat H, Van der Westhuizen C, Jacobs T (2014) Income and non-income inequality in post-apartheid South Africa: what are the drivers and possible policy interventions? Dev South Afr 31(3):461–477

Chen G (2016) The heterogeneity of housing-tenure choice in urban China: a case study based in Guangzhou. Urban Stud 53(5):957–977

Dong H (2018) The impact of income inequality on rental affordability: an empirical study in large American metropolitan areas. Urban Stud 55(10):2106–2122

Drèze J, Sen A (2013) An uncertain glory: India and its contradictions. Princeton University Press.

Dupont V (2004) Socio-spatial differentiation and residential segregation in Delhi: a question of scale? Geoforum 35(2):157–175

Duroudier S (2014) The social and spatial divisions of intermediate cities in the United States. LEspace-Geographique 43(2):134–147. https://doi.org/10.3917/eg.432.0134

Farahi A, Jiao J (2024) Analyzing racial disparities in the United States homeownership: a socio-demographic study using machine learning. Cities 152:105181

Feitosa FF, Camara G, Monteiro AMV, Koschitzki T, Silva MP (2007) Global and local spatial indices of urban segregation. Int J Geogr Inf Sci 21(3):299–323. https://doi.org/10.1080/13658810600911903

Fujita K (2016) Residential segregation in comparative perspective: making sense of contextual diversity. Routledge

Harvey D (1989) The condition of postmodernity, vol 14. Oxford Blackwell

Hulchansky D (2010) The three cities in Toronto: income polarisation among Toronto's neighbourhoods, 1970–2005. University of Toronto, Toronto

Johnston R, Poulsen M, Forrest J (2007) Ethnic and racial segregation in US metropolitan areas, 1980-2000: the dimensions of segregation revisited. Urban Aff Rev 42(4):479–504. https://doi.org/10.1177/1078087406292701

Khondker HH (2011) Globalisation and social inequality in Asia. In: Globalisation and inequality in emerging societies. Springer, pp 31–49

Kundu D, Sharma P, Lahiri B, Sangtani C (2020) Delhi: a city of diversity and disparity. Retrieved from Glasgow

Marcińczak S, Gentile M, Stępniak M (2013) Paradoxes of (post) socialist segregation: metropolitan sociospatial divisions under socialism and after in Poland. Urban Geogr 34(3):327–352

Munir F, Ahmad S, Ullah S, Wang YP (2022) Understanding housing inequalities in urban Pakistan: an intersectionality perspective of ethnicity, income and education. J Race Ethn City 3(1):1–22. https://doi.org/10.1080/26884674.2021.1986442

Musterd S (2017) Socio-economic segregation in European cities: a comparative study of Brussels, Copenhagen, Amsterdam, and Rotterdam. Cities 59:45–53

Nasrabadi MT, Larimian T, Timmis A, Yigitcanlar T (2024) Mapping four decades of housing inequality research: trends, insights, knowledge gaps, and research directions. Sustain Cities Soc 113:105693

Naughton B (2018) The Chinese economy: adaptation and growth. MIT Press

Otsuka K, Yamano T (2005) Green Revolution and regional inequality: implications of Asian experience for Africa. In: The African food crisis: lessons from the Asian Green Revolution. CABI Publishing, Wallingford, pp 239–252

Patel V, Burns JK, Dhingra M, Tarver L, Kohrt BA, Lund C (2018) Income inequality and depression: a systematic review and meta-analysis of the association and a scoping review of mechanisms. World Psychiatry 17(1):76–89

Pryce G (2021) Future directions for research on residential segregation and inequality in China. In: Urban inequality and segregation in Europe and China: towards a new dialogue. Springer, pp 329–361

Reardon SF, O'Sullivan D (2004) Measures of spatial segregation. Sociol Methodol 34(1):121–162. https://doi.org/10.1111/j.0081-1750.2004.00150.x

Roy S, Sowgat T (2024) The neighbourhood effect on mental well-being in the Global South. Cities Health 8:626–652

Roy S, Sowgat T, Ahmed MU, Islam S, Anjum N, Mondal J, Rahman MM (2018) Bangladesh: national urban policies and city profiles for Dhaka and Khulna. GCRF Centre for Sustainable, Healthy and Learning Cities and Neighborhoods (SHLC)

Roy S, Sowgat T, Mondal J (2019) City profile: Dhaka, Bangladesh. Environ Urban ASIA 10(2):216–232. https://doi.org/10.1177/0975425319859126

Sassen S (2013) The global city: New York, London, Tokyo. Princeton University Press

Sengupta U, Shaw A, Kundu D (2022) Urban housing in India. Int J Hous Policy 22(4):467–473. https://doi.org/10.1080/19491247.2022.2133340

Sowgat T, Roy S (2022) Neighborhood segregation in Dhaka. J Urban Aff 46:1371–1391

Spiegel S (2012) Formalisation policies, informal resource sectors and the de-/re-centralisation of power: geographies of inequality in Africa and Asia. CIFOR Report. http://www.cifor.org/fileadmin/subsites/proformal/PDF/RSpiegel1212.pdf (accessed August 1, 2012)

Talen E (2018) Neighborhood. Oxford University Press

Tammaru T, Marcińczak S, Van Ham, M, Musterd S (2014) Socio-economic segregation in European capital cities. Routledge, Abingdon

Tammaru T, Marcińczak S, Aunap R, van Ham M, Janssen H (2020) Relationship between income inequality and residential segregation of socio-economic groups. Reg Stud 54(4):450–461

United Nations Department of Economic and Social Affairs (UNDESA) (2020). World Social Report 2020: Inequality in a Rapidly Changing World. United Nations, New York. ISBN: 978-92-1-130392-6.

UN-Habitat (2020) Global Urban Indicator Database. Retrieved July 26, 2024, from https://data.unhabitat.org/pages/datasets

UN-Habitat (2022) World Cities Report 2022: envisaging the future of cities. United Nations Human Settlement Programme (UN-Habitat), Nairobi, Kenya

United Nations (2020) Inequality—bridging the divide. Un.org. Retrieved July 15, 2024, from https://www.un.org/en/un75/inequality-bridging-divide

Van Ham M, Uesugi M, Tammaru T, Manley D, Janssen H (2020) Changing occupational structures and residential segregation in New York, London and Tokyo. Nat Hum Behav 4(11):1124–1134

Wang YP, Kintrea K, Osborne M, Schweisfurth M, Mitchell R, Kamete A, Yao J, Stewart A, Ahmad S, Young G, Nesterova Y, Everatt D, Lynge H, Abrahams C, Turok I, Scheba A, Visagie J, Manirakiza V, Malonza J, Nduwayezu G, Mugabe L, Nsabimana A, Rutayisire P, Nzayirambaho M, Njunwa JK, Levira F, Moshi I, Msuya I, Kundu D, Sharma P, Debnath T, Roy S, Sowgat T, Islam T, Shakil I, Reyes M, Gamboa M, Rivera R, Caluag A, Manlapas I, Racoma D, Sun T, Zhai L, Li C, Liu Y, Wang C, Zhang L, Sun X, Bhandari R, Baffoe G, Lawson V, Long J (2023) GCRF Centre for Sustainable, Healthy and Learning Cities and Neighbourhoods: household survey and neighbourhood focus group data from seven Asian and African Countries, 2021–2022 [Data Collection]. UK Data Service, Colchester, Essex. https://doi.org/10.5255/UKDA-SN-855998

Wilkinson RG (2020) The impact of inequality: how to make sick societies healthier. Routledge

Wilkinson R, Pickett K (2022) From inequality to sustainability. EARTH4ALL: Deep-Dive Paper, 1, pp 1–14

Wind B, Hedman L (2018) The uneven distribution of capital gains in times of socio-spatial inequality: evidence from Swedish housing pathways between 1995 and 2010. Urban Stud 55(12):2721–2742

Wu L, Bian Y, Zhang W (2019) Housing ownership and housing wealth: new evidence in transitional China. Hous Stud 34(3):448–468

MIX
Papier aus verantwortungsvollen Quellen
Paper from responsible sources
FSC® C105338

If you have any concerns about our products,
you can contact us on
ProductSafety@springernature.com

In case Publisher is established outside the EU,
the EU authorized representative is:
Springer Nature Customer Service Center GmbH
Europaplatz 3, 69115 Heidelberg, Germany

Printed by Libri Plureos GmbH
in Hamburg, Germany